Computational Chemistry and Molecular Modeling

K. I. Ramachandran · G. Deepa · K. Namboori

Computational Chemistry and Molecular Modeling

Principles and Applications

 Springer

Dr. K. I. Ramachandran
Dr. G. Deepa
K. Namboori
Amrita Vishwa Vidyapeetham University
Computational Engineering and Networking
641 105 Ettimadai
Coimbatore
India
ki_ram@ettimadai.amrita.edu
os_deepa@ettimadai.amrita.edu
n_krishnan@ettimadai.amrita.edu

ISBN-13 978-3-540-77302-3   e-ISBN-13 978-3-540-77304-7
DOI 10.1007/978-3-540-77304-7

Library of Congress Control Number: 2007941252

Cover design: KünkelLopka, Heidelberg

Printed on acid-free paper

9 8 7 6 5 4 3 2 1

springer.com

Dedicated to the lotus feet of
Our Beloved Sadguru and Divine Mother
Sri MATA AMRITANANDAMAYI DEVI

Preface

Computational chemistry and molecular modeling is a fast emerging area which is used for the modeling and simulation of small chemical and biological systems in order to understand and predict their behavior at the molecular level. It has a wide range of applications in various disciplines of engineering sciences, such as materials science, chemical engineering, biomedical engineering, etc. Knowledge of computational chemistry is essential to understand the behavior of nanosystems; it is probably the easiest route or gateway to the fast-growing discipline of nanosciences and nanotechnology, which covers many areas of research dealing with objects that are measured in nanometers and which is expected to revolutionize the industrial sector in the coming decades.

Considering the importance of this discipline, computational chemistry is being taught presently as a course at the postgraduate and research level in many universities. This book is the result of the need for a comprehensive textbook on the subject, which was felt by the authors while teaching the course. It covers all the aspects of computational chemistry required for a course, with sufficient illustrations, numerical examples, applications, and exercises. For a computational chemist, scientist, or researcher, this book will be highly useful in understanding and mastering the art of chemical computation. Familiarization with common and commercial software in molecular modeling is also incorporated. Moreover, the application of the concepts in related fields such as biomedical engineering, computational drug designing, etc. has been added.

The book begins with an introductory chapter on computational chemistry and molecular modeling. In this chapter (Chap. 1), we emphasize the four computational criteria for modeling any system, namely stability, symmetry, quantization, and homogeneity. In Chap. 2, "Symmetry and Point Groups", elements of molecular symmetry and point group are explained. A number of illustrative examples and diagrams are given. The transformation matrix for each symmetry operation is included to provide a computational know-how. In Chap. 3, the basic principles of quantum mechanics are presented to enhance the reader's ability to understand the quantum mechanical modeling techniques. In Chaps. 4–10, computational techniques with different levels of accuracy have been arranged. The chapters also cover Huckel's molecular orbital theory, Hartree-Fock (HF) approximation, semi-empirical methods, ab initio techniques, density functional theory, reduced density matrix, and molecular mechanics methods.

Topics such as the overlap integral, the Coulomb integral and the resonance integral, the secular matrix, and the solution to the secular matrix have been included in Chap. 4 with specific applications such as aromaticity, charge density calculation, the stability and delocalization energy spectrum, the highest occupied molecular orbital (HOMO), the lowest unoccupied molecular orbital (LUMO), bond order, the free valence index, the electrophilic and nucleophilic substitution, etc. In the chapter on HF theory (Chap. 5), the formulation of the Fock matrix has been included. Chapter 6 concerns different types of basis sets. This chapter covers in detail all important minimal basis sets and extended basis sets such as GTOs, STOs, double-zeta, triple-zeta, quadruple-zeta, split-valence, polarized, and diffuse. In Chap. 7, semi-empirical methods are introduced; besides giving an overview of the theory and equations, a performance of the methods based on the neglect of differential overlap, with an emphasis on AM1, MNDO, and PM3 is explained. Chapter 8 is on ab initio methods, covering areas such as the correlation technique, the Möller-Plesset perturbation theory, the generalized valence bond (GVB) method, the multi-configurations self consistent field (MCSCF) theory, configuration interaction (CI) and coupled cluster theory (CC).

Density functional theory (DFT) seems to be an extremely successful approach for the description of the ground state properties of metals, semiconductors, and insulators. The success of DFT not only encompasses standard bulk materials but also complex materials such as proteins and carbon nanotubes. The chapter on density functional theory (Chap. 9) covers the entire applications of the theory.

Chapter 10 explains reduced density matrix and its applications in molecular modeling. While traditional methods for computing the orbitals are scaling cubically with respect to the number of electrons, the computation of the density matrix offers the opportunity to achieve linear complexity. We describe several iteration schemes for the computation of the density matrix. We also briefly present the concept of the best n-term approximation.

Chapter 11 is on molecular mechanics and modeling, in which various force fields required to express the total energy term are introduced. Computations using common molecular mechanics force fields are explained.

Computations of molecular properties using the common computational techniques are explained in Chap. 12. In this chapter, we have included a section on a comparison of various modeling techniques. This helps the reader to choose the method for a particular computation.

The need and the possibility for high performance computing (HPC) in molecular modeling is explained in Chap. 13. This chapter explains HPC as a technique for providing the foundation to meet the data and computing demands of Research and Development (R&D) grids. HPC helps in harnessing data and computer resources in a multi-site, multi-organizational context effective cluster management, making use of maximum computing investment for molecular modeling.

Some typical projects/research topics on molecular modeling are included in Chap. 14. This chapter helps the reader to familiarize himself with the modern trends in research connected with computational chemistry and molecular modeling.

Chapter 15 is on basic mathematics and contains an introduction to computational tools such as Microsoft Excel, MATLAB, etc. This helps even a non-mathematics person to understand the mathematics used in the text to appreciate the real art of computing. Sufficient additions have been included as an appendix to cover areas such as operators, HuckelMO hetero atom parameters, Microsoft Excel in the balancing of chemical equations, simultaneous spectroscopic analysis, the computation of bond enthalpy of hydrocarbons, graphing chemical analysis data, titration data plotting, the application of curve fitting in chemistry, the determination of solvation energy, and the determination of partial molar volume.

An exclusive URL (http://www.amrita.edu/cen/ccmm) for this book with the required support materials has been provided for readers which contains a chapterwise PowerPoint presentation, numerical solutions to exercises, the input/output files of computations done with software such as Gaussian, Spartan etc., HTML-based programming environments for the determination of eigenvalues/eigenvectors of symmetrical matrices and interconversion of units, and the step-by-step implementation of cluster computing. A comprehensive survey covering the possible journals, publications, software, and Internet support concerned with this discipline have been included.

The uniqueness of this book can be summarized as follows:

1. It provides a comprehensive background theory for molecular modeling.
2. It includes applications from all related areas.
3. It includes sufficient numerical examples and exercises.
4. Numerous explanatory illustrations/figures are included.
5. A separate chapter on basic mathematics and application tools such as MATLAB is included.
6. A chapter on high performance computing is included with examples from molecular modeling.
7. A chapter on chemical computation using the reduced density matrix method is included.
8. Sample projects and research topics from the area are included.
9. It includes an exclusive web site with required support materials.

With the vast teaching expertise of the authors, the arrangement and designing of the topics in the book has been made according to the requirements/interests of the teaching/learning community. We hope that the reader community appreciates this. Computational chemistry principles extended to molecular simulation are not included in this book; we hope that a sister publication of this book covering that aspect will be released in the near future. We have tried to make the explanations clear and complete to the satisfaction of the reader. However, regarding any queries, suggestions, corrections, modifications and advice, the readers are always welcome to contact the authors at the following email address: n_krishnan@ettimadai.amrita.edu.

The authors would like to take this opportunity to acknowledge the following persons who spend their valuable time in discussions with the authors and helped them to enrich this book with their suggestions and comments:

1. Brahmachari Abhayamrita Chaitanya, the Chief Operating Officer of Amrita University, and Dr. P. Venkata Rangan, the Vice Chancellor of Amrita University, for their unstinted support and constant encouragement in all our endeavours.
2. Dr. C. S. Shastry, Professor of the Department of Science, for his insightful lectures on quantum mechanics.
3. Mr. K. Narayanan Kutty of the Department of Science, for his contribution to the chapter on quantum mechanics.
4. Mr. G. Narayanan Nair of the Systems Department, for his contribution to the section on HPC.
5. Mr. M. Sreevalsan, Mr. P. Gopakumar and Mr. Ajai Narendran of the Systems Department, for their help in making the website for the book.
6. Dr. K. P. Soman, Head of the Centre for Computational Engineering and Networking, for his continuous support and encouragement.
7. Mr. K. R. Sunderlal and Mr. V. S. Binoy from the interactive media group of 'Amrita Vishwa Vidyapeetham-University' for drawing excellent diagrams included in the book.
8. All our colleagues, dear and near ones, friends and students for their cooperation and support.
9. All the officials of Springer-Verlag Berlin Heidelberg and le-tex publishing services oHG, Leipzig for materializing this project in a highly appreciable manner.

Coimbatore, March 2008   K. I. Ramachandran

                                                          Gopakumar Deepa

                                                          Krishnan Namboori P.K.

Contents

Chapter 1
Introduction

1.1 A Definition of Computational Chemistry

Computational chemistry is an exciting and fast-emerging discipline which deals with the modeling and the computer simulation of systems such as biomolecules, polymers, drugs, inorganic and organic molecules, and so on. Since its advent, computational chemistry has grown to the state it is today and it became popular being immensely benefited from the tremendous improvements in computer hardware and software during the last several decades. With high computing power using parallel or grid computing facilities and with faster and efficient numerical algorithms, computational chemistry can be very effectively used to solve complex chemical and biological problems. The major computational requirements are:

1. Molecular energies and structures
2. Geometry optimization from an empirical input
3. Energies and structures of transition states
4. Bond energies
5. Reaction energies and all thermodynamic properties
6. Molecular orbitals
7. Multipole moments
8. Atomic charges and electrostatic potential
9. Vibrational frequencies
10. IR and Raman spectra
11. NMR spectra
12. CD spectra
13. Magnetic properties
14. Polarizabilities and hyperpolarizabilities
15. Reaction pathway
16. Properties such as the ionization potential electron affinity proton affinity
17. Modeling excited states
18. Modeling surface properties and so on

K. I. Ramachandran et al., *Computational Chemistry and Molecular Modeling*
DOI: 10.1007/978-3-540-77304-7, ©Springer 2008

Meeting these challenges could eliminate time-consuming and costly experimentations. Software tools for computational chemistry are often based on empirical information. To use these tools effectively, we need to understand the method of implementation of this technique and the nature of the database used in the parameterization of the method. With this knowledge, we can redesign the tools for specific investigations and define the limits of confidence in results.

In the real modeling procedure of a system, we have to bear in mind the natural criteria associated with the formation of that system and incorporate all these factors to make the model close to the natural system. All natural processes are associated with at least one of the following criteria:

1. *An increase in stability:* Stability is a very broad term comprising structural stability, energy stability, potential stability, and so on. During modeling, the thermodynamic significance (energetics) of stability, is to make the energy of the system as low as possible.
2. *Symmetry:* Nature likes symmetry and dislikes identity. To be more precise, we can say that in nature no two materials are identical, but they may be symmetrical.
3. *Quantization:* This term stands for fixation. For a stable system, everything is quantized. Properties, qualities, quantities, influences, etc. are quantized.
4. *Homogeneity:* A number of natural processes are there such as diffusion, dissolution, etc., which are associated with the reallocation of particles in a homogeneous manner.

The qualitative and quantitative analysis of molecules on the basis of these criteria are the main objectives of computational chemistry and molecular modeling. Now we shall familiarize ourselves with some of the computational terms.

1.2 Models

A scientific method of explaining anything involves a hypothesis, theory and laws. A hypothesis is just an educated guess or logical conclusion from known facts. The hypothesis is then compared with all available data and the details are developed. If the hypothesis is found to be consistent with known facts it is called a theory and is usually published. Most of the theories explain observed phenomena, predict the results of future experiments, and can be presented in mathematical form. When a theory is found to be always correct for a long time, it is eventually referred to as a scientific law. This process is very useful; however, we often use some *constructs*, which do not fit in the scheme of the scientific method. However, a construct is a very useful tool, and can be used to communicate in science. One of the most commonly used constructs is a *model*. A model is a simple way of describing and predicting scientific results. Models may be simple mathematical descriptions or completely non-mathematical visuals. Models are very useful because they allow us to predict and understand phenomena without performing the complex mathemati-

Fig. 1.1 The Lewis representation of the oxygen atom

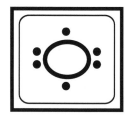

cal manipulations dictated by a rigorous theory. A model, in fact, is simpler than the system it mimics. It is a subset or subsystem of the original system. Experienced researchers continue to use models that were taught in the introductory level; however, they realize that there will always be exceptions to the rules of these models.

A simple model, which we consider at an elementary level, is the Lewis dot (electron dot) representation. For example, the Lewis Dot Structure of the oxygen atom is given in Fig. 1.1. Electron dot formulation (also referred to as the Lewis Dot formula) seeks to designate the atom as a symbol representing what is called the "core" which includes the part of the atom other than the valence electrons.

This model is not a complete description of the system, since it does not provide the kinetic energies of the particles or Coulombic interactions between the electrons and nuclei and so on. The theory of quantum mechanics, which accounts correctly for all these properties, needs to be included. The Lewis model accounts for the pairing of electrons keeping opposite spin and for the number of energy levels available to the electrons under normal temperature and pressure. The Lewis model is able to predict chemical bonding patterns and give some indication of the strength of the bonds (single bonds, double bonds, etc.). However, none of the quantum mechanics equations are used in applying this technique.

1.3 Approximations

Approximations are other types of constructs that are often seen. Even though a theory may give a rigorous mathematical description of chemical phenomena, the mathematical complexities might be so great that it is just not feasible to solve a problem exactly. If a quantitative result is desired, the best technique is often to do only part of the work. One of the techniques applied in approximation is to completely leave out the complex part of the calculation. Another type of approximation is to use an average rather than an exact mathematical description. Some other common approximation methods are variations, perturbations, simplified functions, and fitting parameters to reproduce experimental results.

Quantum mechanics gives a mathematical description of the behavior of electrons, which has never been found to be wrong. However, the quantum mechanical equations have never been solved exactly for any chemical system other than for the hydrogen atom. Thus, the entire field of computational chemistry is built around approximate solutions. Some of these solutions are very crude, and others are more accurate than any experiment that has yet been designed. There are several implications of this situation. Firstly, computational chemists require knowledge of each approximation being used in the computation and the level of computational accuracy that can be expected. Secondly, to get very accurate results, we require extremely powerful computers. Thirdly, if the equations could be solved exactly, much of the work now done on supercomputers could be done faster and more accurately on a PC.

1.4 Reality

There are certain things known to us exactly. For example, the quantum mechanical description of the hydrogen atom matches the observed spectrum as accurately as any experimental result. If an approximation is used, one must ask how accurate an answer must be. Computations of energetics of molecules and reactions often attempt to achieve what is called "chemical accuracy," meaning an error less than about 1 kcal/mol, since this is sufficient to describe van der Waals interactions, the weakest interaction possible between molecules. Most of the computational scientists do not have any interest in results more accurate than this, as even biological modeling such as drug designing can be done within that limit. A student of computational chemistry must realize that theories, models, and approximations are powerful tools for understanding and achieving research goals. But one should remember that results obtained from none of these tools are perfect. This may not be an ideal situation, but it is the best that the scientific community can offer.

The term *theoretical chemistry* may be defined as the mathematical description of chemistry. Very few aspects of chemistry can be computed exactly, but almost every aspect of chemistry has been described in a qualitative or approximate quantitative computational scheme. The biggest mistake that a computational chemist may make is to assume that any computed number is exact. However, just as not all spectra are perfectly resolved, often a qualitative or approximate computation can give useful insight into chemistry if you understand what it tells you and what it does not.

1.5 Computational Chemistry Methods

Computational chemistry is comprised of a theoretical (or structural) modeling part, known as *molecular modeling,* and a modeling of processes (or experimentations) known as *molecular simulation.* The former alone is the topic of this book. Depending upon the level of theory that we observe in a computation, the following methods have been identified.

1.5.1 Ab Initio Calculations

The term Ab initio is the Latin term meaning "from the beginning." This name is given to computations which are derived directly from theoretical principles (such as the Schrödinger equation), with no inclusion of experimental data. This method, in fact, can be seen as an approximate quantum mechanical method. The approximations made are usually mathematical approximations, such as using a simpler functional form for a function, or getting an approximate solution to a differential equation.

The most common type of ab initio calculation is called a Hartree Fock calculation (HF), in which the primary approximation is called the central field approximation. This method does not include Coulombic electron-electron repulsion in the calculation. However, its net effect is included in the calculation. This is a variational calculation, meaning that the approximate energies calculated are all equal to or greater than the exact energy. The energies calculated are usually in units called Hartrees (1 Hartree = 27.2114 eV – An HTML-based GUI for energy conversion is made available in the text URL). Because of the central field approximation, the energies from HF calculations are always greater than the exact energy and tend to a limiting value called the Hartree Fock limit.

The second approximation in HF calculations is that the wavefunction must be described by some functional form, which is only known exactly for a few one-electron systems. The functions used most often are linear combinations of Slater type orbitals (e^{-ax}) or Gaussian type orbitals $\left(e^{\left(-ax^2\right)}\right)$, abbreviated as, respectively, STO and GTO. The wavefunction is formed from linear combinations of atomic orbitals, or more often from linear combinations of basis functions. Because of this approximation, most HF calculations give a computed energy greater than the Hartree Fock limit. The exact set of basis functions used is often specified by an abbreviation, such as STO-3G or 6-311++g**.

Most of these computations begin with a HF calculation, followed by further corrections for the explicit electron-electron repulsion, referred to as correlations. Some of these methods are the Möller-Plesset perturbation theory (MPn, where n is the order of correction), the Generalized Valence Bond (GVB) method, Multi-Configurations Self Consistent Field (MCSCF), Configuration Interaction (CI) and Coupled Cluster theory (CC). As a group, these methods are referred to as *correlated calculations*.

A method, which avoids making the HF mistakes in the first place, is called Quantum Monte Carlo (QMC). There are several flavors of QMC, namely variational, diffusion, and Green's functions. These methods work with an explicitly correlated wavefunction and evaluate integrals numerically using a Monte Carlo integration. These calculations can be very time-consuming, but they are probably the most accurate methods known today.

An alternative ab initio method is the Density Functional Theory (DFT), *in which the total energy is expressed in terms of the total electron density, rather than the wavefunction.* In this type of calculation, there is an approximate Hamiltonian and an approximate expression for the total electron density.

The favorable aspect of ab initio methods is that they eventually converge to the exact solution, once all the approximations are made sufficiently small in magnitude. However, this convergence is not monotonic. Sometimes, the smallest calculation gives the best result for a given property.

The unfavorable aspect of ab initio methods is that they are expensive. These methods often take enormous amounts of computer CPU time, memory, and disk space. The HF method scales as N^4, where N is the number of basis functions, so a calculation twice as big takes 16 times as long to complete. Correlated calculations often scale much worse than this. *In practice, extremely accurate solutions are obtainable only when the molecule contains half a dozen electrons or less.*

In general, ab initio calculations give very good qualitative results and can give increasingly accurate quantitative results as the molecules in question become smaller.

1.5.2 Semiempirical Calculations

Semiempirical calculations are set up with the same general structure as a HF calculation. Within this framework, certain pieces of information, such as two electron integrals, are approximated or completely omitted. In order to correct for the errors introduced by omitting part of the calculation, the method is parameterized, by curve fitting in a few parameters or numbers, in order to give the best possible agreement with experimental data. The *merit* of semiempirical calculations is that they are much faster than the ab initio calculations. The *demerit* of semiempirical calculations is that the results can be slightly defective. If the molecule being computed is similar to molecules in the database used to parameterize the method, then the results may be very good. If the molecule being computed is significantly different from anything in the parameterization set, the answers may be very poor.

Semiempirical calculations have been very successful in the description of organic chemistry, where there are only a few elements used extensively and the molecules are of moderate size. However, semiempirical methods have been devised specifically for the description of inorganic chemistry as well.

1.5.3 Modeling the Solid State

The electronic structure of an infinite crystal is defined by a band structure plot, which gives energies of electron orbitals for each point in k-space, called the *Brillouin zone*. Since ab initio and semiempirical calculations yield orbital energies, they can be applied to band structure calculations. However, if it is time-consuming to calculate the energy for a molecule, it is even more time-consuming to calculate energies for a list of points in the Brillouin zone.

Band structure calculations have been done for very complicated systems; however, the software is not yet automated enough or sufficiently fast enough that anyone does band structures casually.

1.5.4 Molecular Mechanics

If a molecule is too big to effectively use a semiempirical treatment, it is still possible to model its behavior by totally avoiding quantum mechanics. The methods, referred to as *molecular mechanics*, set up a simple algebraic expression for the total energy of a compound, with no necessity to compute a wavefunction or total electron density [2]. *The energy expression consists of simple classical equations, such as the harmonic oscillator equation in order to describe the energy associated with bond stretching, bending, rotation, and intermolecular forces, such as van der Waals interactions and hydrogen bonding. All of the constants in these equations must be obtained from experimental data or an ab initio calculation.*

In a molecular mechanics method, the database of compounds used to parameterize the method (a set of parameters and functions is called a force field) is crucial to its success. The molecular mechanics method may be parameterized against a specific class of molecules, such as proteins, organic molecules, organo-metallics, etc. Such a force field would only be expected to have any relevance to describing other proteins.

Molecular mechanics allows the modeling of very large molecules, such as proteins and segments of DNA, making it the primary tool of computational biochemists. The defect of this method is that there are many chemical properties that are not even defined within the method, such as electronic excited states. In order to work with extremely large and complicated systems, often most of the molecular mechanics software packages will have highly powerful and easy to use graphical interfaces.

1.5.5 Molecular Simulation

Molecular simulation is a computational experiment conducted on a molecular model. This can be set up in different levels of accuracy. A number of simulation techniques have been designed such as the Monte Carlo simulation (MC), the Conformational Biased Monte Carlo (CBMC) simulation, the Molecular Dynamics (MD) simulation, the Car-Parrinello Molecular Dynamics (CPMD) simulation, and so on [3].

1.5.6 Statistical Mechanics

Statistical mechanics is the mathematical means to extrapolate the thermodynamic properties of bulk materials from a molecular description of the material. Statistical mechanics computations are often tacked onto the end of ab initio calculations for gas phase properties. For condensed phase properties, often molecular dynamics calculations are necessary in order to do a computational experiment.

1.5.7 Thermodynamics

Thermodynamics is one of the most well-developed mathematical chemical descriptions. Very often, any thermodynamic treatment is left for trivial pen and paper work, since many aspects of chemistry are so accurately described with very simple mathematical expressions.

1.5.8 Structure-Property Relationships

Structure-property relationships are qualitatively or quantitatively empirically defined empirical relationships between molecular structure and observed properties. In some cases this may seem to duplicate statistical mechanical results; however, structure-property relationships need not be based on any rigorous theoretical principles.

The simplest case of structure-property relationships are qualitative thumb rules. For example, an experienced polymer chemist may be able to predict whether a polymer will be soft or brittle based on the geometry and bonding of the monomers.

When structure-property relationships are mentioned in the current literature, it usually implies a quantitative mathematical relationship. These relationships are most often derived by using curve fitting software to find the linear combination of molecular properties, which best reproduces the desired property. The molecular properties are usually obtained from molecular modeling computations. Other molecular descriptors, such as molecular weight or topological descriptions, are also used.

When the property being described is a physical property, such as the boiling point, this is referred to as a Quantitative Structure-Property Relationship (QSPR). When the property being described is a type of biological activity (such as a drug activity), this is referred to as a Quantitative Structure-Activity Relationship (QSAR).

1.5.9 Symbolic Calculations

Symbolic calculations are performed when the system is just too large for an atom-by-atom description to be viable at any level of approximation. An example might be the description of a membrane by describing the individual lipids as some representative polygon with some expression for the energy of interaction. This sort of treatment is used for computational biochemistry and even microbiology.

1.5.10 Artificial Intelligence

Techniques invented by computational scientists concerned with artificial intelligence (AI) have been applied mostly to drug design in recent years. These methods are also known as De Novo or rational drug design. The general scenario is that some functional site will be identified, and it is desirable to come up with a structure for a molecule that will interact (dock) with that site in order to hinder its functionality. Rather than making trials with hundreds or thousands of possibilities, the molecular mechanics is built into an AI program, which tries enormous numbers of "reasonable" possibilities in an automated fashion. The number of techniques for describing the "intelligent" part of this operation is so diverse that it is impossible to make any generalization about how this is implemented in the program.

1.5.11 The Design of a Computational Research Program

When we are using computational chemistry to answer a chemical question, the obvious requirement is to know how to use the software. Moreover, we need to assess how good the answer is going to be. Normally, a computational chemist should preliminarily answer the following questions before getting into any research activity.

1. What do we need to recognize from computations?
2. Why do we stick to computational tools?
3. What should be the permissible accuracy level?

In analytical chemistry, we do a number of identical measurements, then work out the error from a standard deviation. With computational experiments, repeating the same experiment should always give exactly the same result. The way that we estimate our error is to compare a number of similar computations to the experimental answers. If none exist, we may have to guess which method should be reasonable, based on its assumptions, for which we may have to study the computational results with known systems and make a proper standardization of the technique before applying the same computational techniques to unknown systems. Regarding the level of computation, often ab initio calculations would be the most reliable. However, it is time-consuming, and sometimes we would take a decade to do a single calculation even with a high performance computing facility. If we need to scale a computation, we need to do the simplest possible calculations, then use the scaling equation to estimate the possible time required to complete the required computation.

1.5.12 Visualization

Data visualization is the process of displaying information in any sort of pictorial or graphical representation. A number of computer programs are now available to apply a colorization scheme to data or to work with three-dimensional representations [1].

1.6 Journals and Book Series Focusing on Computational Chemistry

The following is a list of common journals and book series focusing on computational chemistry:

1. Advances in Molecular Modeling
2. Chemical Informatics Letters
3. Chemical Modelling: Applications and Theory
4. Computational and Theoretical Polymer Science
5. Computers and Chemistry
6. International Journal of Quantum Chemistry
7. Journal of Biomolecular Structure and Dynamics
8. Journal of Chemical Information and Computer Science
9. Journal of Chemometrics
10. Journal of Computational Chemistry
11. Journal of Computer-Aided Materials Design
12. Journal of Computer-Aided Molecular Design
13. Journal of Mathematical Chemistry
14. Journal of Molecular Graphics and Modelling
15. Journal of Molecular Modeling
16. Journal of Molecular Structure
17. Journal of Molecular Structure: THEOCHEM
18. Macromolecular Theory and Simulations
19. Molecular Simulation
20. Quantitative Structure-Activity Relationships
21. Reviews in Computational Chemistry
22. SAR and QSAR in Environmental Research
23. Structural Chemistry
24. Theoretical Chemistry Accounts: Theory, Computation, and Modeling (Formerly Theoretica Chimica Acta)

1.7 Journals and Book Series
Often Including Computational Chemistry

 1. Advances in Chemical Physics
 2. Advances in Drug Research
 3. Annual Review of Biochemistry
 4. Annual Review of Biophysics and Bioengineering
 5. Annual Review of Biophysics and Biomolecular Structure
 6. Annual Review of Physical Chemistry
 7. Biochemistry
 8. Biophysical Journal
 9. Biopolymers
10. Chemical Reviews
11. Chemometrics and Intelligent Laboratory Systems
12. Computer Applications in the Biosciences
13. Current Opinions in Biotechnology
14. Current Opinions in Structural Biology
15. Drug Design and Discovery
16. Drug Discovery Today
17. Journal of Chemical Physics
18. Journal of Mathematical Biology
19. Journal of Medicinal Chemistry
20. Journal of Molecular Biology
21. Journal of Organic Chemistry
22. Journal of Physical Chemistry
23. Journal of the American Chemical Society
24. Journal of Theoretical Biology
25. Modern Drug Discovery
26. Perspectives in Drug Discovery and Design
27. Protein Engineering
28. Protein Science
29. Proteins: Structure, Function, and Genetics
30. Reviews in Modern Physics

1.8 Common Reference Books Available
on Computational Chemistry

Since the advent of computers into the world of science and technology, scientists have started seeking the help of computers in their computational works. Hence, large number of books are available today in this area, starting from the very beginning to the present day. Some of the relevant reference books are listed below, arranged in chronological order.

1. Peter Lykos and Isaiah Shavitt, Supercomputers in Chemistry, in ACS Symposium Series 173, American Chemical Society, Washington, DC, 1981.
2. E. Stuper, W. Brugger, and P. Jurs, Computer-Aided Analysis of the Relation Between Chemical Structure and Biological Activity, Mir, Moscow, 1982.
3. Klaus Ebert and Hanns Ederer, Computers. Use in Chemistry, Mir, Moscow, 1988.
4. S. R. Heller and R. Potenzone Jr., Computer Applications in Chemistry, Proceedings of the 6th International Conference on Computers in Chemical Research and Education, in Analytical Chemistry Symposium Series, Vol. 15, Elsevier, Amsterdam, The Netherlands, 1983.
5. V. D. Maiboroda, S. G. Maksimova, and Yu. G. Orlik, Solution of Problems in Chemistry Using Programmable Microcalculators, Izd. Universitetskoe, Minsk, USSR, 1988.
6. Kenneth L. Ratzlaff, Introduction to Computer-Assisted Experimentations, Wiley-Interscience, New York, 1988.
7. K. Ebert, H. Ederer, and T. L. Isenhour, Computer Applications in Chemistry. An Introduction for PC Users, With Two Diskettes in BASIC and PASCAL, VCH, Weinheim, 1989.
8. Josef Brandt and Ivar K. Ugi, Computer Applications in Chemical Research and Education, Huethig Verlag, Heidelberg, 1989.
9. G. Gauglitz, Software-Development in Chemistry 3. Proceedings of the 3rd Workshop on Computers in Chemistry, Tuebingen, November 16–18, 1988, Springer-Verlag, Berlin, 1989.
10. Russell F. Doolittle, Molecular Evolution: Computer Analysis of Protein and Nucleic Acid Sequences, in Methods in Enzymology, Vol. 183, Academic Press, San Diego, 1990.
11. Uwe Harms, Supercomputer and Chemistry 2, Debis Workshop 1990, Ottobrunn, November 19–20, 1990, Springer, Berlin, 1991.
12. Juergen Gmehling, Computers in Chemistry, Proceedings of the 5th Workshop in Software Development in Chemistry, Oldenburg, November 21–23, 1990, Springer, Berlin, 1991.
13. Ludwig Brand and Michael L. Johnson, Numerical Computer Methods, in Methods Enzymol., Vol. 210, Academic Press, San Diego, 1992.
14. Mototsugu Yoshida, Computer Aided Chemistry: Introduction to New Method for Chemistry Research, Tokyo Kagaku Dozin, Tokyo, 1993.
15. Rogers, Computational Chemistry Using the PC, 2nd ed., VCH, Weinheim, 1995.
16. W. J. Hehre, Practical Strategies for Electronic Structure Calculations, Wavefunction, Inc., Irvine, CA, 1995.
17. Guy H. Grant and W. Graham Richards, Computational Chemistry, Oxford University Press, Oxford, UK, 1995.
18. G. W. Robinson, S. Singh, and M. W. Evans, Water in Biology, Chemistry and Physics: Experimental Overviews and Computational Methodologies, World Scientific, Singapore, 1996.

19. Peter C. Jurs, Computer Software Applications in Chemistry, 2nd ed., Wiley, New York, 1996.
20. W. J. Hehre, A. J. Shusterman, and W. W. Huang, A Laboratory Book of Computational Organic Chemistry, Wavefunction, Inc., Irvine, CA, 1996.
21. Jane S. Murray and Kalidas Sen, Molecular Electrostatic Potentials: Concepts and Applications, in Theor. Comput. Chem., Vol. 3, Elsevier, Amsterdam, The Netherlands, 1996.
22. S. Wilson and G. H. F. Diercksen, Problem Solving in Computational Molecular Science: Molecules in Different Environments, Proceedings of the NATO Advanced Study Institute held 12–22 August 1996, in Bad Windsheim, Germany, in NATO ASI Ser., Ser. C, Vol. 500, Kluwer, Dordrecht, 1997.
23. Jerzy Leszczynski, Computational Chemistry: Reviews of Current Trends, Vol. 3, World Scientific, Singapore, 1999.
24. Frank Jensen, Introduction to Computational Chemistry, Wiley, Chichester, 1999.
25. K. Ohno, K. Esfarjan, and Y. Kawazoe, Computational Materials Science: From Ab Initio to Monte Carlo Methods, Springer, Berlin, 1999.

1.9 Computational Chemistry on the Internet

A number of resources are available on the Internet for computational chemistry and molecular modeling. Some of them are included here for your information:

1. ACCVIP Australian Computational Chemistry via the Internet Project
   (http://www.chem.swin.edu.au/)
2. WWW Computational Chemistry Resources
   (http://www.chem.swin.edu.au/chem_ref.html)
3. Some resources on computational chemistry
   (http://www.zyvex.com/nanotech/compChemLinks.html)
4. Internet Resources for Science and Mathematics Education,
   collected by Tom O'Haver
   (http://www.towson.edu/csme/mctp/Technology/Chemistry.html)
5. Chemistry (and some other) Internet Resources
   (http://www.technion.ac.il/technion/chemistry/links/chem_resources.html)
6. Intute Science, Engineering and Technology
   (http://www.intute.ac.uk/sciences//)
7. NIST ChemistryWebBook (http://webbook.nist.gov/chemistry/)
8. Chemcyclopedia (http://www.chemcyc.org/ME2/Default.asp)
9. Computational Chemistry List (CCL) a mailing list of computational chemists
   (http://www.ccl.net/)
10. ChemFinder.com (http://chemfinder.cambridgesoft.com/)

1.10 Some Topics of Research Interest Related to Computational Chemistry

At present, computational chemistry has entered into all areas of research, so that an awareness of this discipline becomes essential for all advanced research activities. Some of the areas of research interest are given below:

1. Drug discovery and materials research imaging of a computer rendering of molecular systems
2. Computational drug designing
3. Computational study of new chemical compounds and materials such as pharmaceuticals, plastics, microprocessors, glass, metal, paint, aerospace, and automobiles
4. Study of free energy surfaces to guide the improvement of models for biomolecular simulations
5. Introduction of multi-scale methods for examining macromolecular systems
6. Modeling protein-mediated oxidation of small molecules
7. Investigating statistical scoring functions
8. Modeling of electrostatics of proteins in solvent continua
9. Free energy calculations on biomolecules such as ribosomes
10. Mesoscopic simulations of actin filaments, lipid vesicles, and nanoparticles
11. Modeling of "membrane proteins" in action
12. Multiscale modeling of photoactive liquid crystalline systems
13. Protein dynamics: from nanoseconds to microseconds and beyond
14. Photochemistry and non-adiabatic quantum dynamics: multiconfigurational methods and effective-mode models for large systems
15. Study of hydrogen bonding pathways and hydrogen transfer in biochemical processes
16. Modeling of bio-motors
17. Study of hydrogen bonding interactions of water on hydroxylated silica surfaces
18. Electronic structure calculations on the adsorption and reaction of molecules at catalyst surfaces
19. High-performance computing and the design of chemical software for parallel computers
20. Structure, bonding, and reactivity in main-group, organometallic and organic chemistry
21. Modeling of solvation and transport properties of pharmaceutical compounds
22. Computational study of chiral surfaces used in chromatography
23. Calculation of penetrant solubilities in polymers, in particular, investigating the effects of specific polymer-penetrant interactions, which are difficult to access by experimental probes
24. Modeling penetrant-induced plasticization of glassy polymers

References

1. Gund P, Barry DC, Blaney JM, and Cohen NC (1988) Guidelines for Publications in Molecular Modeling Related to Medicinal Chemistry. J Med Chem 31:2230
2. Boyd DB, Lipkowitz KB, eds. (2000) Reviews in Computational Chemistry, History of the Gordon Conferences on Computational Chemistry. Wiley-VCH, New York, p 399–439.
3. Schleyer PVR, Allinger NL, Clark T, Gasteiger J, Kollman P and Schaefer III HF, eds. (1998) Encyclopedia of Computational Chemistry, Vols. 1–5. Wiley, Chichester

Chapter 2
Symmetry and Point Groups

2.1 Introduction

Symmetry plays a vital role in the analysis of the structure, bonding, and spectroscopy of molecules. We will explore the basic symmetry *elements* and *operations*
and their use in determining the symmetry classification (point group) of different
molecules. The symmetry of objects (and molecules) may be evaluated through certain tools known as the elements of symmetry.

2.2 Symmetry Operations and Symmetry Elements

A symmetry operation is defined as *an operation performed on a molecule that
leaves it apparently unchanged*. For example, if a water molecule is rotated by $180°$
around a line perpendicular to the molecular plane and passing through the central oxygen atom, the resulting structure is indistinguishable from the original one
(Fig. 2.1). A symmetry element can be defined *as the point, line or plane with respect to which a symmetry operation is performed*. The symmetry element associated with the rotation drawn above is the line, or rotation axis, around which the
molecule was rotated. The water molecule is said to possess this symmetry element.
Table 2.1 includes the types of symmetry elements, operations and their symbols [2].

Fig. 2.1 Water molecule undergoing rotation by $180°$

K. I. Ramachandran et al., *Computational Chemistry and Molecular Modeling*
DOI: 10.1007/978-3-540-77304-7, ©Springer 2008

Table 2.1 Types of symmetry elements, operations, and their symbols

Element	Operation	Symbol
Symmetry plane	Reflection through the plane	σ
Inversion center	Inversion: Every point x, y, z translated into $-x, -y, -z$	i
Proper axis	Rotation about the axis by $360/n$	C_n
Improper axis	1. Rotation by $360/n$ degrees	S_n
	2. Reflection through the plane perpendicular to the rotation axis	

2.3 Symmetry Operations and Elements of Symmetry

2.3.1 The Identity Operation

Every molecule possesses at least one symmetry element, the identity. The identity operation amounts to doing nothing to a molecule or a rotation of the molecule by $360°$ and so leaving the molecule completely unchanged. The symbol of the identity element is E and the corresponding operation is designated as $\hat{E}$. Let us assign the coordinates (x_1, y_1, z_1) to any atom of the molecule. The identity operation does not alter these coordinates. If the coordinates after the operation are designated as (x_2, y_2, z_2), then we get the following equations:

$$x_2 = 1x_1 + 0y_1 + 0z_1 \tag{2.1}$$

$$y_2 = 0x_1 + 1y_1 + 0z_1 \tag{2.2}$$

$$z_2 = 0x_1 + 0y_1 + 1z_1 . \tag{2.3}$$

Or, the identity operation matrix can be represented as:

$$\begin{bmatrix} x_2 \\ y_2 \\ z_2 \end{bmatrix} = \begin{bmatrix} 1 & 0 & 0 \\ 0 & 1 & 0 \\ 0 & 0 & 1 \end{bmatrix} \begin{bmatrix} x_1 \\ y_1 \\ z_1 \end{bmatrix} \tag{2.4}$$

Or, the transformation matrix (T) corresponding to E becomes:

$$T = \begin{bmatrix} 1 & 0 & 0 \\ 0 & 1 & 0 \\ 0 & 0 & 1 \end{bmatrix} \tag{2.5}$$

The identity operation will get the same representation as Eq. 2.4 for a molecule belonging to any point group. We can take internal coordinates of all the atoms (e.g. water) of the molecule for determining the transformation matrix corresponding to the identity operation as shown in Fig. 2.2.

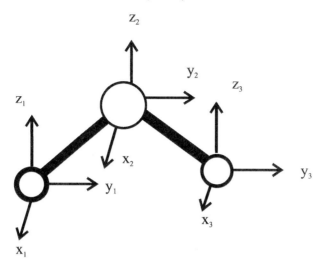

Fig. 2.2 Identity operation of three atoms of water

The transformation matrix for E will be a 9×9 diagonal matrix as shown in Eq. 2.6.

$$T = \begin{bmatrix} 1 & 0 & 0 & 0 & 0 & 0 & 0 & 0 & 0 \\ 0 & 1 & 0 & 0 & 0 & 0 & 0 & 0 & 0 \\ 0 & 0 & 1 & 0 & 0 & 0 & 0 & 0 & 0 \\ 0 & 0 & 0 & 1 & 0 & 0 & 0 & 0 & 0 \\ 0 & 0 & 0 & 0 & 1 & 0 & 0 & 0 & 0 \\ 0 & 0 & 0 & 0 & 0 & 1 & 0 & 0 & 0 \\ 0 & 0 & 0 & 0 & 0 & 0 & 1 & 0 & 0 \\ 0 & 0 & 0 & 0 & 0 & 0 & 0 & 1 & 0 \\ 0 & 0 & 0 & 0 & 0 & 0 & 0 & 0 & 1 \end{bmatrix} \qquad (2.6)$$

2.3.2 Rotation Operations

This symmetry operation, denoted by the symbol $\hat{C}_n$, corresponds to the rotation about an axis by $(360°/n)$. When the molecule is rotated with respect to an axis by $360°$, if n-times symmetrical structures are obtained, then the axis is said to be a C_n axis or n-fold axis. The water molecule is left unchanged by a rotation of $180°$ or twice symmetrical structures are obtained by rotation of $360°$. The operation is said to be a two-fold or $\hat{C}_2$ rotation and the symmetry element is a C_2 rotation axis. Another example is the plane triangular BF_3 molecule. It is left unchanged by a rotation of $120°$ around an axis perpendicular to the molecular plane. Hence here, the operation is a threefold or $\hat{C}_3$ rotation. The symmetry element is a C_3 rotation axis. Actually, two different types of rotations are possible about this axis: clockwise and anti-clockwise rotations (Figs. 2.3 and 2.4). It can be seen that these rotations result in different spatial arrangements.

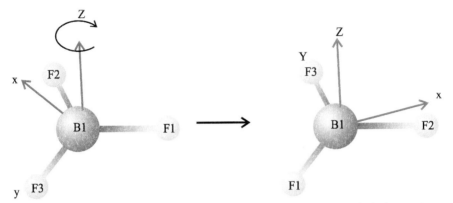

Fig. 2.3 Symmetry operation, rotation by 120° on a boron tri fluoride (BF$_3$)-clockwise rotation

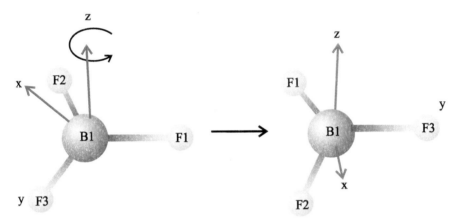

Fig. 2.4 Symmetry operation, rotation by 120° on a boron tri fluoride (BF$_3$)-anticlockwise rotation

The matrix representation of C_n depends on the group. We shall consider a general case of a rotation of a molecule through θ about the z-axis (Fig. 2.5). By inserting the appropriate value of θ, the matrix representation on C_n group can be determined. Atom A has coordinates (x_1, y_1, z_1). On rotating the atom through θ about the z-axis, it reaches the point $B(x_2, y_2, z_2)$. The z coordinate remains the same, i.e. $(z_2 = z_1)$. Hence, the rotation can be considered as a 2D rotation by an angle θ. The initial position of the vector (x_1, y_1) can be written in polar coordinates as follows:

$$(x_1, y_1) = (r\cos\phi, r\sin\phi) \tag{2.7}$$

$$(x_2, y_2) = [r\cos(\phi + \theta), r\sin(\phi + \theta)]$$
$$= [(r\cos\phi\cos\theta - r\sin\phi\sin\theta), (r\sin\phi\cos\theta + r\cos\phi\sin\theta)]$$
$$= [(x_1\cos\theta - y_1\sin\theta), (y_1\cos\theta + x_1\sin\theta)]$$

$$(x_2, y_2) = [(x_1\cos\theta - y_1\sin\theta), (x_1\sin\theta + y_1\cos\theta)] \tag{2.8}$$

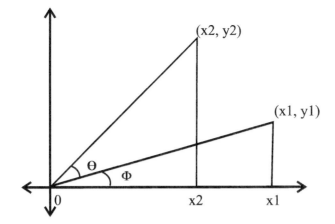

Fig. 2.5 C_n-representation by rotation through an angle θ

Hence:

$$x_2 = x_1 \cos\theta - y_1 \sin\theta + 0z_1 \tag{2.9}$$

$$y_2 = x_1 \sin\theta + y_1 \cos\theta + 0z_1 \tag{2.10}$$

$$z_2 = 0x_1 + 0y_1 + 1z_1 \tag{2.11}$$

In matrix notation, C_n can be written as:

$$\begin{bmatrix} x_2 \\ y_2 \\ z_2 \end{bmatrix} = \begin{bmatrix} \cos\theta & -\sin\theta & 0 \\ \sin\theta & \cos\theta & 0 \\ 0 & 0 & 1 \end{bmatrix} \begin{bmatrix} x_1 \\ y_1 \\ z_1 \end{bmatrix} \tag{2.12}$$

Hence, the transformation matrix for C_n will be as follows:

$$T = \begin{bmatrix} \cos\theta & -\sin\theta & 0 \\ \sin\theta & \cos\theta & 0 \\ 0 & 0 & 1 \end{bmatrix} \tag{2.13}$$

As, for example, in the water molecule, the symmetry operation C_n is C_2 as the rotation of the molecule by $180°$ produces identical configurations. Hence, the transformation matrix for the water molecule will be as follows:

$$T = \begin{bmatrix} -1 & 0 & 0 \\ 0 & -1 & 0 \\ 0 & 0 & 1 \end{bmatrix} \tag{2.14}$$

The BF_3 molecule possesses three C_2 axes and two C_3 axes, as illustrated in Figs. 2.3, 2.4, and 2.6. The axis with the highest value is considered as the principal rotation axis. Hence, the three fold C_3 is considered as the principal rotation axis for BF_3.

Fig. 2.6 Three C_2 axes (1, 2 and 3) for BF_3

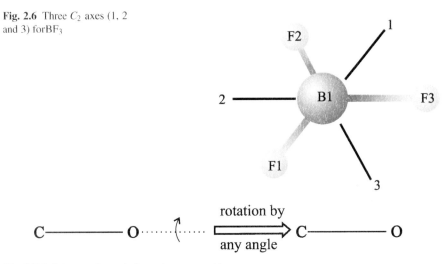

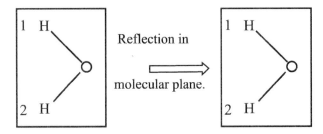

Fig. 2.7 Infinite rotation axis for carbon monoxide

In linear molecules such as CO_2 and CO rotation by any angle with respect to the molecular axis returns the molecule unchanged. Hence, such molecules possess C_∞, the infinite rotation axis. For CO, such an infinite rotation axis is shown in Fig. 2.7.

If a molecule keeps a number of axis of symmetry, then the axis providing maximum symmetry by the operation $\hat{C}_n$ or the axis with maximum value of n is known as the principal axis.

2.3.3 Reflection Planes (or Mirror Planes)

The reflection operation, denoted by the symbol σ, corresponds to the reflection in a mirror plane. The water molecule possesses two distinct mirror planes, labeled as σ_v and $\sigma_{v'}$, the reflection in the plane of the molecule, and the reflection in a plane perpendicular to the molecule, as given in Figs. 2.8 and 2.9.

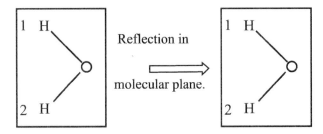

Fig. 2.8 Water molecule reflection in the molecular plane

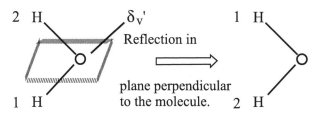

Fig. 2.9 Water molecule reflection perpendicular to the molecular axis

These mirror planes are given the subscript label "v" to indicate that they are "vertical" mirror planes. To understand this notation, consider the C_2 axis of the water molecule. If a molecule is bisected by a plane and each atom in one half of the bisected molecule is reflected through the plane and encounters a similar atom in the other half, the molecule has a plane of symmetry. Both the element and the operator are designated by sigma. Every planar molecule has at least one plane of symmetry, the molecular plane. BF_3 has, in addition, three vertical planes of symmetry, each containing one B−F bond and bisecting the angle between the other two B−F bonds. Reflection in a plane always results in a change of sign of the coordinates perpendicular to this plane. Coordinates parallel to this plane are unchanged. Thus, σ_{xy} changes (x,y,z) to $(x,y,-z)$, σ_{yz} changes (x,y,z) into $(-x,y,z)$, and σ_{xz} changes (x,y,z) to $(x,-y,z)$. If the plane is normal to the principal axis of symmetry, then the plane of symmetry is horizontal (σ_h). It is σ_v if it contains the principal rotation axis and is a vertical plane. It is considered as σ_d if is a dihedral plane (containing the principal axis and bisecting a pair of C_2 axes).

An atom A at (x_1, y_1, z_1) on the reflection in xz plane changes to $B(x_2, y_2, z_2)$. The reflection on the xz plane does not change the x and z coordinates, but changes the sign of the y. Thus:

$$x_2 = x_1 + 0y_1 + 0z_1 \tag{2.15}$$
$$y_2 = 0x_1 - y_1 + 0z_1 \tag{2.16}$$
$$z_2 = 0x_1 + 0y_1 + z_1 \tag{2.17}$$

The matrix representation for the symmetry operation is:

$$\begin{bmatrix} x_2 \\ y_2 \\ z_2 \end{bmatrix} = \begin{bmatrix} 1 & 0 & 0 \\ 0 & -1 & 0 \\ 0 & 0 & 1 \end{bmatrix} \begin{bmatrix} x_1 \\ y_1 \\ z_1 \end{bmatrix} \tag{2.18}$$

The transformation matrix for the symmetry operation σ_{xz} is $\begin{bmatrix} 1 & 0 & 0 \\ 0 & -1 & 0 \\ 0 & 0 & 1 \end{bmatrix}$. Similarly,

the transformation matrix for the symmetry operation σ_{xy} is $\begin{bmatrix} 1 & 0 & 0 \\ 0 & 1 & 0 \\ 0 & 0 & -1 \end{bmatrix}$ and the transformation matrix for the symmetry operation σ_{yz} is $\begin{bmatrix} -1 & 0 & 0 \\ 0 & 1 & 0 \\ 0 & 0 & 1 \end{bmatrix}$. In BF$_3$ both

mirror planes are coming *vertically* out of the plane of the paper. The molecular plane of the BF$_3$ molecule is a "horizontal" mirror plane, labeled as σ_h (Fig. 2.10).

Again, the labeling can be understood by viewing the molecule through a rotation axis as shown in Fig. 2.11. BF$_3$ possesses a C_3 and 2 C_2 axes where the C_3 axis is the principal axis. The labeling refers to the relationship between the plane and the principal axis. The mirror plane lies horizontally, in the plane of the paper. *Note that the "v" and h labeling refers to the relationship between the planes and the principal rotation axis, not to the plane of the molecule.* The mirror plane, dihedral, or σ_d planes bisect two C_2 axes. The principal rotation axis of the benzene molecule is a C_6 axis running perpendicular to the molecule (Fig. 2.12). It also possesses 3 C_2 axes running through opposite carbon atoms. Benzene possesses three types of mirror plane, a plane perpendicular to the principal (C_6) axis, σ_h plane. The other two types of mirror planes both lie vertically with respect to the C_6 axis. However, the one on the right cuts between two C_2 axes and is called a dihedral plane, σ_d. *It is to be noted that molecules keeping at least one mirror plane are not "chiral".*

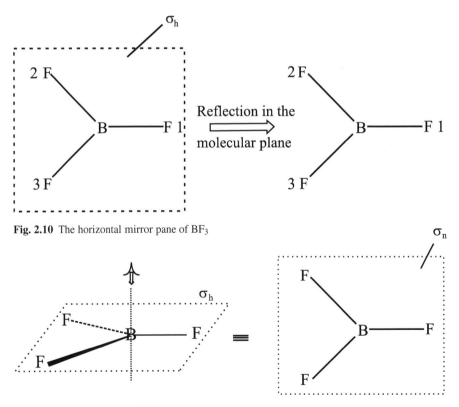

Fig. 2.10 The horizontal mirror pane of BF$_3$

Fig. 2.11 Illustration of σ_h of BF$_3$

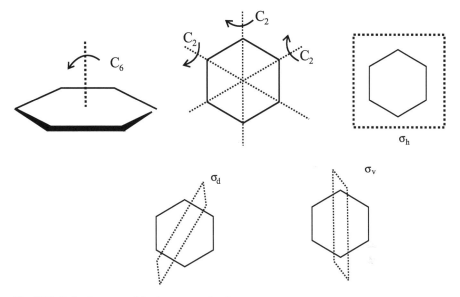

Fig. 2.12 Reflection axes of the benzene molecule

2.3.4 Inversion Operation

In this operation every atom is moved in a straight line to the center of the molecule and then moved out (extrapolated to) the same distance on the other side. If symmetry is observed by this operation, then the molecule is said to be keeping the center of inversion. This symmetry operation is called inversion and is denoted by $\hat{\imath}$. The inversion operation can be considered as a two-fold rotation, followed by the reflection in the horizontal plane. Or:

$$\hat{\imath} = \hat{C}_n \hat{\sigma}_h \tag{2.19}$$

An octahedral molecule is unchanged by inversion through the center of the molecule, as shown in the hypothetical molecule of the type (MF_6) as illustrated in Fig. 2.13. An example of such a molecule is sulphur hexafluoride. The center of the molecule is called the *center of inversion*.

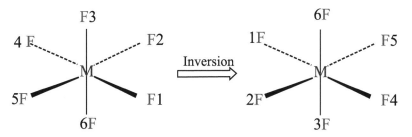

Fig. 2.13 Inversion operation on octahedral molecules

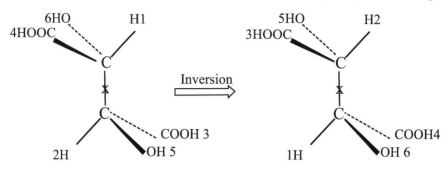

Fig. 2.14 Center of inversion in meso-tartaric acid

The center of inversion needs not coincide with an atom as in the meso-tartaric acid molecule (Fig. 2.14). *Molecules possessing a center of inversion are not chiral.* Both of the carbon atoms in meso-tartaric acid are bonded to four different groups (asymmetric); still, the molecule is not chiral as it is keeping a center of inversion.

2.3.5 Improper Rotations

Improper rotations consist of two separate operations, an n-fold rotation (rotation by $360°/n$) about an axis followed by reflection in a plane perpendicular to that axis. The symbol for an improper rotation is S_n. The improper rotation operation can be considered as:

$$\hat{S}_n = \hat{C}_n \hat{\sigma}_h \qquad (2.20)$$

Improper axes are often the most difficult symmetry elements to locatem as, for example, methane possesses an S_4 axis, though it is not keeping any C_4 axis. In methane, rotation by $90°$ followed by the reflection in a perpendicular plane restores the structure as is shown in Fig. 2.15.

2.4 Consequences for Chirality

A chiral molecule is one which cannot be superimposed on its mirror image. A generalization for chirality can be deduced from the symmetry elements of a molecule. *A chiral molecule should not possess an S_n axis.* It should also not possess any reflection plane. But, the reflection plane is the same as S_1 improper rotation, i.e., rotation by $360/1 = 360°$ followed by a reflection. Similarly, chiral molecules should not possess a center of inversion. In fact, an inversion is the same as S_2 improper rotation.

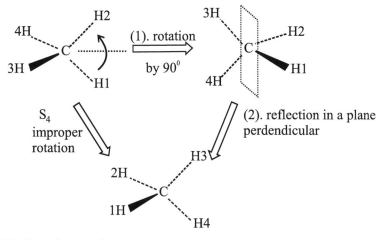

Fig. 2.15 S_4-axis in the methane molecule

2.5 Point Groups

The symmetry of a molecule can be completely specified by listing all the symmetry elements (E, C_n, σ, i and S_n) it possesses. Every element is characterized by a set of symmetry elements. If chemically different molecules possess precisely the same set of symmetry elements, they are symmetrically related and must be classified together. Thus, phenanthrene and water go together. E, C_2, σ_{xz}, and σ_{yz} together form a mathematical group. Since each of the operations leaves at least one point (the center of mass) unchanged, they are said to constitute a symmetry point group. All the known molecules can be classified into 32 symmetry point groups which are given Schoenflies symbols that convey essential information about the symmetry of the molecule. Types of point groups with their characteristics and suitable examples are given in Tables 2.2, 2.3, and 2.4.

Table 2.2 General types of point groups

Sl. no.	Point group	Characteristic symmetry elements
1	C_s	E and only one σ
2	C_i	E and a center of inversion
3	C_n	E and one C_n axis
4	C_{nv}	E one C_n axis and n σ_v
5	C_{2h}	E one C_n axis and one σ_h plane
6	D_{nh}	E one C_n axis, nC_2 axes and $n\sigma_v$ planes
7	D_{nd}	E one C_n axis nC_2 axes and $n\sigma_v$ planes

Table 2.3 Special types of point groups

Sl. no.	Point group	Characteristic symmetry elements
1	$D_{\infty v}$	Linear molecules with center of inversion
2	$C_{\infty v}$	Linear molecules without center of inversion
3	Td	Tetrahedron
4	Oh	Octahedron
5	Ih	Icosahedron

Table 2.4 Point group examples

Point group	Shape	Molecule
Oh	Octahedral	SF_6, $Co(NH_3)_6^{3+}$
Td	Tetrahedral	CH_4, $Ni(CO)_4$
D_{6h}	Hexagonal	Benzene
D_{4h}	Square planar Oh	$Ni(CN)_4^{2-}$, $PtCl_4^{2-}$ Sp-, trans-$Co(NH_3)_4Cl_2^+ - Oh$
D_{3h}	Trigonal planar	BF_3, CO_3^{2-} NO_3^-
D_{2h}	Square planar	Trans-$Pt(NH_3)_2Cl_2$
C_{4v}	Distorted octahedral	SF_5Cl
C_{3v}	Pyramidal or distorted Td	$NH_3 -$ py, $CHCl_3$, $POCl_3$-Td
C_{2v}	Oh, v-shaped, square planar or Td	Cis-$Pt(NH_3)_2Cl_2^{2+} - Oh$, $H_2O - V-$, Cis-$Pt(NH_3)_2Cl_2 -$ sp$-$, $Co(py)_2Cl_2$-Td

2.6 The Procedure for Determining the Point Group of Molecules

The general procedure for finding the point group of any molecule is given as follows:

1. In the first step, identify all the symmetry elements of the molecule.
2. Look for the highest rotation axis.
3. If the axis is C_∞ (the molecule is linear) look for the presence of a center of symmetry i. If the molecule has i, then it will be definitely keeping σ_h and it belongs to $D_{\infty h}$. If it does not have i, it belongs to $C_{\infty v}$.
4. If the highest axis is C_3, C_4, or C_5, check for other axes of the same order. (a) Six five fold axes: If it has 15 planes it belongs to Ih, otherwise to I. (b) Three 4-fold axes: If the molecule also has 9 planes, it belongs to Oh, otherwise to O. (c) Four 3-fold axes: if the molecule has neither i, nor any planes of symmetry, it belongs to T. Planes but no $i - Td$: center i, then Th.
5. If the molecule has only one C_n axis with $n > 2$, or if the highest axis is C_2, look for n two-fold axes perpendicular to the principal axis. If there are any, look for planes of symmetry. (a) No plane of symmetry: D_n (b) n-vertical planes but no horizontal plane: D_{nd}(c) n vertical planes and a horizontal plane: D_{nh} (d) Principal axis C_2 and there are two C_2 axes perpendicular to it and if the molecule has i: D_{2d}.

6. Has n-fold axis, C_n: look for S_{2n} (a) S_{2n} exists: point group is S_{2n} (b) No S_{2n}: look for planes, no planes: C_n; n-vertical planes but no horizontal plane: C_{nv} (c) A horizontal plane but no vertical planes: C_{nh}.
7. If there are no axes (other than C_1), look for a plane and a center. (a) One plane: C_s (b) Center i: Ci (c) Neither i nor planes: C_1.

A flow chart for finding the point group of molecules is included in Fig. 2.16. We shall illustrate the procedure with the help of a few examples.

1. *Water* (H_2O)

   a. It does not belong to any special group. Hence, the point group is not Oh, Ih, and Td.
   b. It is non-linear. Hence, the absence of $C_{\infty v}$ and $D_{\infty h}$.
   c. The principal axis is C_2 and has no S_4 axis. No C_2 axis perpendicular to the principal axis. Hence, D and S groups are ruled out.

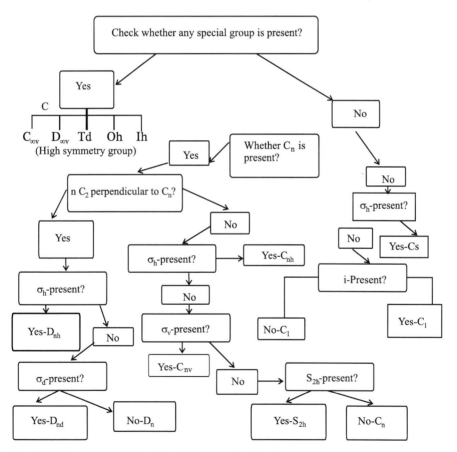

Fig. 2.16 Flow chart for finding the point groups of molecules

   d. The molecule belongs to C_2, C_{2h}, or C_{2v}.
   e. There is a horizontal plane of symmetry (σ_h) and a vertical plane of symmetry (σ_v).
   f. Hence, the point group of water is C_{2v}.

2. *Ammonia* (NH_3)

   a. It does not belong to a special group.
   b. There is a C_3 axis.
   c. There are no other C_2 axes.
   d. There is no σ_h plane.
   e. There is a σ_v plane.
   f. Hence, the point group is C_{3v}.

3. *Boron trifluoride* (BF_3)

   a. It does not belong to a special group.
   b. There is a C_3 axis.
   c. There are 3 C_2 axes perpendicular to C_3.
   d. There is a σ_h plane.
   e. Hence, the point group is D_{3h}.

4. *Trans-dichloro ethene* ($C_2H_2Cl_2$)

   a. It does not belong to a special group.
   b. There is a C_2 axis.
   c. There are no other C_2 axes.
   d. There is a σ_h plane.
   e. Hence, the point group is C_{2h}.

2.7 Typical Molecular Models

Some typical molecular geometries and their point group are depicted below.

1. *Tetrahedral* (Td): Methane, elemental phosphorous, and B_4Cl_4 are examples, as given in Figs. 2.17 and 2.18.

Fig. 2.17 Tetrahedral (Td) structure

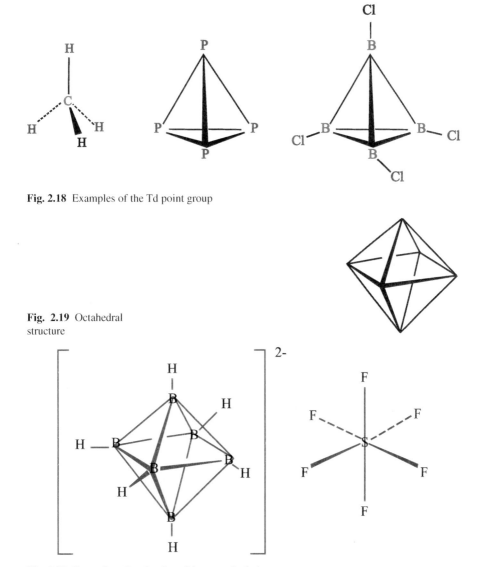

Fig. 2.18 Examples of the Td point group

Fig. 2.19 Octahedral structure

Fig. 2.20 Examples of molecules with an octahedral structure

2. *Octahedral (Oh)*: A diagrammatic representation of the octahedral structure is shown in Fig. 2.19. The structures of molecules B_6H_6 and SF_6 are included in Fig. 2.20.

3. *Icosahedron (Ih)*: The molecular structure is included in Fig. 2.21. $[B_{12}H_{12}]^{2-}$ and elemental boron are examples of this point group.

Fig. 2.21 Icosahedral
structure

2.8 Group Representation of Symmetry Operations

Group theory is the mathematical study of symmetry, as embodied in the structures known as groups [1]. These are sets with a closed binary operation satisfying the following three properties:

1. The operation must be associative.
2. There must be an identity element.
3. Every element must have a corresponding inverse element.

Let us consider the symmetry operation of the C_{2v} point group. We have already seen the transformation matrices for identity, rotation, and reflection operators. These matrices obey the group multiplication table and are representations of the group.

The water molecule, for example, possesses four elements of symmetry, E, $C_2(z)$, $\sigma_v(xz)$, and $\sigma_v(yz)$.

It can be proven that the product of any two operations gives rise to one of the operations in the group. This is illustrated in the multiplication table for the symmetry operation of water ($C_{2v\,point\,group}$). The matrix representation for the symmetry operation of the C_{2v} point group obeying the group multiplication table is called the representation of the group (Table 2.5). For example, we have seen that $\sigma_{xz}C_2 = \sigma_{yz}$. Using matrices:

$$\begin{bmatrix} 1 & 0 & 0 \\ 0 & -1 & 0 \\ 0 & 0 & 1 \end{bmatrix} \begin{bmatrix} -1 & 0 & 0 \\ 0 & -1 & 0 \\ 0 & 0 & 0 \end{bmatrix} = \begin{bmatrix} -1 & 1 & 0 \\ 0 & 1 & 0 \\ 0 & 0 & 1 \end{bmatrix} \qquad (2.21)$$

Table 2.5 Multiplication table for the symmetry operation of water

C_{2v}	E	$C_2(z)$	$\sigma_v(xz)$	$\sigma_v(yz)$
E	E	$C_2(z)$	$\sigma_v(xz)$	$\sigma_v(yz)$
$C_2(z)$	$C_2(z)$	E	$\sigma_v(yz)$	$\sigma_v(xz)$
$\sigma_v(xz)$	$\sigma_v(xz)$	$\sigma_v(yz)$	E	$C_2(z)$
$\sigma_v(yz)$	$\sigma_v(yz)$	$\sigma_v(xz)$	$C_2(z)$	E

2.9 Irreducible Representations

We have seen that for water in the equilibrium geometry, four symmetry operations are possible, $\hat{E}$, $\hat{C}_2$, $\hat{\sigma}_{(x,z)}$, and $\hat{\sigma}_{(y,z)}$. By Mulliken's convention (the standard convention), the molecular plane is assigned the (y,z) plane. As the symmetry elements constitute a group, the corresponding operations follow commutative law.

Hence, the electronic wavefunctions can be considered as simultaneous eigenfunctions of all four symmetry operators. Since $\hat{E}$ is a unit operator, $\hat{E}\psi_{(electron)} = \psi_{(electron)}$. For the remaining symmetry operators, $\hat{O}^2 = 1$ providing two eigenvalues, ± 1. Hence, each electronic wavefunction of water is an eigenfunction of $\hat{E}$ with eigenvalue $+1$ and an eigenfunction of the remaining three symmetry operators ($\hat{C}_2$, $\hat{\sigma}_{(x,z)}$, $\hat{\sigma}_{(y,z)}$) with eigenvalues ± 1. We may propose eight possible sets of eigenvalues as given in Table 2.6.

All these eight eigenvalues are not possible for the water molecule. Symmetry operators multiply in the same manner as symmetry operations, as is shown in Table 2.6. From this table we can rule out some of the eigenvalues. We know that $\hat{C}_2 \times \hat{\sigma}_{(x,z)} = \hat{\sigma}_{(y,z)}$. Hence, eigenvalues not satisfying this equation are not possible for water, which limits the possible eigenvalues to be four, as given in Table 2.7.

Symmetry eigenvalues for the higher order (with positive $\hat{C}_n$ or $\hat{S}_n$) are designated as A and the lower one is designated as B. Each possible set of eigenvalues is called an irreducible set or symmetry species set. The species with all the eigenvalues positive is called the totally symmetric species (here, A_1).

Table 2.6 Eigenvalues corresponding to the symmetry operations

$\hat{E}$	$\hat{C}_2$	$\hat{\sigma}_{(x,z)}$	$\hat{\sigma}_{(y,z)}$
1	1	1	1
1	1	1	-1
1	1	-1	1
1	1	-1	-1
1	-1	1	1
1	-1	1	-1
1	-1	-1	1
1	-1	-1	-1

Table 2.7 Irreducible representation of C_{2v} of water

	$\hat{E}$	$\hat{C}_2$	$\hat{\sigma}_{(x,z)}$	$\hat{\sigma}_{(y,z)}$
A_1	1	1	1	1
A_2	1	1	-1	-1
B_1	1	-1	1	-1
B_2	1	-1	-1	1

Table 2.8 Designation of orbitals on the basis of degeneracy

Degeneracy	1	2	3	4	5
Designation	A,B	E	T	G	H

2.10 Labeling of Electronic Terms

Along with the symmetry labeling of orbitals, spin multiplicity $(2S+1)$ is also included, where S is the electronic spin. For example, the electronic state of water with one electron unpaired and with the electronic wavefunction unchanged by the symmetry operators can be designated as $^2A_1$. Based on the orbital degeneracy the following labeling is shown (Table 2.8):

For molecules with a center of symmetry and having an eigenvalue of $+1$, then the subscript g is added, while if the eigenvalue is -1, then the subscript u is included. For example, the possible symmetry elements of a D_{6h} molecule are A_{1g}, A_{2g}, B_{1g}, B_{2g}, E_{1g}, E_{2g}, A_{1u}, A_{2u}, B_{1u}, B_{2u}, E_{1u}, and E_{2u}.

2.11 Exercises

2.11.1 Questions

1. Determine the point group for the following molecules:
   NCl_3, CCl_4, $CH_2 = CH_2$, $CF_2 = CH_2$, SO_3, PCl_5, SnF_4, SeF_4 and PCl_3.
2. Find the point groups of the following species:
   SO_4^{2-}, SiF_6^{2-}, and BrF_4^-.
3. Identify the symmetry elements and find the point group of the following:
   NH_2Cl, CO_3^{2-}, SiF_4, HCN, $SiFClBrI$, and BF_4^-.
4. Write the irreducible representation of the C_{3v} point group.
5. Identify the point groups of molecules producing polar molecules.
6. Identify the point groups of molecules producing optically active molecules.
7. List the symmetry operations possible for a. NH_3 b. $HOCl$ c. CH_2F_2.
8. Find the eigenvalues of $\hat{O}_{C_4}$.
9. Find the order (number of elements of symmetry in a group) of
   a. C_{3v} b. D_{3h} c. C_s.

2.11.2 Answers to Selected Questions

1. $NCl_3 - C_{3v}$, $CCl_4 - Td$, $CH_2 = CH_2 - D_{2h}$, $CF_2 = CH_2 - C_{2v}$, $SO_3 - D_{3h}$, $PCl_5 - D_{3h}$, $SnF_4 - Td$, $SeF_4 - C_{2v}$ and $PCl_3 - C_{3v}$.
2. $SO_4^{2-} - Td$, $SiF_6^{2-} - Oh$, $BrF_4^- - D_{4h}$

3. $NH_2Cl - E, \sigma : C_s$,
   $CO_3^{2-} - E, C_3, C_2, \sigma_h, \sigma_v, S_3 : D_{3h}$,
   $SiF_4 - E, C_3, C_2, \sigma_d, S_4 : Td$,
   $HCN - E, C_\infty, C_2, \sigma_v : C_{\infty v}$,
   $SiFClBrI - E : C_1$
   and $BF_4^- - E, C_3, C_2, \sigma_d, S_4 : Td$.

References

1. Cotton FA (1990) Chemical Applications of Group Theory, 3rd ed. Wiley, New York
2. Kettle SFA (1995) Structure and Symmetry (Readable Group Theory for Chemists), 2nd ed. John Wiley and Sons, Chichester

Chapter 3
Quantum Mechanics: A Brief Introduction

*I think it is safe to say that no one understands quantum
mechanics. Do not keep saying to yourself, if you can possibly
avoid it, "But how can it be like that?" because you will get
"down the drain" into a blind alley from which nobody has yet
escaped. Nobody knows how it can be like that.*

– Richard Feynman (1918–1988)

3.1 Introduction

The development of quantum mechanics was initially provoked by two main observations that established the inadequacy of classical physics. They are called the
ultraviolet catastrophe and the *photoelectric effect*.

3.1.1 The Ultraviolet Catastrophe

A blackbody is a unique object which absorbs and emits all frequencies of electromagnetic radiations incident on it. Classical physics can be used to derive an equation which describes the intensity of blackbody radiation as a function of frequency
for different temperatures. This generalization is known as the Rayleigh-Jeans law.
Let us look at the spectrum in detail. When an iron block is heated, the color of the
metal is gray at a low temperature, bright red at about 1270 K and dazzling white
at 1770 K. This feature is described in Fig. 3.1. Although the Rayleigh-Jeans law
works for low frequencies, it diverges at higher ones. This divergence at higher frequencies is called the ultraviolet catastrophe.

Max Planck [1] gave an explanation to the blackbody spectrum in the year 1900
by assuming that the energies of the oscillations of electrons which gave rise to the
radiation must be proportional to integral multiples of the frequencies. Using statistical mechanics, Planck derived an equation similar to the Rayleigh-Jeans equation,
but with the adjustable parameter h. Planck found that for $h = 6.626 \times 10^{-34}$ Js
(Planck's constant), the experimental data could be reproduced to its finest detail.
This famous revolutionary relation is given by Eq. 3.1.

$$E = nh\vartheta \qquad (3.1)$$

K. I. Ramachandran et al., *Computational Chemistry and Molecular Modeling*
DOI: 10.1007/978-3-540-77304-7, ©Springer 2008

Fig. 3.1 Intensity of radiation of heated iron against frequency. The values corresponding to the Rayleigh-Jeans relationship are represented by a *dashed curve*. It fits well to experimental data at low frequencies, but becomes departing at higher frequencies

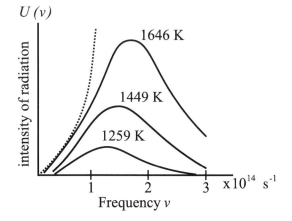

Where n is a positive integer, ϑ is the frequency of the oscillator, and E is the energy. But Planck could not offer a good justification for his assumption of energy quantization. Scientists did not take this energy quantization idea seriously until Einstein invoked a similar assumption to explain the photoelectric effect.

3.1.2 The Photoelectric Effect

Heinrich Hertz in 1887 discovered that irradiation by ultraviolet light would cause electrons to be ejected from a metal surface. According to the classical wave theory of light, the intensity of the light determines the amplitude of the wave, and so a greater intensity of light should cause the electrons on the metal to oscillate more violently and to be ejected with a greater kinetic energy. In contrast, the experiment showed that the kinetic energy of the ejected electrons depended only on the frequency of the light. On the other hand, the intensity of light affects only the number of ejected electrons and not their kinetic energies.

Einstein explained the problem of the photoelectric effect in 1905. Instead of assuming that the electronic oscillators had energies given by Planck's equation (Eq. 3.1), Einstein assumed that the radiation itself consisted of packets of energy E, which are now called photons. Einstein successfully explained the photoelectric effect by using this assumption, and he calculated a value of h close to that obtained by Planck.

Two years later, Einstein showed that, like light, atomic vibrations were also quantized. Classical physics predicts that the molar heat capacity at a constant volume (C_v) of a crystal is $3R$, where R is the molar gas constant. This works well for high temperatures, but for low temperatures C_v actually falls to zero. Einstein was able to explain this result by assuming that the oscillations of atoms about their equilibrium positions are quantized according to Eq. 3.1 – Planck's quantization condition for electronic oscillators. This confirmed that the energy quantization concept was important even for a system of atoms in a crystal, which could be well-modeled by a system of masses and springs (i.e., by classical mechanics).

3.1.3 The Quantization of the Electronic Angular Momentum

Rutherford proposed a classical atomic structure in which the electrons are considered as revolving round the nucleus of atom. One problem with this model is that orbiting electrons experience a centripetal acceleration. Such accelerating charges should lose energy by radiation making stable electronic orbits classically forbidden. Bohr proposed stable electronic orbits with the electronic angular momentum quantized as:

$$l = mvr = n\hbar \qquad (3.2)$$

where m is the mass of the electron, v its velocity, and r the radius of the orbit, $\hbar = h/2\pi$, $n = 1,2,3\dots$ The quantization of angular momentum leads to discretization of radius as well as the energy of the orbit. Bohr's atom model could explain the atomic spectrum of the hydrogen atom. Bohr assumed that the discrete lines seen in the spectrum of the hydrogen atom were due to transitions of electrons from one allowed orbit/energy level to another. He further assumed that the energy of a transition is acquired or released in the form of a photon as proposed by Einstein, such that:

$$\Delta E = h\vartheta \qquad (3.3)$$

This is known as the *Bohr frequency condition*. This condition, along with Bohr's expression for the allowed energy levels, gives a good match to the observed hydrogen atom spectrum. However, it works only for atoms with one electron. It could not explain the fine spectrum even for the hydrogen atom.

3.1.4 Wave-Particle Duality

Einstein had shown that the momentum of a photon is:

$$p = \frac{h}{\lambda} \qquad (3.4)$$

This can be easily shown as follows. Assuming $E = hv$ for a photon and $\lambda v = c$ for an electromagnetic wave, we obtain:

$$E = \frac{hc}{\lambda} \qquad (3.5)$$

Now we use the result of Einstein's special theory of relativity, $E = mc^2$ to get:

$$\lambda = \frac{h}{mc} \tag{3.6}$$

This is equivalent to Eq. 3.4. Here, m refers to the relativistic mass, not the rest mass. Note that the rest mass of a photon is zero. Light can behave both as a wave (it exhibits properties such as diffraction, interference, and polarization, and it has a wavelength), and as a particle (it contains packets of energy hv). De Broglie established a similar relationship in 1924 for material particles by proposing a dual nature for matter, and particles as well as waves [2]. He proposed an equation for finding the wave length (λ – the de Broglie wave length) included in Eq. 3.7, which is similar to Eq. 3.6. Here, m is mass, and v is the velocity of the particle.

$$\lambda = \frac{h}{mv} \tag{3.7}$$

In 1927, Davisson and Germer observed diffraction patterns by bombarding metals with electrons, confirming de Broglie's proposition.

De Broglie's equation offers a justification for Bohr's assumption (Eq. 3.2). According to Bohr's atom model, only those circular orbits in which the angular momentum of the electron, an integral multiple of $\hbar = \frac{h}{2\pi}$ is permitted.

$$mvr = n\hbar = n\frac{h}{2\pi} \tag{3.8}$$

According to de Broglie, the electrons have a wave character also. For the waves to be completely in phase, the circumference of the orbit should be an integral multiple of wavelength. Therefore:

$$2\pi r = n\lambda \tag{3.9}$$

Where, r is the radius of the orbit. Substituting λ from Eq. 3.7:

$$mvr = n\hbar = n\frac{h}{2\pi} \tag{3.10}$$

This is identical with Bohr's equation (Eq. 3.3).

Heisenberg showed that the wave-particle duality leads to the famous uncertainty principle:

$$\Delta x \times \Delta p \geq \frac{h}{4\pi} \tag{3.11}$$

where Δx is the uncertainty in position and Δp is the uncertainty in momentum. One result of the uncertainty principle is that if the orbital radius of an electron in an atom r is known exactly, then the angular momentum must be uncertain. The problem with Bohr's model is that it specifies r exactly and it also ensures that the orbital angular momentum must be an integral multiple of $\hbar = \frac{h}{2\pi}$. Thus, the stage was set for a new quantum theory, which was consistent with the uncertainty principle. The first principle in quantum theory stands for Schrödinger equation. Modeling molecules from the first principle is generally referred to as *ab initio modeling* [3].

3.2 The Schrödinger Equation

In 1925, Erwin Schrödinger and Werner Heisenberg independently developed the new quantum theory. Schrödinger method involves partial differential equations, whereas Heisenberg's method employs matrices; however, a year later the two methods were shown to be mathematically equivalent. Schrödinger equation seems to have a better physical interpretation via the classical wave equation. Indeed, the Schrödinger equation can be viewed as a form of the wave equation applied to matter waves.

3.2.1 The Time-Independent Schrödinger Equation

We start with the one-dimensional classical wave equation:

$$\frac{\partial^2 u}{\partial x^2} = \frac{1}{v^2}\frac{\partial^2 u}{\partial t^2} \tag{3.12}$$

where v is velocity.

By introducing the separation of variables:

$$u(x,t) = \psi(x)f(t) \tag{3.13}$$

we obtain:

$$f(t)\frac{d^2\psi(x)}{dx^2} = \frac{1}{v^2}\psi(x)\frac{d^2 f(t)}{dt^2} \tag{3.14}$$

If we introduce one of the standard wave equation solutions for $f(t)$ such as $e^{i\omega t}$ (the constant can be taken care of later in the normalization), we obtain:

$$\frac{d^2\psi(x)}{dx^2} = \frac{-\omega^2}{v^2}\psi(x) \tag{3.15}$$

Now we have an ordinary differential equation describing the spatial amplitude of the matter wave as a function of position. The energy of a particle is the sum of kinetic and potential parts:

$$E = \frac{p^2}{2m} + V(x)$$

(3.16)

which can be solved for the momentum, p, to obtain:

$$p = \{2m[E - V(x)]\}^{1/2}$$

(3.17)

Now we can use the de Broglie formula (Eq. 3.4) to get an expression for the wavelength:

$$\lambda = \frac{h}{p} = \frac{h}{\{2m[E - V(x)]\}^{1/2}}$$

(3.18)

The term ω^2/v^2 in Eq. 3.15 can be rewritten in terms of λ if we recall that $\omega = 2\pi\vartheta$ and $\vartheta\lambda = v$, where ω is the angular momentum, λ is the wavelength and ϑ is the frequency:

$$\frac{\omega^2}{v^2} = \frac{4\pi^2\vartheta^2}{v^2} = \frac{4\pi^2}{\lambda^2} = \frac{2m[E - V(x)]}{\hbar^2}$$

(3.19)

(where $\hbar = h/2\pi$). When this result is substituted into Eq. 3.15 we obtain the famous *time-independent Schrödinger equation* [4]:

$$\frac{d^2\psi(x)}{dx^2} + \frac{2m}{\hbar^2}[E - V(x)]\,\psi(x) = 0$$

(3.20)

which is almost always written in the form:

$$-\frac{\hbar^2}{2m}\frac{d^2\psi(x)}{dx^2} + V(x)\psi(x) = E\psi(x)$$

(3.21)

This single-particle one-dimensional equation can easily be extended to the case of three dimensions, where it becomes:

$$-\frac{\hbar^2}{2m}\nabla^2\psi(r) + V(r)\psi(r) = E\psi(r)$$

(3.22)

A two-body problem can also be treated by this equation if the mass m is replaced with a reduced mass.

It is important to point out that this analogy with the classical wave equation only goes so far. We cannot, for instance, derive the time-*dependent* Schrödinger equation in an analogous fashion (for instance, that equation involves the partial first derivative with respect to time instead of the partial second derivative). In fact, Schrödinger (see Fig. 3.2) presented his time-independent equation first, and then went back and postulated the more general time-dependent equation.

Fig. 3.2 Erwin Schrödinger
(1887–1961)

A careful analysis of the process of observation in atomic physics has shown that the sub-atomic particles have no meaning as isolated entities, but can only be understood as inter-connections between the preparation of an experiment and the subsequent measurement.

– Erwin Schrödinger

3.2.2 The Time-Dependent Schrödinger Equation

We are now ready to consider the time-dependent Schrödinger equation. Although we were able to derive the single-particle time-independent Schrödinger equation starting from the classical wave equation and the de Broglie relation, the time-dependent Schrödinger equation cannot be derived using elementary methods and is generally given as a postulate of quantum mechanics. The single-particle three-dimensional time-dependent Schrödinger equation is:

$$i\hbar \frac{\partial \psi(r,t)}{\partial t} = -\frac{\hbar^2}{2m}\nabla^2 \psi(r,t) + V(r)\psi(r,t) \tag{3.23}$$

where V is assumed to be a real function and represents the potential energy of the system. *Wave mechanics* is the branch of quantum mechanics with Eq. 3.23 as its dynamical law. Note that Eq. 3.23 does not yet account for spin or relativistic effects.

Of course the time-dependent equation can be used to derive the time-independent equation. If we write the wavefunction as a product of spatial and temporal terms, $\psi(r,t) = \psi(r)f(t)$, then Eq. 3.23 becomes

$$\psi(r)i\hbar \frac{\mathrm{d}f(t)}{\mathrm{d}t} = f(t)\left[-\frac{\hbar^2}{2m}\nabla^2 + V(r)\right]\psi(r) \tag{3.24}$$

$$\text{Or}: \frac{i\hbar}{f(t)}\frac{\mathrm{d}f}{\mathrm{d}t} = \frac{1}{\psi(r)}\left[-\frac{\hbar^2}{2m}\nabla^2 + V(r)\right]\psi(r) \tag{3.25}$$

Since the left-hand side is a function of t only and the right hand side is a function of r only, the two sides must equal a constant. If we tentatively designate this constant E (since the right-hand side clearly must have the dimensions of energy), then we extract two ordinary differential equations, namely:

$$\frac{1}{f(t)}\frac{df(t)}{dt} = -\frac{iE}{\hbar} \tag{3.26}$$

and

$$\left[-\frac{\hbar^2}{2m}\nabla^2\psi(r)+V(r)\psi(r)\right] = E\psi(r) \tag{3.27}$$

$$\left[-\frac{\hbar^2}{2m}\nabla^2 +V(r)\right]\psi(r) = E\psi(r) \tag{3.28}$$

where the term in square bracket on the LHS is called the Hamiltonian operator.

The latter equation is once again the time-independent Schrödinger equation. The former equation is easily solved to yield:

$$f(t) = e^{-iEt/\hbar} \tag{3.29}$$

The Hamiltonian in Eq. 3.27 is a Hermitian operator, and the eigenvalues of a Hermitian operator must be real, so E is real. This means that the solutions $f(t)$ are purely oscillatory, since $f(t)$ never changes in magnitude (recall Euler's formula $e^{\pm i\theta} = \cos\theta \pm i\sin\theta$)

Thus, if:

$$\psi(r,t) = \psi(r)e^{-iEt/\hbar} \tag{3.30}$$

then the total wavefunction $\psi(r,t)$ differs from $\psi(r)$ only by a phase factor of a constant magnitude. There are some interesting consequences of this. Firstly, the quantity $\psi(r,t)^2$ is time independent, as we can easily show:

$$\left|\psi(r,t)^2\right| = \psi^*(r,t)\psi(r,t) = e^{iEt/\hbar}\psi^*(r)e^{-iEt/\hbar}\psi(r) \quad = \psi^*(r)\psi(r) \tag{3.31}$$

Secondly, the expectation value for any time-independent operator is also time-independent, if $\psi(r,t)$ satisfies Eq. 3.30. By the same reasoning applied above,

$$\langle A\rangle = \int \psi^*(r,t)\hat{A}\psi(r,t) = \int \psi^*(r)\hat{A}\psi(r) \tag{3.32}$$

For these reasons, wavefunctions of the type in Eq. 3.30 are called *stationary states*. The state $\psi(r,t)$ is qutstationary, but the particle it describes is not!

Of course, Eq. 3.30 represents a particular solution to Eq. 3.23. The general solution to Eq. 3.23 will be a linear combination of these particular solutions, i.e.:

$$\psi(r,t) = \sum_i c_i e^{-iE_i t/\hbar}\psi_i(r) \tag{3.33}$$

3.3 The Solution to the Schrödinger Equation

Solutions to Schrödinger equation are called wavefunctions. Out of various solutions to the Schrödinger equation, those satisfying the following conditions are listed here: [5]

1. ψ must be continuous. The wavefunction and its derivative must be continuous.
2. ψ must be finite everywhere.
3. It must approach zero at infinite distance.
4. ψ must be single-valued.

Solutions that do not satisfy these properties do not generally correspond to physically realizable circumstances. These permitted solutions to the equation are called eigenfunctions. Each permitted solution corresponds to a definite energy state and is known as orbital. The electron orbitals in atoms are called atomic orbitals, while those in a molecule are called molecular orbitals.

A typical quantum mechanical problem consists of the following steps:

1. Writing the Schrödinger equation for the system under study.
2. Solving the equation and finding the eigenvalues corresponding to the equation.
3. Characterizing the system based on the solutions.

Please refer to the Appendix to learn more about operators.

3.4 Exercises

3.4.1 Question 1

What should be the range values of the work function of a metal in order to be useful in a photo cell for detecting visible light?

3.4.2 Answer 1

A wave length (λ) of visible light is 4000–7000 Å. Here:

$$h = 6.626 \times 10^{-34} \text{J S}$$

$$c = 3 \times 10^{8} \text{m s}^{-1}$$

$$\text{and} \quad 1 \text{ Joule} = \frac{1}{1.602 \times 10^{-19}} \text{eV}$$

Energy corresponding to 4000 Å:

$$= \frac{hc}{\lambda} = \frac{6.626 \times 10^{-34} \times 3 \times 10^8}{4000 \times 10^{-10} \times 1.602 \times 10^{-19}} = 3.102 \, \text{eV}$$

Similarly energy corresponding to 7000 Å

$$= \frac{6.626 \times 10^{-34} \times 3 \times 10^8}{7000 \times 10^{-10} \times 1.602 \times 10^{-19}} = 1.77 \, \text{eV}$$

Therefore, any metal with work function between 1.77 eV and 3.10 eV are the probable candidates for detecting visible light.

3.4.3 Question 2

Calculate the potential difference that must be applied to stop the fastest photo electrons emitted by a surface when irradiated by an electromagnetic radiation of frequency 1.5×10^{15} Hz. (The work function is 4 eV.)

3.4.4 Answer 2

$$\text{Energy of photon} = h\nu = (6.626 \times 10^{-34}) \times (1.5 \times 10^{15}) \text{J}$$
$$= \frac{(6.626 \times 10^{-34}) \times (1.5 \times 10^{15})}{(1.602 \times 10^{-19})} \text{eV} = 6.204 \, \text{eV}$$

Therefore, the energy for the fastest photo electron is $6.204 - 4 = 2.204$ eV. Or, the potential difference to be applied is 2.204 volts.

3.4.5 Question 3

An electron is accelerated through a potential difference of 400 V. Determine its de Broglie wave length.

3.4.6 Answer 3

Kinetic energy gained by the electron (non-relativistic), $T = \dfrac{p^2}{2m} = 400 \, \text{eV}$

$$\therefore P = \sqrt{2mT}$$
$$\text{Mass of the electron} = 9.11 \times 10^{-31} \, \text{kg}$$
$$\text{Charge of the electron} = 1.602 \times 10^{-19} \, \text{Coulombs}$$

Hence, the linear momentum,

$$p = \left[\left(400 \times 1.602 \times 10^{-19}\,\mathrm{J} \right) \times \left(2 \times 9.11 \times 10^{-31}\,\mathrm{kg.} \right) \right]^{1/2}$$
$$= 10.798 \times 10^{-24}\,\mathrm{kg.ms}^{-1}$$

de Broglie wave length $\quad \lambda = \dfrac{h}{p} = \dfrac{6.626 \times 10^{-34}}{10.798 \times 10^{-24}} = 0.6132 \times 10^{-10}\,\mathrm{m}$

$$= 0.6132\,\text{Å}$$

3.4.7 Question 4

The energy of certain X-rays is found to be equal to that of a 1 KeV electron. Compare their wave lengths.

3.4.8 Answer 4

The Kinetic energy is:

$$T = \frac{p^2}{2m} = 1000\,\mathrm{eV} = 1.602 \times 10^{-19} \times 10^3\,\mathrm{J} = 1.602 \times 10^{-16}\,\mathrm{J}$$

According to de Broglie, the wave length of an electron is:

$$\lambda = \frac{h}{p} = \frac{h}{\sqrt{2mT}} = \frac{6.626 \times 10^{-34}\,\mathrm{J.s}}{[2(9.11 \times 10^{-31}\,\mathrm{kg}) \times (1.602 \times 10^{-16}\,\mathrm{J})]^{1/2}}$$
$$= 0.39 \times 10^{-10}\,\mathrm{m} = 0.39\,\text{Å}$$

Energy of X-rays: $\quad E = h\nu = \dfrac{hc}{\lambda}$

$$\text{Or}: \quad \lambda = \frac{hc}{E} = \frac{(6.626 \times 10^{-34}\,\mathrm{J.s})(3 \times 10^8\,\mathrm{m.s}^{-1})}{1.602 \times 10^{-16}\,\mathrm{J}} = 12.408\,\text{Å}$$

Hence, $\quad \dfrac{\text{wave length of X-rays}}{\text{de Broglie wave length of electron}} = \dfrac{12.408}{0.39} = 31.85$

The wave length of the X-rays is 31.85 times the de Broglie wavelength of the electron.

3.4.9 Question 5

The speed of an electron is found to be $1\,km.s^{-1}$ within an accuracy of 0.02%. Calculate the uncertainty in its position.

3.4.10 Answer 5

The momentum of the electron: $p = mv = (9.11 \times 10^{-31}\,kg)(1000\,m\,s^{-1})$

$$\% \text{ accuracy} = \frac{\Delta P \times 100}{P} = 0.02\%$$

$$\Delta P = \frac{0.02 \times 9.11 \times 10^{-31} \times 1000}{100} = 1.822 \times 10^{-31}\,kg.m\,s^{-1}$$

$$\Delta x; \frac{h}{4\pi\Delta P} = \frac{6.626 \times 10^{-34}\,J.s}{4\pi \times 1.822 \times 10^{-31}\,kg.m\,s^{-1}} = 2.894 \times 10^{-4}\,m$$

3.4.11 Question 6

In a hydrogen atom, the electron in the $n = 2$ excited state remains there for 10^{-8} seconds on an average before making a transition to the ground state ($n = 1$). (a) Calculate the uncertainty in energy of the excited state. (b) What is the fraction of the transition energy? (c) Compute the width of wave length corresponding to this.

3.4.12 Answer 6

a)

$$\Delta E \times \Delta t \geq h$$

$$\Delta E \geq \frac{h}{\Delta t} = \frac{6.626 \times 10^{-34}\,J.s}{10^{-8}\,s} = 6.626 \times 10^{-26}\,J$$

$$Or \quad \frac{6.626 \times 10^{-26}}{1.602 \times 10^{-19}} eV = 4.14 \times 10^{-7}\,eV$$

b)

$$\text{Energy of } n = 2 \rightarrow n = 1 \text{ transition is } -13.6\,eV\left(\frac{1}{2^2} - \frac{1}{1^2}\right) = 10.2\,eV$$

$$\text{Fraction of energy} = \frac{\Delta E}{E} = \frac{4.14 \times 10^{-7}}{10.2} = 4.06 \times 10^{-8}$$

c)

$$\lambda = \frac{hc}{E} = \frac{\left(6.626 \times 10^{-34}\,\text{J.s}\right)\left(3 \times 10^{8}\,\text{m.s}^{-1}\right)}{10.2 \times 1.602 \times 10^{-19}\,\text{J}} = 1218\,\text{Å}$$

The spectral line width of this line $= \dfrac{\Delta\lambda}{\lambda} = \dfrac{\Delta v}{v} = \dfrac{\Delta E}{E}$

$$\Delta\lambda = \frac{\Delta E \times \lambda}{E} = 4.06 \times 10^{-8} \times 1.218 \times 10^{-7} = 4.95 \times 10^{-7}\,\text{Å}$$

3.4.13 Question 7

Write down the normalized wavefunction if $\psi(x) = A\exp\left(-kx^2\right)$, where k and A are real constants over the entire domain.

3.4.14 Answer 7

$$\psi(x) = A\exp(-kx^2)$$

For the normalized wavefunction:

$$\int_{-\infty}^{+\infty} (A\psi^*)(A\psi)\,dx = 1$$

$$A^2 \int_{-\infty}^{+\infty} \exp\left(-2kx^2\right)dx = 1$$

But we know that $\displaystyle\int_{-\infty}^{+\infty} \exp\left(-2kx^2\right)dx = \sqrt{\dfrac{\pi}{2k}}$

$$\text{Or} \quad A = \left(\frac{2k}{\pi}\right)^{1/4}$$

$$\psi(x) = \left(\frac{2k}{\pi}\right)^{1/4} \exp(-kx^2)$$

3.4.15 Question 8

Given $\psi(x) = A\sin(kx)$. Find the eigenvalues of the operator $\hat{O} = \dfrac{\partial^2}{\partial x^2}$. Find out whether $\hat{O} = \dfrac{\partial}{\partial x}$ is an eigenoperator.

3.4.16 Answer 8

$$\frac{\partial \psi(x)}{\partial x} = \frac{\partial [A\sin(kx)]}{\partial x} = Ak\cos(kx).$$ This is not of the form $\hat{k}\psi(x) = k\psi(x)$

$\therefore \dfrac{\partial}{\partial x}$ is not an eigenoperator for the function.

$$\frac{\partial^2 \psi(x)}{\partial x^2} = \frac{\partial^2 [A\sin(kx)]}{\partial x^2} = -k^2 A\sin(kx)$$

$\therefore \dfrac{\partial^2}{\partial x^2}$ is an eigenoperator with an eigenvalue of $(-k^2)$ for the function.

3.4.17 Question 9

Find the voltage with which electrons in an electron microscope have to be accelerated to get a wavelength of 1 Å.

3.4.18 Answer 9

Let V be the voltage to be applied on electrons. Then the kinetic energy gained $= eV$ Joules. $(e = 1.602 \times 10^{-19}\,\text{Coulombs})$. The de Broglie wavelength can be calculated from the relation:

$$\lambda = \frac{h}{p}. \tag{3.34}$$

Now the kinetic energy $\dfrac{p^2}{2m} = eV$. Or:

$$p = \sqrt{2\,meV} \tag{3.35}$$

From Eqs. (3.34) and (3.35), the de Broglie wavelength is:

$$\lambda = \frac{h}{\sqrt{2\,meV}} = 1\,\text{Å}$$

Hence:

$$V = \frac{h^2}{2me\lambda^2} = \frac{\left(6.626 \times 10^{-34}\,\text{J.s}\right)^2}{(1.602 \times 10^{-19}) \times 2 \times (9.11 \times 10^{-31}\,\text{kg.}) \times (1 \times 10^{-10}\,\text{m})^2}$$

$$= 150\,\text{V}$$

3.4.19 Question 10

Calculate the minimum energy of an electron inside a hydrogen atom whose radius is 0.53 Å using the uncertainty principle.

3.4.20 Answer 10

$$\Delta x = 5.3 \times 10^{-11}\,\text{m}$$

$$\therefore \Delta p = \frac{\eta}{2\Delta x} \geq 9.9 \times 10^{-25}\,\text{kg} \cdot \text{m}\,\text{s}^{-1}$$

$$\text{Kinetic energy of electron} \quad = \frac{(\Delta p)^2}{2 \times 2\pi m} = \frac{\left(9.9 \times 10^{-25}\,\text{kg} \cdot \text{m}\,\text{s}^{-1}\right)^2}{2 \times 9.11 \times 10^{-31}\,\text{kg.}}$$

$$= 5.4 \times 10^{-19}\,\text{J}$$

$$= 3.37\,\text{eV}$$

3.5 Exercises

1. Calculate the wavelength of an electron that has been accelerated through a potential of 100 million volts.
2. An electron has a speed of $500\,\text{m}\,\text{s}^{-1}$ with an uncertainty of 0.02%. What is the uncertainty in locating its position?
3. The ionization energy of a hydrogen atom in the ground state is $1312\,\text{kJmol}^{-1}$. Calculate the wavelength of radiation emitted when the electron in the hydrogen atom makes a transition from principal quantum level, $n = 2$ to $n = 1$.

4. Calculate the de Broglie wavelength for an electron traveling at 1 percent of the speed of light.
5. Find the eigenfunctions of the momentum operator assuming that $P'\phi = p\phi$, where p is the momentum.
6. Assume that the Hamiltonian operator is invariant under time reversal. Prove that the wavefunction for a spin less non-degenerate system at any given instant of time can always be real.
7. The Hamiltonian operator for a spin 1 system is given by $H = \alpha S_z^2 + \beta \left(S_x^2 - S_y^2 \right)$. Solve this equation to find the normalized energy states and eigenvalues. Is this Hamiltonian invariant under time reversal? How do the normalized eigenstates transform under time reversal?
8. Using uncertainty principle show that an electron can not be confined to the nucleus of the atom. (The typical radius of a nucleus $= 10^{-15}$ m).

References

1. Kragh H (2000) Max Planck: The Reluctant Revolutionary. Phys Wor Dec
2. Jackson JD (2006) Mathematics for Quantum Mechanics: An Introductory Survey of Operators, Eigenvalues, and Linear Vector Spaces. http://www.amazon.com
3. Moore W (1989) Schrödinger: Life and Thought. Cambridge
4. Eisberg R, Resnick R (1985) Quantum Physics of Atoms, Molecules, Solids, Nuclei, and Particles. Wiley, New York
5. Bohm A (1994) Quantum Mechanics Foundations and Applications, 3rd ed. Springer, New York

Chapter 4
Hückel Molecular Orbital Theory

4.1 Introduction

Quantum mechanical computation is based on solving the Schrödinger equation, $\hat{H}\psi = E\psi$, where $\hat{H}$ is the Hamiltonian energy operator, and ψ is an amplitude function, which is the eigenfunction with E as the eigenvalue. Perhaps the great disappointment of quantum chemistry is that, while the Schrödinger equation is powerful enough to describe almost all properties of systems, it is too complex to solve for all but the simplest of systems. The equation is unique for each system as the Hamiltonian for different systems are different. The Schrödinger equation for only a few systems can be solved accurately like particles in a one-dimensional box, the hydrogen atom, and the hydrogen molecule ion. In such cases the equation of the system is separated into different uncoupled equations involving only one space variable (the dimension). These separated equations are solved and corresponding energies (eigenvalues) are calculated. The total wavefunction of the system is the product of wavefunctions of the separated ones. But in most cases, the exact equation cannot be separated into uncoupled equations. One approach for overcoming the problem is by introducing some approximations that permit us to separate the function into uncoupled space variables. Three major approximations are widely used to separate the Schrödinger equation into a set of smaller equations before carrying out Hückel calculations [2]:

1. The Born-Oppenheimer approximation
2. The independent particle approximation
3. The π-electron separation approximation

4.2 The Born-Oppenheimer Approximation

The Born-Oppenheimer approximation is an efficient approximation resulting in energies close to the actual energy of the system. The masses of the nuclei are much

K. I. Ramachandran et al., *Computational Chemistry and Molecular Modeling*
DOI: 10.1007/978-3-540-77304-7, ©Springer 2008

greater than the electrons, hence the electrons can respond almost instantaneously to any change in the nuclear positions. Thus, to a high-quality approximation, we can consider the electrons as moving in a field of fixed nuclei. This helps us to separate the Schrödinger equation into two parts, one for the nuclei and the other for electrons. Moreover, within this approximation, the nuclear kinetic energy term can be neglected and the nuclear–nuclear repulsion term can be taken as a constant. We retain the inter-nuclear repulsion terms, which can be calculated from the nuclear charges and the inter-nuclear distances. In this approximation, we retain all terms involving electrons, including the potential energy terms due to attractive forces between the nuclei and electrons and those due to repulsive forces among electrons.

For example, the helium atom consists of a nucleus of a charge $+2e$ surrounded by two electrons (Fig. 4.1). Let the nucleus lie at the origin of the Cartesian coordinate system, let the position vectors of the two electrons be r_1 and r_2, respectively, and let the distance between the electrons be r_{12}. Applying the Born-Oppenheimer approximation, the Hamiltonian of the system takes the form of Eq. 4.1.

$$\hat{H} = -\frac{\hbar^2}{2m_e}\left(\nabla_1^2\right) + -\frac{\hbar^2}{2m_e}\left(\nabla_2^2\right) - \frac{Ze^2}{4\pi\xi_0}\left(\frac{1}{r_1}\right) - \frac{Ze^2}{4\pi\xi_0}\left(\frac{1}{r_2}\right) + \frac{1}{4\pi\xi_0}\left(\frac{1}{r_{12}}\right)$$

$$(4.1)$$

Here we have neglected reduced mass effects. The terms in the above expression represent the kinetic energy of the first electron, the kinetic energy of the second electron, the electrostatic attraction between the nucleus and the first electron, the electrostatic attraction between the nucleus and the second electron, and the electrostatic repulsion between the two electrons, respectively. It is the final term which results in difficulties as it requires measuring the distance between the two moving electrons, which is not possible by Heisenberg's uncertainty principle. There is a very convenient and simple way of writing the Hamiltonian operator for atomic and molecular systems as given below.

The kinetic energy term: $\frac{\hbar^2}{2m_e}\nabla^2 = \frac{1}{2}\nabla^2$. From the potential energy term, $\frac{1}{4\pi\xi_0}$ is dropped. With these simplifications, the Hamiltonian for the helium atom (Nuclear

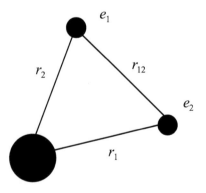

Fig. 4.1 Helium atom showing two electrons, e_1 and e_2

charge $= 2$) takes the form of Eq. 4.2.

$$\hat{H} = -\frac{1}{2}\nabla_1^2 - \frac{1}{2}\nabla_2^2 - \frac{2}{r_1} - \frac{2}{r_2} + \frac{1}{r_{12}} \qquad (4.2)$$

The Schrödinger equation for helium (Eq. 4.3) atom can be formulated as follows;

$$\hat{H}\psi = E\psi$$

But,

$$\hat{H}\psi = \left[-\frac{1}{2}\nabla_1^2 - \frac{1}{2}\nabla_2^2 - \frac{2}{r_1} - \frac{2}{r_2} + \frac{1}{r_{12}} \right]\psi$$

Hence,

$$\left[-\frac{1}{2}\nabla_1^2 - \frac{1}{2}\nabla_2^2 - \frac{2}{r_1} - \frac{2}{r_2} + \frac{1}{r_{12}} \right]\psi = E\psi \qquad (4.3)$$

Where ∇^2 is an operator given by: $\nabla^2 = \dfrac{\partial^2}{\partial x^2} + \dfrac{\partial^2}{\partial y^2} + \dfrac{\partial^2}{\partial z^2}$ spaced in Cartesian axes.

The Schrödinger equation can be suited in spherical coordinates. Let r be the distance of the radius vector making an angle θ with the reference (z) axis and ϕ be the angle of the image of the vector on the xy plane with the x-axis. The relationship between polar coordinates (r, θ, ϕ) and Cartesian coordinates (x, y, z) is illustrated as follows (Fig. 4.2).

$$x = r\sin\theta\cos\phi, \quad y = r\sin\theta\sin\phi, \quad z = r\cos\theta \text{ and } x^2 + y^2 + z^2 = r^2.$$

The solution to the Schrödinger equation based on polar coordinates takes the form of $\psi = R(r).\Theta(\theta).\Phi(\phi)$ where $R(r)$ is the radial function while $\Theta(\theta)$ and

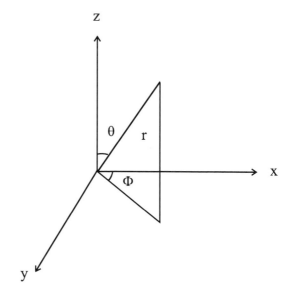

Fig. 4.2 Polar (spherical coordinates)

$\Phi(\phi)$ are angular functions. It may be noted that $R(r)$ depends on the principal quantum number (n) and azimuthal quantum number (l), $\Theta(\theta)$ depends on azimuthal (l) and magnetic (m_l) quantum numbers while $\Phi(\phi)$ depends on magnetic quantum number (m_l).

The Hamiltonian for many-electron systems will have a kinetic energy operator sum $\left(\sum -\frac{1}{2}\nabla_i^2\right)$ and the potential energy operator sum $\left(\sum V_i\right)$. The kinetic energy term is always negative as it is associated with a decrease in energy. Potential energy can be positive (if it is due to repulsion, the electron–electron repulsion) or negative (if it is due to attraction, the electron–nucleus repulsive). The electron–electron repulsive term $\left(\frac{1}{2}\sum \frac{1}{r_{ij}}\right)$ is multiplied by $\frac{1}{2}$ to avoid the double counting of terms. Nuclear repulsive terms are avoided in the Born-Oppenheimer approximation. The Hamiltonian of such a system takes the form of $\hat{H} = -\sum \left[\frac{1}{2}\nabla_i^2 + V_i\right] + \frac{1}{2}\sum \frac{1}{r_{ij}}$, where the first sum term is attractive while the second sum term is repulsive.

4.3 Independent Particle Approximation

In predicting molecular electronic structure one of the solutions is the Linear Combination of Atomic Orbitals model (LCAO). Here molecular orbital (MO) behavior is approximated as the resultant of the linear combination of atomic orbitals. If ψ is the molecular orbital function formed from atomic orbitals with functions, let $\phi_1, \phi_2, \phi_3, \ldots .\phi_n$ and $c_1, c_2, c_3, \ldots .c_n$ be their respective contributions, then $\psi = c_1\phi_1 + c_2\phi_2 + c_3\phi_3 + \ldots . + c_n\phi_n$. Or, $\psi = \sum_n c_n\phi_n$. In the MO treatment of H_2^+, two molecular orbitals are obtained by the linear combination of atomic orbitals ($1s$), σ_{1s} is the bonding molecular (lower energy and more probable) orbital and σ_{1s}^* is the antibonding molecular (the higher energy and less probable) orbital. An energy level diagram of H_2^+ is given in Fig. 4.3.

In the bonding molecular orbital the electron probability density is relatively high between the nuclei and in the antibonding molecular orbital, there is a node (zero probability plane) in density between the nuclei as illustrated in Fig. 4.4.

In the Hamiltonian of many electron systems (molecules), all the electrons have to be considered providing an expression of the form of Eq. 4.4.

$$\hat{H}_{molecule} = \hat{H}_{(1+2+3+\ldots .+n)} \tag{4.4}$$

Here also the Columbic repulsive terms between electrons in the Hamiltonian make an actual solution to the Schrödinger equation difficult. Independent particle approximation is one of the methods to overcome the above difficulty.

The principle of independent particle approximation is at the heart of many methods such as the Hartree-Fock (HF) theory, the density functional theory and in the Hückel MO theory, which are very popular methods to solve the electronic

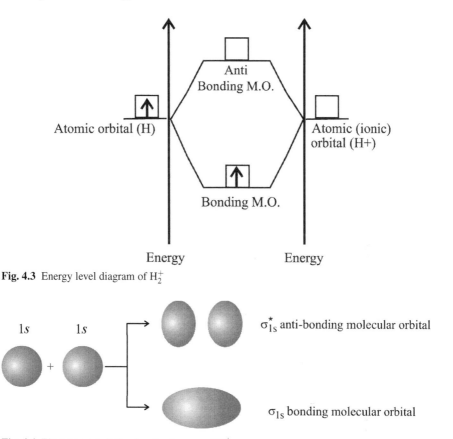

Fig. 4.3 Energy level diagram of H_2^+

Fig. 4.4 Electron probability density diagram of H_2^+

Schrödinger. In this approximation each particle (electron) is considered as independent, i.e., each particle is assumed to be in a different orbital, so that we can write the wavefunction of the system as a product of wavefunctions of constituents (Eq. 4.5):

$$\phi(r_1, r_2, \ldots, r_n) = \eta_1(r_1) . \eta_2(r_2) \ldots \eta_n(r_n) \qquad (4.5)$$

The system is considered as having n orbitals and n electrons, $\eta_n(r_n)$ is the wavefunction corresponding n^{th} electron at a distance of (r_n). The approximate form of the wavefunction represented in Eq. 4.4 is often known as the Hartree product (see Chap. 5). In this approximation, an average potential function $(V^*(i))$ is introduced which covers the potential due to the nucleus and all the electrons other than the specified electron. Hence, the Hamiltonian for the ith electron can be written as:

$$\hat{H}(i) = -\frac{1}{2}\nabla^2(i) + V^*(i) \qquad (4.6)$$

The Hamiltonian for all electrons can be similarly written. The Schrödinger equation for each electron can be written as:

$$\hat{H}(i)\psi(i) = E(i)\psi(i) \tag{4.7}$$

4.4 π-Electron Approximation

In unsaturated molecules (molecules keeping multiple bonds between the same atoms), the bonds are formed by two different modes of overlapping of atomic orbitals. The end-on or coaxial overlapping results in a sigma (σ) bond while lateral or side-wise overlapping results in a pi (π) bond. Most of the properties of such molecules will be due to the presence of π-bond. As for example alkenes and alkynes are characterized by organic addition reactions distinctive of the presence of π-electrons. Hence in such molecules, sigma bond and pi-bond contributions can be separated and the required π-bond contribution can be characterized. This type of approximation is known as π-electron approximation.

For unsaturated systems, refinement of Hamiltonian expression can be done through π-electron approximation. This method is unique to Hückel's generalization. In such cases the Hamiltonian for sigma and pi electrons of the molecule are separated and the sigma contribution is neglected. Hamiltonian for each π-electron is calculated and sum of these functions makes the molecular Hamiltonian as given is Eq. 4.8.

$$\hat{H}(\pi) = \sum_{1}^{n}\left(-\frac{1}{2}\nabla^2(i) + V_\pi(i)\right) + \frac{1}{2}\sum\frac{1}{r_{ij}} \tag{4.8}$$

Here, n is the number of π-electrons of kinetic energy $-\frac{1}{2}\nabla^2(i)$ and the potential energy term $V_\pi(i)$ represents the potential energy of a single pi-electron in the average field of the framework of nuclei and all electrons except electron i. In an alkenic double bond, each carbon keeps a single π-electron while in an alkynic triple bond, each carbon carries two π-electrons.

4.5 Hückel's Calculation

In alkenes and alkynes the pi-electrons are present in the unhybridized p-orbitals, which are considered as independent of the sigma framework of hybrid orbitals and sigma electrons. Molecular orbital wavefunction ψ is given by Eq. 4.9:

$$\psi = a_1\phi_1 + a_2\phi_2 + \ldots\ldots + a_i\phi_i \tag{4.9}$$

where a_i is the contribution of the electronic wavefunction ϕ_i. As only p-electrons are contributing to the wavefunction, the above equation can be written as in Eq. 4.10:

$$\psi = a_1 p_1 + a_2 p_2 + \ldots + a_i p_i \tag{4.10}$$

For ethene, each carbon atom keeps a p-electron. Let p_1 and p_2 be the two pi-electrons present in carbon atoms 1 and 2. Let their respective contributions be a_1 and a_2. For the unhybridized p-electrons, molecular orbitals are formed by the LCAO of p_1 and p_2. Overlapping between atomic orbitals can be either in a symmetric manner or in an unsymmetrical manner, with the respective wavefunctions ψ^+ (resulting in a bonding molecular orbital) and ψ^- (resulting in an antibonding molecular orbital). Hence:

$$\psi^+ = a_1 p_1 + a_2 p_2 \tag{4.11}$$

and

$$\psi^- = a_1 p_1 - a_2 p_2 . \tag{4.12}$$

As p_1 and p_2 are atomic orbitals and the wavefunction ψ is for molecular orbital, the exact MO solution is not provided from the above expressions.

4.6 The Variational Method and the Expectation Value

Taking back the Schrödinger equation $\hat{H}\psi = E\psi$ and pre-multiplying both sides by ψ, we get $\psi\hat{H}\psi = \psi E \psi$. Energy E being a scalar value, $\psi\hat{H}\psi = \psi^2 E$. For many electron systems a similar expression is obtained by integrating both sides in a volume $d\tau$: $\int \psi\hat{H}\psi \, d\tau = E \int \psi^2 \, d\tau$

Or, energy:

$$E = \frac{\int \psi\hat{H}\psi \, d\tau}{\int \psi^2 \, d\tau} \tag{4.13}$$

When the Hamiltonian involved is exact, energy calculated from Eq. 4.13 will also be exact. In the Hamiltonian each interaction term leads into a decrease in energy. When the entire interactions are included, the corresponding Hamiltonian will also be exact and minimum. But in all experiments, the calculated Hamiltonian will be higher than the actual one due to the dropping or skipping of some unimportant interaction terms. Once we get the approximate energy, we can repeat the experiment by modifying the Hamiltonian. It is a fundamental postulate of quantum mechanics that E in Eq. 4.12 is the expectation value of the energy and will be higher than the actual energy. By repeating the experiment, we will be generating a number of expectation energies, out of which higher ones must be farther from the true value than the lower one, so they are discarded. The identification of the energy value close to the actual one involves a minimization process of calculated energy from a set of basis functions. This principle is called the variational method. The

ψ-value can further be modified by taking criteria other than energy. Note that in all these criteria the variational principle is applied.

4.7 The Expectation Energy and the Hückel MO

From the LCAO possible in ethene, the ψ-value corresponding to Eq. 4.10 produces an expectation energy value, E, given by Eq. 4.14.

$$E = \frac{\int (a_1 p_1 + a_2 p_2)\hat{H}(a_1 p_1 + a_2 p_2)\,d\tau}{\int (a_1 p_1 + a_2 p_2)^2\,d\tau} \tag{4.14}$$

$$E = \frac{\int \left[a_1^2(p_1\hat{H}p_1) + a_1 a_2(p_1\hat{H}p_2) + a_2 a_1(p_2\hat{H}p_1) + a_2^2(p_2\hat{H}p_2) \right]\,d\tau}{\int \left[a_1^2 p_1 p_1 + 2a_1 a_2 p_1 p_2 + a_2^2 p_2 p_2 \right]\,d\tau} \tag{4.15}$$

The integrals included in Eq. 4.14 can be simplified as follows:

$$\int (p_1\hat{H}p_1)\,d\tau = \alpha, \qquad \text{known as the Coulomb integral.}$$

$$\int (p_1\hat{H}p_2)\,d\tau = \int (p_2\hat{H}p_1)\,d\tau = \beta, \qquad \begin{array}{l}\text{known as the exchange integral} \\ \text{or resonance integral.}\end{array}$$

$$\int p_1 p_1\,d\tau = S_{11} = \int p_2 p_2\,d\tau = S_{22} \qquad \text{and}$$

$$\int p_1 p_2\,d\tau = S_{12} = \int p_2 p_1\,d\tau = S_{21}$$

known as the overlap integral. With these simplified notations, the energy expression can be written as Eq. 4.16:

$$E = \frac{a_1^2 \alpha + 2a_1 a_2 \beta + a_2^2 \alpha}{a_1^2 S_{11} + 2a_1 a_2 S_{12} + a_2^2 S_{22}} \tag{4.16}$$

By knowing α, β, and S, the energy can be calculated. Setting the minimization criterion with respect to some minimization parameters:

$$\frac{\partial E}{\partial a_1} = \frac{\partial E}{\partial a_2} = 0 \tag{4.17}$$

Here, instead of varying the trial function to find the minimum value of E, we need to vary the linear coefficients. This is a relatively straightforward case of searching for the minimum of a function. If N is the numerator, D is the denominator in the energy expression, N' is the first derivative of numerator and D' is the first derivative of the denominator, then:

$$\frac{\partial E}{\partial a_1} = \frac{N'D - ED'}{D^2} = \frac{N' - ED'}{D} = 0 \tag{4.18}$$

Or,

$$N' - ED' = 0$$

$$\frac{\partial E}{\partial a_1} = \frac{N'D - ED'}{D^2} = \frac{N' - ED'}{D} = 0$$

$$N' - ED' = 0$$

$$N' = ED' \tag{4.19}$$

We get Eqs. 4.20 and 4.21:

$$a_1\alpha + a_2\beta = E(a_1S_{11} + a_2S_{12}) \tag{4.20}$$

$$a_1\beta + a_2\alpha = E(a_1S_{12} + a_2S_{22}) \tag{4.21}$$

From Eq. 4.20:

$$a_1\alpha - Ea_1S_{11} + a_2\beta - Ea_2S_{12} = 0$$

Or:

$$a_1(\alpha - ES_{11}) + a_2(\beta - ES_{12}) = 0 \tag{4.22}$$

From Eq. 4.21:

$$a_1\beta - Ea_1S_{12} + a_2\alpha - Ea_2S_{22} = 0$$

Or:

$$a_1(\beta - ES_{12}) + a_2(\alpha - ES_{22}) = 0 \tag{4.23}$$

Moreover, it is assumed that wavefunctions p_1 and p_2 retain the orthonormality condition even in the molecular state, i.e.:

$$\int p_1 p_2 \, d\tau = \int p_2 p_1 \, d\tau = 0,$$

Or

$$S_{12} = S_{21} = 0$$

and:

$$\int p_1 p_1 \, d\tau = \int p_2 p_2 \, d\tau = 1$$

or:

$$S_{11} = S_{22} = 1.$$

Substituting these approximations in Eqs. 4.22 and 4.23, we get:

$$a_1(\alpha - E) + a_2\beta = 0 \tag{4.24}$$

$$a_2(\alpha - E) + a_1\beta = 0 \tag{4.25}$$

These orthonormal equations are called secular equations. The coefficient matrix of these equations is represented by Eq. 4.26:

$$\begin{bmatrix} (\alpha - E) & \beta \\ \beta & (\alpha - E) \end{bmatrix} \tag{4.26}$$

From this matrix, the solution to E is computationally simple as it is the eigenvalue of the secular coefficient matrix. For ethene containing two sp^2-hybridized carbon atoms and two π-electrons, a 2×2 matrix of the form of Eq. 4.26 is obtained. In general, for a conjugated system keeping alternate double and single bonds containing n carbon atoms, an $n \times n$ coefficient matrix is obtained. Such an equation yields n eigenvalues corresponding to n energy levels, known as the spectrum of energy levels.

4.8 The Overlap Integral (S_{ij})

The overlap integral is given by the expression, $S_{ij} = \int p_i p_j \, d\tau$. If $i = j$, the overlap integral, $S_{ii} = \int p_i p_i \, d\tau = 1$ for the normalized atomic orbitals. If $i \neq j$, the overlap integral, $S_{ij} = \int p_i p_j \, d\tau = 0$ for the orthogonal atomic orbitals. It is obvious that the value of the overlap integral varies from zero to unity and is a measure of the non-orthogonality of the orbitals. Orthogonal p-functions are independent functions. Since p-functions of orbitals are widely separated in space and are independent; these functions are expected to be orthogonal. The closer the centers of the p-functions, the larger is the overlap integral. In this sense, S_{ij} is called the overlap integral since it is a measure of an overlapping of the orbitals i and j. In the usual "zeroth" approximation of the LCAO method, $S_{ij} = 0$ when, $i \neq j$. This simplifies the computation to a large extent. A variation of S_{ij} of different types of carbon atoms is shown in Fig. 4.5.

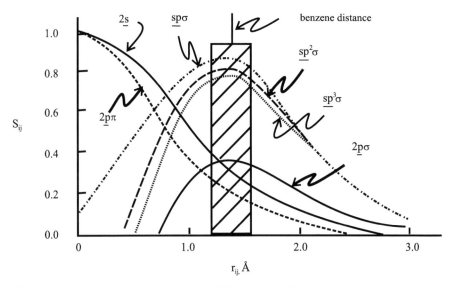

Fig. 4.5 Variation of the overlap integral with different types of c-atoms

4.9 The Coulomb Integral (α)

The Coulomb integral is: $\alpha = \int p_i \hat{H} p_i \, d\tau$. To a zeroth-order approximation, α, is the Hamiltonian for the Coulomb energy of an electron with a wavefunction p_i in the field of atom i and influenced by its nucleus and is unaffected by any other nuclei farther away. This approximation, of course, will be most valid where the surrounding atoms have no net electrical charges. The Coulomb integral α is a function of the nuclear charge and the type of orbital. As it involves attraction, it is a negative number.

4.10 The Resonance (Exchange) Integral (β)

The resonance (exchange) integral is a measure of the resonance or exchange and it amounts to the energy of an electron in the fields of atoms i and j, involving the wavefunctions p_i and p_j. It is a function of the atomic number, the orbital types, and the degree of overlap. As it is a function of the degree of overlap, it is also a function of the internuclear distance. In the zeroth order approximation, β_{ij} is neglected if i and j are not in the customary bond forming distance.

4.11 The Solution to the Secular Matrix

Eq. 4.25 is the coefficient matrix for the secular equations. Dividing elements of the matrix by β, we get $\begin{bmatrix} (\alpha - E)/\beta & 1 \\ 1 & (\alpha - E)/\beta \end{bmatrix}$. If $(\alpha - E)/\beta$ is put as x, the matrix takes the form of:

$$\begin{bmatrix} x & 1 \\ 1 & x \end{bmatrix} \tag{4.27}$$

The solution to the above matrix can be made by expanding the corresponding determinant (secular determinant) or by finding the eigenvalues and eigenvectors of the matrix (secular matrix). For the equation set to be linearly dependent, the secular determinant must be zero. Hence:

$$\begin{vmatrix} x & 1 \\ 1 & x \end{vmatrix} = 0 \tag{4.28}$$

Expanding the determinant:

$$x^2 - 1 = 0$$
$$x^2 = 1$$
$$x = \pm 1$$

The eigenvalues of this matrix can be computed using any scientific environment such as MATLAB or MATHEMATICA. Working out the problem with MATLAB is as follows:

```
>> syms x;
>> eig([x 1;1 x])

ans =
x-1
x+1
```

We get two eigenvalues to the secular coefficient matrix of ethene, $(x = +1)$ and $(x = -1)$ where $x = (\alpha - E)/\beta$.
Taking the first eigenvalue:

$$\frac{\alpha - E}{\beta} = x = 1 \tag{4.29}$$

$$(\alpha - E) = \beta$$
$$E = (\alpha - \beta) \tag{4.30}$$

Similarly, from the second eigenvalue:

$$\frac{\alpha - E}{\beta} = x = -1$$

$$(\alpha - E) = -\beta$$
$$E = (\alpha + \beta) \tag{4.31}$$

On fixing the reference point of energy as α, we get energy eigenvalues of π-electrons of ethene as one greater than β (antibonding) and the other less than β (bonding). The energy level diagram of the π-MO of ethene is given in Fig. 4.6.

4.12 Generalization

The method can be generalized to conjugated systems of any size. The dimension of the matrix is the number of atoms in the π-conjugated system. Label the carbon atoms from one end if it is an open chain compound. Otherwise, labeling can be started from anywhere and be continued until the cycle is completed. Let us take the three-carbon system allyl $[CH_2=CH-CH_2-]$ as our next example. With labeling, the system can be represented as $\left[\underset{1}{C}H_2 = \underset{2}{C}H - \underset{3}{C}H_2 - \right]$. Here, we get a 3×3 matrix as the secular coefficient matrix. Elements in the matrix are based on the following rules:

1. Each period stands for the connectivity of the corresponding atom.
2. In each period the reference atom is labeled as x ($i = j$ positions of the matrix).

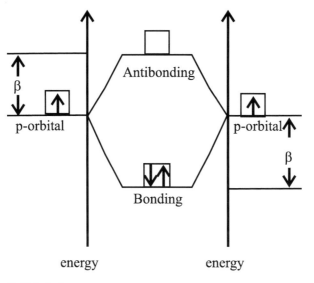

Fig. 4.6 Hückel's MO of ethene

3. If $i \neq j$, and if the corresponding atom is connected to the respective reference atom, put 1 as the element.
4. If $i \neq j$, and if the corresponding atom is not connected to the respective reference atom, put 0 as the element.

For the allyl system, the secular matrix will be as follows:

$$\begin{bmatrix} x & 1 & 0 \\ 1 & x & 1 \\ 0 & 1 & x \end{bmatrix} \tag{4.32}$$

The determination of the eigenvalues of the matrix suggests three MOs for the allyl system with energy values, $E = \alpha, E = \alpha + \beta\sqrt{2}$ and $E = \alpha - \beta\sqrt{2}$, in which the lowest energy level will be occupied with two electrons obtained from unhybridized orbitals of two carbon atoms.

Now, let us take 1,3-butadiene $(CH_2{=}CH{-}CH{=}CH_2)$. The molecule can be labeled as $\underset{1}{C}H_2 = \underset{2}{C}H - \underset{3}{C}H = \underset{4}{C}H_2$. The secular coefficient matrix of the molecule is a 4×4 matrix as given in Eq. 4.33:

$$\begin{bmatrix} x & 1 & 0 & 0 \\ 1 & x & 1 & 0 \\ 0 & 1 & x & 1 \\ 0 & 0 & 1 & x \end{bmatrix} \tag{4.33}$$

Eigenvalues of the matrix are calculated to get the spectrum of energies. Four eigenvalues are obtained for the matrix with $x = (\alpha - E)/\beta$ values $-1.6180, -0.6180, +0.6180, +1.6180$.

4.13 The Eigenvector Calculation of the Secular Matrix

The expansion of any molecular orbital over a basis set ϕ_k, $\psi = \sum_k a_k \phi_k$ leads to a set of arbitrary expansion coefficients a_k, which we optimize by imposing the conditions of optimization, $\dfrac{\partial E}{\partial a_1} = \dfrac{\partial E}{\partial a_2} = \dfrac{\partial E}{\partial a_3} = \ldots = \dfrac{\partial E}{\partial a_k} = \ldots = \dfrac{\partial E}{\partial a_n} = 0$, to find the energy minimum in an n-dimensional vector space by calculating the eigenvector. The eigenvector calculation using MATLAB is quite simple. The entries are given as follows:

```
>> A=[0 1 0;1 0 1;0 1 0];
>> [V,D] = eig(A)


V =
         0.5000      -0.7071       0.5000
        -0.7071       0.0000       0.7071
         0.5000       0.7071       0.5000


D =
        -1.4142             0             0
              0       -0.0000             0
              0             0        1.4142
```

The elements of the diagonal in the d matrix correspond to the eigenvalues. The eigenvector of the matrix with -1.414 as the eigenvalue is:

$$\begin{bmatrix} 0.5000 \\ -0.7071 \\ 0.5000 \end{bmatrix}$$

The eigenvector of the matrix with 0 as the eigenvalue is:

$$\begin{bmatrix} -0.7071 \\ 0.0000 \\ 0.7071 \end{bmatrix}.$$

4.14 The Chemical Applications of Hückel's MOT

The Hückel results show some interesting features for conjugated hydrocarbons keeping alternate double and single bonds [2, 3]:

1. The orbital energies are in pairs of equal magnitude and opposite signs. This means that if there is an odd number of orbitals, there must be an orbital energy of zero (a non-bonding orbital) that pairs with itself. (An example is the benzyl radical in Table 4.1).

Table 4.1 Benzyl radical with electrons in the molecular orbital

Number	Orbital	Energy	Number of electrons	Electronic energy
1	MO-1	2.101	2	4.202
2	MO-2	1.258	2	2.518
3	MO-3	1.000	2	2.000
4	MO-4	0.000	1	0.000
5	MO-5	− 1.000	0	0.000
6	MO-6	− 1.259	0	0.000
7	MO-7	− 2.101	0	0.000
		Total energy		8.720

2. For the pairs of orbitals the coefficients are also paired. For a given atomic orbital, the coefficients in the two molecular orbitals are equal in magnitude. For one set of atoms ("starred" or "non-starred") the coefficients are equal; for the other set of atoms ("non-starred" or "starred") the coefficients are of opposite signs.
3. The charge densities are all unity, so the Hückel theory predicts that conjugated hydrocarbons are nonpolar.
4. The spin densities in the output refer to the density of the odd electron in the +ve (carbocation) and −ve (carbanion) ions formed by removing or adding an electron to the molecule. The spin densities of the +ve and −ve ions for hydrocarbons are equal, since they are just the squares of the coefficients in the highest occupied and lowest unoccupied molecular orbitals which are a pair of orbitals. To a first approximation, the electron spin resonance (ESR) spectrum depends on these spin densities, so the Hückel theory predicts that the ESR spectrum of the +ve and −ve ions of conjugated hydrocarbons are equal. Experimentally, they are very similar.

4.15 Charge Density

Eigenvectors can be transformed into derived quantities that give us a better, intuitive sense of how HMO calculations relate to the physical properties of molecules. One of these quantities is the charge density. The magnitude of the coefficient of an orbital a_{ij} at a carbon atom c_i gives the relative amplitude of the wavefunction at that atom. The square of the wavefunction is a probability function; hence, the square of the eigenvector coefficient gives a relative probability of finding the electron within an orbital j near a carbon atom i. This is a measure of the relative charge density, too, because a point in the molecule at which there is a high probability of finding electrons is a point of a large negative charge density and a portion of the molecule at which electrons are not likely to be found is positively charged, relative to the rest of the molecule. There may be one or two electrons in an orbital ($N = 1, N = 2$).

Unoccupied (virtual) orbitals make, of course, no contribution to the charge density. To obtain the total charge density q_i at atom c_i we must sum over all occupied or partially occupied orbitals and subtract the result from 1.0, the π-charge density of the carbon atom alone:

$$q_i = \left(1 - \sum Na_i^2\right) \tag{4.34}$$

Where $\sum Na_i^2$ is the total electron density at c_i. As, for example, in allyl carbocation $(CH_2=CH-CH_2^\oplus)$, the Hückel MOT was followed to generate the charge density. The eigenvalues and eigenvectors of the system can be computed using MATLAB as follows:

```
>> A=[0 1 0;1 0 1;0 1 0];
>> [V,D]=eig(A)

V =

     0.5000    -0.7071     0.5000
    -0.7071     0.0000     0.7071
     0.5000     0.7071     0.5000


D =

    -1.4142          0          0
         0    -0.0000          0
         0          0     1.4142
```

From the above output data, the eigenvector corresponding to the eigenvalue of 1.4142 is: $\begin{bmatrix} 0.5000 \\ 0.7071 \\ 0.5000 \end{bmatrix}$.

With these values, the system can be labeled as follows:

$$\left(\underset{0.5000}{C}\ H_2 = \underset{0.7071}{C}\ H - \underset{0.5000}{C}\ H_2^\oplus\right)$$

The charge density in each atom can be calculated based on Eq. 4.35. The probabilities of the charge (due to two electrons) to be on the three carbon atoms are calculated as follows:

$$q_1 = 1 - 2(0.5)^2 = 1 - 0.5 = 0.5 \tag{4.35}$$

$$q_2 = 1 - 2(0.7071)^2 = 1 - 0.9999 = 0.0000 \tag{4.36}$$

$$q_3 = 1 - 2(0.5)^2 = 1 - 0.5 = 0.5 \tag{4.37}$$

The energy level spectrum of the cation is included in Fig. 4.7. Remember that $\beta-$ is a negative energy and that is why $(\alpha+1.414\beta)$ level is the lowest energy level. The two end carbon atoms carry equal charges. This suggests that the positive charge is delocalized between the two end carbon atoms. This sort of delocalization of charge

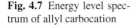

Fig. 4.7 Energy level spectrum of allyl carbocation

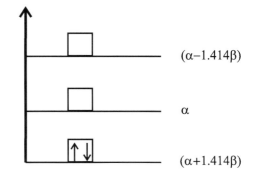

$(\alpha - 1.414\beta)$

α

$(\alpha + 1.414\beta)$

is an autostabilization technique. Similarly allyl carbanion, $(CH_2=CH-CH_2^-)$, has 4 electrons to be arranged in the same energy levels. Hence:

$$q_i = q_3 = 1 - 2(0.5)^2 - 2(0.7071)^2 = 1 - 0.5 - 1 = -0.5$$

and

$$q_2 = 1 - 2(0.7071)^2 - (0.000)^2 = 0.0000 \,.$$

4.16 The Hückel (4n + 2) Rule and Aromaticity

Recall that the interaction (overlap) of two atomic orbitals leads to a more stable (lower E) bonding MO and a less stable (higher E) antibonding MO, compared to the energies of original atomic orbitals [4]. The number of new molecular orbitals is equal to the number of atomic orbitals involved (the linear combination). The relative stability or energies of the molecular orbital in a fully conjugated, cyclic, planar polyene can be effectively predicted with the Hückel MOT. A stable species should have closed shell p-electron configurations, that is, no unpaired molecular orbital. This concept can be extended to predict the stability of such a species, as, for example, the stability of benzene can be predicted as follows. A Frost diagram and a comparison of stability is included in Fig. 4.8. Entries in MATLAB to generate eigenvalues and eigenvectors of benzene are given below:

```
>> A= [0 1 0 0 0 1;1 0 1 0 0 0;0 1 0 1 0 0;0 0 1 0 1 0;0 0 0 1 0 1;
       1 0 0 0 1 0];
>> [V,D]=eig(A)

V =
        0.4082    -0.2887    -0.5000     0.5000     0.2887    -0.4082
       -0.4082    -0.2887     0.5000     0.5000    -0.2887    -0.4082
        0.4082     0.5774          0          0    -0.5774    -0.4082
       -0.4082    -0.2887    -0.5000    -0.5000    -0.2887    -0.4082
        0.4082    -0.2887     0.5000    -0.5000     0.2887    -0.4082
       -0.4082     0.5774          0          0     0.5774    -0.4082
```

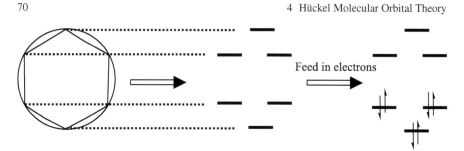

Fig. 4.8 Frost diagram and stability of benzene

$$
D =
\begin{array}{cccccc}
-2.0000 & 0 & 0 & 0 & 0 & 0 \\
0 & -1.0000 & 0 & 0 & 0 & 0 \\
0 & 0 & -1.0000 & 0 & 0 & 0 \\
0 & 0 & 0 & 1.0000 & 0 & 0 \\
0 & 0 & 0 & 0 & 1.0000 & 0 \\
0 & 0 & 0 & 0 & 0 & 2.0000
\end{array}
$$

We can correlate π-electron energy and stability by the following procedure:

1. If, on the ring closure, the π electron energy of an open chain polyene (alternating single and double bonds) decreases (increases in terms of β as it is negative) the molecule is classified as aromatic: refer to Fig. 4.9 and Table 4.2. From the table it is obvious that the ring closure of 1,3,5-hexatriene is favoured, and the corresponding cyclic molecule (benzene) is aromatic.
2. If, on the ring closure, the π electron energy increases, (decreases in terms of β) the molecule is classified as antiaromatic (Fig. 4.10). The computed values suggest that the ring closure of 1,3-butadiene is associated with an increase in energy (Table 4.3) or the corresponding cyclic compound is nonaromatic.

Fig. 4.9 Ring closure of
1,3,5-hexatriene

Table 4.2 Comparisons of computed energies associated with ring closure of 1,3,5-hexatriene

Number	Orbital	Energy		Number of electrons		Electronic energy	
		Cyclic	Open chain	Cyclic	Open chain	Cyclic	Open chain
1	MO-1	2.000	1.802	2	2	4.000	3.604
2	MO-2	1.000	1.247	2	2	2.000	2.494
3	MO-3	1.000	0.445	2	2	2.000	0.890
4	MO-4	−1.000	−0.445	0	0	0.000	0.000
5	MO-5	−1.000	−1.247	0	0	0.000	0.000
6	MO-6	−2.000	−1.802	0	0	0.000	0.000
	Total energy in terms of beta negative energy					8.000	6.988

Table 4.3 Comparison of computed energies associated with the ring closure of 1,3-butadiene

Number	Orbital	Energy		Number of electrons		Electronic energy	
		Cyclic	Open chain	Cyclic	Open chain	Cyclic	Open chain
1	MO-1	2.000	1.618	2	2	4.000	3.236
2	MO-2	0.000	0.618	1	2	0.000	1.236
3	MO-3	0.000	−0.618	1	0	0.000	0.000
4	MO-4	−2.000	−1.618	0	0	0.000	0.000
	Total energy in terms of beta negative energy					4.000	4.472

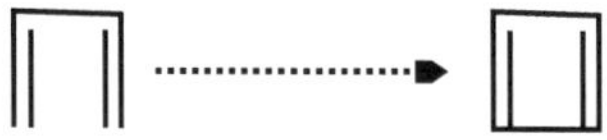

Fig. 4.10 Ring closure of
1,3-butadiene

3. If, on the ring closure, the π electron energy remains the same, the molecule is classified as nonaromatic, e.g., 1,3,5,7-cyclooctatetraene (C_8H_8−COT).

4.17 The Delocalization Energy

We have seen that localized ethene provides a ground level energy of $E = 2\alpha + 2\beta$. The next higher homologue propene (allyl radical) can be considered as an sp^3-hybridized carbon connected to the radical obtained from ethene. If we assume a localized double bond in propene, π-electron energy of propene will be the same as that of ethene. With the delocalization of the double bond between three carbon atoms, another π-electron energy is obtained. The difference in energy is the delocalization energy. The delocalization of π-electrons stabilizes the molecule as is evident from the energy values. The secular matrix for propene, neglecting the possibility for delocalization, will be:

$$\begin{bmatrix} x & 1 \\ 1 & x \end{bmatrix} \tag{4.38}$$

Putting $x = 0$:

$$\begin{bmatrix} 0 & 1 \\ 1 & 0 \end{bmatrix} \tag{4.39}$$

The secular matrix for propene (Table 4.5) providing the possibility for delocalization, will be:

$$\begin{bmatrix} x & 1 & 0 \\ 1 & x & 1 \\ 0 & 1 & x \end{bmatrix} . \tag{4.40}$$

Table 4.4 Pi-electron energy calculation of propene without delocalization

Number	Orbital	Energy	Number of electrons	Electronic energy
1	MO-1	1.000	2	2.000
2	MO-2	−1.000	0	0.000
	Total energy in terms of beta negative energy			2.000

Table 4.5 Pi-electron energy calculation of propene with delocalization

Number	Orbital	Energy	Number of electrons	Electronic energy
1	MO-1	1.414	2	2.828
2	MO-2	0.000	0	0.000
3	MO-3	−1.414	0	0.000
	Total energy in β			2.828

For localized and delocalized 1,3-butadiene are given below and the corresponding energies are tabulated in Table 4.6.

Table 4.6 Delocalization energy of 1,3-butadiene*

Number	Orbital	Energy		Number of electrons		Electronic energy	
		Localized	Delocalized	Localized	Delocalized	Localized	Delocalized
1	MO-1	1	1.618	2	2	2.000	3.236
2	MO-2	1	0.618	2	2	2.000	1.236
3	MO-3	−1	−0.618	0	0	0.000	0.000
4	MO-4	−1	−1.618	0	0	0.000	0.000
		Total energy (in terms of β-negative energy)				4.000	4.472

* Obviously, the delocalization energy of 1,3-butadiene is 0.472 β.

Putting $x = 0$:

$$\begin{bmatrix} 0 & 1 & 0 \\ 1 & 0 & 1 \\ 0 & 1 & 0 \end{bmatrix} \qquad (4.41)$$

A summary of π-electron energy calculation is given in Table 4.4. The difference in these two, $0.828\ \beta$, is the pi-electron delocalization energy of propene. Similarly, we can find the delocalization energy of 1,3-butadiene.

$$\begin{bmatrix} x & 1 & 0 & 0 \\ 1 & x & 0 & 0 \\ 0 & 0 & x & 1 \\ 0 & 0 & 1 & x \end{bmatrix} \qquad \begin{bmatrix} x & 1 & 0 & 0 \\ 1 & x & 1 & 0 \\ 0 & 1 & x & 1 \\ 0 & 0 & 1 & x \end{bmatrix}$$

Localized   Delocalized

4.18 Energy Levels and Spectrum

Hückel's MOT is a convenient method of expressing the energy levels generated by the p-orbitals of carbon atoms. Energies will be in units of β relative to α. The energy of α can be arbitrarily standardized as zero. From this the lowest unoccupied molecular orbital (LUMO) and highest occupied molecular orbital (HOMO) can be identified. The molecular energy level with the same energy as α is known as the nonbonding molecular orbital, the molecular energy level with a higher energy than α is known as the antibonding molecular orbital, and the molecular energy level with a lower energy than α is known as the bonding molecular orbital. The energy level diagram obtained is sometimes referred to as an energy level spectrum. From the energy level diagram, probable spectral lines caused by $\pi \rightarrow \pi^*$ electronic transitions can be predicted. Usually it is the transition from the HOMO to LUMO, is most often of interest. In the case of butadiene, this process is depicted in Fig. 4.11. As can be seen, the energy difference between the HOMO and the LUMO is:

$$\alpha - 1.414\beta - (\alpha + 1.414\beta) = -2.828\beta \ .$$

But by Planck's equation

$$\Delta E = h\upsilon = \frac{hc}{\lambda} \tag{4.42}$$

Or, wavelength:

$$\lambda = \frac{hc}{\Delta E} \tag{4.43}$$

Assuming the value of $\beta-$ to be $-2.7\,\text{eV} = \left(-2.7 \times 1.602 \times 10^{-19}\right)\text{J.}$, the wavelength of the transition is expected to be: ($\beta-$ can be taken as equal to $-2.7\,\text{eV} = \left(-2.7 \times 1.602 \times 10^{-19}\right)\text{J.}$):

$$\lambda = \frac{hc}{\Delta E} = \frac{6.626 \times 10^{-34} \times 3 \times 10^8}{-2.828\beta} = \frac{6.626 \times 10^{-34} \times 3 \times 10^8}{-2.828 \times \left(-2.7 \times 1.602 \times 10^{-19}\right)}$$
$$= 1.6286 \times 10^{-7}\,\text{m} = 162.8 \times 10^{-9}\,\text{m} = 162.8\,\text{nm}$$

Thus, the value of the lowest energy absorption allyl group is predicted to lie in the vacuum UV [5]; a very energetic photon would be necessary to excite this electron. Unfortunately, the correct answer is closer to 400 nm, but the fact that we can get this close is pretty amazing. Also, it is highly dependent on the method used to determine β.

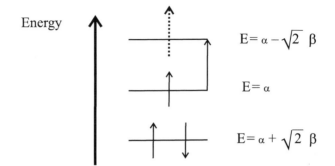

Fig. 4.11 $\pi \rightarrow \pi^*$ electronic transition spectrum of 1,3-butadiene

4.19 Wave Functions

The following illustrative example clearly shows the application of Hückel's method to determine the wavefunction of a conjugated system. 1,3-butadiene is taken as the example.

4.19.1 Step 1: Writing the Secular Matrix

The molecule is $C\underset{1}{H_2}=C\underset{2}{H}-C\underset{3}{H}=C\underset{4}{H_2}$. Hence, the secular matrix can be expressed as follows:

$$\begin{bmatrix} x & 1 & 0 & 0 \\ 1 & x & 1 & 0 \\ 0 & 1 & x & 1 \\ 0 & 0 & 1 & x \end{bmatrix} \tag{4.44}$$

4.19.2 Step 2: Solving the Secular Matrix

4.19.2.1 Method 1

Find the eigenvalue expressions and solve them by putting them as zero to get the values of x. Using MATLAB, the problem is worked out as follows:

```
>> syms x;
>> A= [ x 1 0 0 ;1 x 1 0;0 1 x 1; 0 0 1 x];
>> eig (A)
```

ans =
```
    -1/2*5^(1/2)+1/2+x
    1/2+1/2*5^(1/2)+x
    -1/2*5^(1/2)-1/2+x
    -1/2+1/2*5^(1/2)+x
```

On solving these equations, we get x as $x = \pm 1.61804$ and $x = \pm 0.61804$. Eigenvectors corresponding to these eigenvalues represent the coefficients.

4.19.2.2 Method 2

This system of equations has a nontrivial solution only if its determinant is equal to zero, which leads to the HMO determinantal equation for butadiene:

$$\begin{vmatrix} x & 1 & 0 & 0 \\ 1 & x & 1 & 0 \\ 0 & 1 & x & 1 \\ 0 & 0 & 1 & x \end{vmatrix} = 0 \tag{4.45}$$

$$\Rightarrow x \begin{vmatrix} x & 1 & 0 \\ 1 & x & 1 \\ 0 & 1 & x \end{vmatrix} - \begin{vmatrix} 1 & 1 & 0 \\ 0 & x & 1 \\ 0 & 1 & x \end{vmatrix} \Rightarrow x(x^3 - 2x) - (x^2 - 1)$$

$$= x^4 - 3x^2 + 1 = 0 \tag{4.46}$$

This quartic equation can be converted into a quadratic equation by putting $u = x^2$. Then equation becomes:

$$u^2 - 3u + 1 = 0 \tag{4.47}$$

Hence:

$$u = \frac{3 \pm \sqrt{5}}{2} = 2.618 \,\&\, 0.382 \,.$$

$$\text{Or } x = \pm\sqrt{\frac{3 + \sqrt{5}}{2}} \,\&\, \pm\sqrt{\frac{3 - \sqrt{5}}{2}}$$

$$x = \pm 0.61804 \text{ and } x = \pm 1.61804 \,.$$

The Hückel molecular orbital energy scheme for butadiene is given in Fig. 4.12.
The delocalized wavefunctions of butadiene can be represented as follows:

$$\psi_{\text{butadiene}} = c_1 p_1 + c_2 p_2 + c_3 p_3 + c_4 p_4 \tag{4.48}$$

To calculate c_n values we can proceed as follows. We obtain the ratios $\dfrac{c_n}{c_1}$ by Eqs. 4.49 and 4.50:

$$\frac{c_n}{c_1} = \frac{(\text{cofactor})_n}{(\text{cofactor})_1} \,; \quad n = \text{odd} \,. \tag{4.49}$$

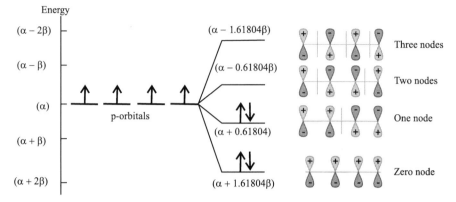

Fig. 4.12 HMO energy scheme for butadiene

$$\frac{c_n}{c_1} = -\frac{(\text{cofactor})_n}{(\text{cofactor})_1} ; \quad n = \text{even} \tag{4.50}$$

For butadiene, the cofactor ratios are determined as follows:

$$\frac{c_1}{c_1} = +\frac{\begin{vmatrix} x & 1 & 0 \\ 1 & x & 1 \\ 0 & 1 & x \end{vmatrix}}{\begin{vmatrix} x & 1 & 0 \\ 1 & x & 1 \\ 0 & 1 & x \end{vmatrix}} = 1 ; \qquad \frac{c_2}{c_1} = -\frac{\begin{vmatrix} 1 & 1 & 0 \\ 0 & x & 1 \\ 0 & 1 & x \end{vmatrix}}{\begin{vmatrix} x & 1 & 0 \\ 1 & x & 1 \\ 0 & 1 & x \end{vmatrix}} = -\frac{(x^2 - 1)}{(x^3 - 2x)} ;$$

$$\frac{c_3}{c_1} = +\frac{\begin{vmatrix} 1 & x & 0 \\ 0 & 1 & 1 \\ 0 & 0 & x \end{vmatrix}}{\begin{vmatrix} x & 1 & 0 \\ 1 & x & 1 \\ 0 & 1 & x \end{vmatrix}} = \frac{x}{(x^3 - 2x)} = \frac{1}{(x^2 - 2)} ; \qquad \frac{c_4}{c_1} = -\frac{\begin{vmatrix} 1 & x & 1 \\ 0 & 1 & x \\ 0 & 0 & 1 \end{vmatrix}}{\begin{vmatrix} x & 1 & 0 \\ 1 & x & 1 \\ 0 & 1 & x \end{vmatrix}} = -\frac{1}{(x^3 - 2x)} .$$

Substituting the values of x, the cofactor ratios can be computed. The calculations are tabulated in Table 4.7. Here, c_n is obtained as the quotient of (c_n/c_1) divided by $\sqrt{\sum (c_n/c_1)^2}$. The denominator is 2.6900 for $x = -1.61804$ and 1.6625 for $x = -0.61804$. Wavefunctions can be written directly from the cofactor values:

$$\psi_1 = 0.3717 p_1 + 0.6015 p_2 + 0.6015 p_3 + 0.3717 p_4 \tag{4.51}$$
$$\psi_2 = 0.6015 p_1 + 0.3717 p_2 - 0.3717 p_3 - 0.6015 p_4 \tag{4.52}$$

Similarly, from the other two values of x,s two more wavefunctions are obtained:

$$\psi_3 = 0.6015 p_1 - 0.3717 p_2 - 0.3717 p_3 + 0.6015 p_4 \tag{4.53}$$
$$\psi_4 = 0.3717 p_1 - 0.6015 p_2 + 0.6015 p_3 - 0.3717 p_4 . \tag{4.54}$$

Table 4.7 Cofactor ratio computation

n	(C_n/C_1)		$(C_n/C_1)^2$		C_n	
	$x = -1.61804$	$x = -0.61804$	$x = -1.61804$	$x = -0.61804$	$x = -1.61804$	$x = -0.61804$
1	1.0000	1.0000	1.0000	1.0000	0.3717	0.6015
2	1.6180	0.6180	2.6180	0.3819	0.6015	0.3717
3	1.6180	-0.6180	2.6180	0.3819	0.6015	-0.3717
4	1.0000	-1.0000	1.0000	1.0000	0.3715	-0.6015

Table 4.8 The coefficients of wavefunctions of butadiene

MO	Atom 1	Atom 2	Atom 3	Atom 4
MO 1	0.3717	0.6015	0.6015	0.3717
MO 2	0.6015	0.3717	-0.3717	-0.6015
MO 3	0.6015	-0.3717	-0.3717	0.6015
MO 4	0.3717	-0.6015	0.6015	-0.3717

Total energy of pi-electrons:

$$2(\alpha + 0.61804\beta) + 2(\alpha + 1.61804\beta) = (4\alpha + 4.47216\beta) . \qquad (4.55)$$

4.20 Bond Order

The pi-bond order is a measure of pi-electron density between carbon atoms in a compound. It is the number (quantity) of pi-bonds established between the atoms. If C_j and C_k are the connecting carbon atoms, N is the number of electrons in a single orbital (1 or 2), a_{ij} and a_{ik} are the coefficients (eigenvectors) then bond orders:

$$P_{jk} = \sum N a_{ij} a_{ik} . \qquad (4.56)$$

The bond order thus calculated is known as the mobile bond order or the Coulson bond order. As an example, The coefficients of the wavefunction of butadiene are given in Table 4.8. Only the first two molecular orbitals are occupied (with 2 electrons each). Hence, the bond order in the molecule $\underset{1}{C}H_2{=}\underset{2}{C}H{-}\underset{3}{C}H{=}\underset{4}{C}H_2$ can be computed as follows.

The pi-bond order between carbon atoms 1 and 2 $= P_{12}$:

$$= (2 \times 0.3717 \times 0.6015) + (2 \times 0.6015 \times 0.3717)$$
$$= (0.4472 + 0.4472) = 0.8944$$

Similarly, the pi-bond order between carbon atoms 2 and 3 $= P_{23}$

$$= (2 \times 0.6015 \times 0.6015) + (2 \times 0.3717 \times -0.3717)$$
$$= (0.7236 - 0.2763) = 0.4473$$

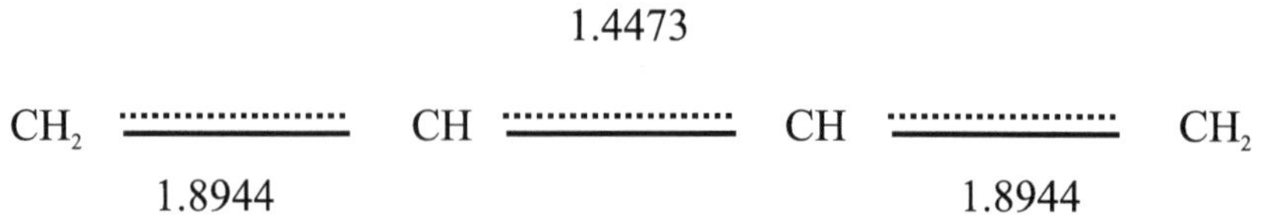

1.4473

CH$_2$ ⋯⋯⋯⋯⋯⋯ CH ⋯⋯⋯⋯⋯⋯ CH ⋯⋯⋯⋯⋯⋯ CH$_2$

1.8944   1.8944

Fig. 4.13 Bond order representation of butadiene

And the pi-bond order between carbon atoms 3 and 4 $= P_{34}$

$$= (2 \times 0.6015 \times 0.3717) + (2 \times -0.3717 \times -0.6015)$$
$$= (0.4472 + 0.4472) = 0.8944 \,.$$

If we take σ-bond order between carbon atoms to be one each, the bond order representation of butadiene can be represented as follows (Fig. 4.13).

The Coulson pi-bond order calculation helps to make a check on the calculated pi-bond energy. The pi-bond energy is as follows:

$$E_\pi = 2\beta \left(\sum P_{ij} \right) + N\alpha \,. \tag{4.57}$$

For 1, 3-butadiene, the pi-bond energy is:

$$E_\pi = 2\beta \left(2 \times 0.8944 + 0.4473 \right) + 4\alpha = \left(4\alpha + 4.4722\beta \right) \,.$$

This value is in close agreement with the calculated pi-bond energy in the previous section.

4.21 The Free Valence Index

The free valence index is a measure of chemical reactivity. The measurement of the free valence index involves the determination of the degree that the atoms in a molecule are bonded to adjacent atoms relative to their theoretical maximum bonding power. Coulson [1] defines the free valence index, F_r as follows:

$$F_r = \left(N_{\text{max}}, \text{maximum possible bonding power of } i\text{th atom} \right) - \sum P_{ij} \,. \tag{4.58}$$

Where $\sum P_{ij}$ is the sum of the bond orders of all bonds to the ith atom including σ-bonds. In a trimethylene methane system (Fig. 4.14) with the central carbon sp^2-hybridized, the Coulson is calculated N_{max} as the sum of the sigma bond order and the pi-bond order and is equal to $(3 + \sqrt{3}) = 4.732$. For butadiene (Fig. 4.15), each carbon atom makes use of 3 sigma bonds; the pi-bond orders for different carbon atoms have been calculated earlier. With these values the free valence index of different carbon atoms can be computed. These values are tabulated in Table 4.9. From these values we can presume that butadiene could well be more reactive to neutral

Fig. 4.14 Trimethylene
methane

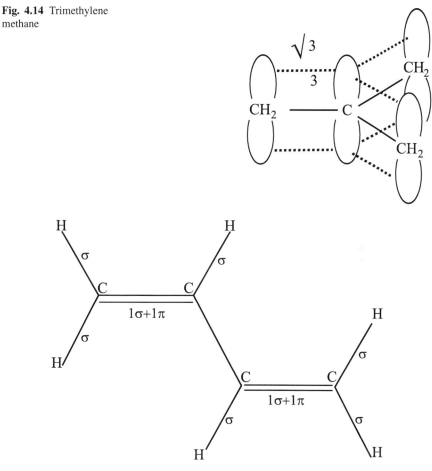

Fig. 4.15 Nature of bonding in butadiene

Table 4.9 Free valence index calculation of butadiene

Carbon	sigma	pi	Total $\left(\sum P_{ij}\right)$	$F_r = \left(4.732 - \sum P_{ij}\right)$
1	3	0.8944	3.8944	0.8376
2	3	1.3417	4.3417	0.3903
3	3	1.3417	4.3417	0.3903
4	3	0.8944	3.8944	0.8376

nonpolar reagents, such as free radicals, at the 1 and 4 carbons, than at the 2 and
3 carbons. Neutral nonpolar reagents are specified here so as to avoid charge distribution effects. The free valence index values of some free radicals and alkenes are
included in Fig. 4.16.

Fig. 4.16 Free valence index
of alkenes and organic radicals

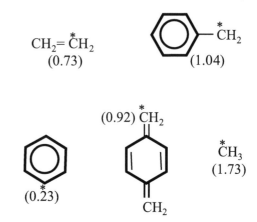

$$CH_2 = \overset{*}{C}H_2$$
$$(0.73)$$

4.22 Molecules with Nonbonding Molecular Orbitals

A conjugated system carrying an odd number of π-electron centers will be keeping
nonbonding molecular orbitals (NBMOs). The NBMO coefficients determine the
calculated distribution of the odd electron in the radical and the charges in the cation
and anion intermediates that could be developed. We shall illustrate this application
by taking a benzyl radical (Fig. 4.17). The calculated energy values for all molecular
orbitals are tabulated in Table 4.10.

The coefficients of orbitals of the NBMO (Table 4.11) clearly show that the odd
electron is delocalized. The squares of the coefficients give the electron density. If

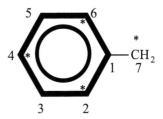

Fig. 4.17 Benzyl radical

Table 4.10 Energy values of molecular orbitals of the benzyl radical

Orbital	Electrons	Energy
MO-1	2	$\alpha + 2.101\beta$
MO-2	2	$\alpha + 1.259\beta$
MO-3	2	$\alpha + \beta$
MO-4	1	α
MO-5	0	$\alpha - \beta$
M0-6	0	$\alpha - 1.259\beta$
M0-7	0	$\alpha - 2.101\beta$
Total	7	$\alpha + 8.721\beta$

Table 4.11 Electron density calculation of the NBMO of the benzyl radical

	Atom-1	Atom-2	Atom-3	Atom-4	Atom-5	Atom-6	Atom-7
Coefficients NBMO	0.000	0.378	0.000	−0.378	0.000	0.378	−!0.756
Square of coefficients	0.000	0.143	0.000	0.143	0.000	0.143	0.572
% electron density	0.000	14.300	0.000	14.300	0.000	14.300	57.200

an electron is added to get an anion, or an electron is removed to get the cation, the effect remains the same as the changes take place only to the NBMO.

This can very effectively predict the directive property of the monosubstituents like ortho-para or the metadirecting and activation or deactivation effect associated with substitution.

4.23 The Prediction of Chemical Reactivity

The Hückel theory can be used to make predictions regarding electrophilic and nucleophilic substitution reaction possibilities. An electrophile is a species in search of electron density. The Hückel theory can tell us to identify the carbon atom in a molecule with the most accessible electron density. The highest energy level is the most accessible, and the corresponding electrons will be found in the HOMO. We must remember that the electrons in an orbital are spread across all of the atoms in the molecule in proportion to the square of the coefficients multiplying their respective atomic orbitals. *Therefore, the carbon atom P orbital with the largest squared coefficient in the HOMO will be the atom most likely to undergo electrophilic aromatic substitution.* On the other hand, nucleophilic substitution involves the donation of electron density to the molecule by a nucleophile. The corresponding election density will most likely to be in the empty MO of the lowest energy, the LUMO. *The carbon atom with the largest squared coefficient in the LUMO, once again, will be the site best able to accept the donated electron density and will therefore be the site of nucleophilic substitution.* The coefficients and squares of the coefficients of the HOMO and LUMO Hückel molecular orbitals are recorded in Table 4.12. Hence, in this molecule (Fig. 4.18), positions 1,4,5 and 8 are more susceptible for electrophilic substitution as well as for nucleophilic substitution.

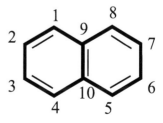

Fig. 4.18 Naphthaleine

Table 4.12 LUMO and HOMO coefficients and electron densities of naphthaleine

	Atom-1	Atom-2	Atom-3	Atom-4	Atom-5	Atom-6	Atom-7	Atom-8	Atom-9	Atom-10
Coefficients HOMO	0.425	0.263	−0.263	−0.425	0.425	0.263	−0.263	−0.425	0.000	0.000
Coefficients LUMO	0.425	−0.263	−0.263	0.425	−0.425	0.263	0.263	−0.425	0.000	0.000
Square of coefficients of HOMO	0.181	0.069	0.069	0.181	0.181	0.069	0.069	0.181	0.000	0.000
Square of coefficients of LUMO	0.181	0.069	0.069	0.181	0.181	0.069	0.069	0.181	0.000	0.000

4.24 The HMO and Symmetry

In symmetric molecules, the HMOs will also keep well defined symmetry properties. If two atoms, 1 and 2, are symmetrically equivalent, then the coefficients for the $2p\pi$ atomic orbitals on these atoms are related as:

$$C_1 = \pm C_2 \tag{4.59}$$

For example, *trans*-butadiene belongs to the $C2h$ point group with symmetry elements: E, $C2$, i, σh. In the molecule (Fig. 4.15), atoms 1 and 4 are symmetrically related. Similarly, atoms 2 and 3 are also related. When we compare their coefficients (Table 4.8), Eq. 4.59 can be visualized. Moreover, the respective bond orders are also equal. HMO wavefunctions can be written as:

$$\phi_1 = 0.372\chi_1 + 0.602\chi_2 + 0.602\chi_3 + 0.372\chi_4 \tag{4.60}$$
$$\phi_2 = 0.602\chi_1 + 0.372\chi_2 - 0.372\chi_3 - 0.602\chi_4 \tag{4.61}$$
$$\phi_3 = 0.602\chi_1 - 0.372\chi_2 - 0.372\chi_3 + 0.602\chi_4 \tag{4.62}$$
$$\phi_4 = 0.372\chi_1 - 0.602\chi_2 + 0.602\chi_3 - 0.372\chi_4 \tag{4.63}$$

where χ_i is the atomic orbital wavefunction of the ith atom. By group theory, each HMO belongs to a definite irreducible representation of the point group of the molecule. Let us verify this on the example of *trans*-butadiene. Firstly, we need to establish the results of the action of the $C2h$ symmetry elements on all the atomic orbitals. The effect of various symmetry operations to the $2p\pi$-HMO system can be studied. Let us see the effect of identity operation on the function (Fig. 4.19).

From the figure it is clear that $\hat{E}\phi_1 = \phi_1$. When the molecule is subjected to $\hat{C}_2$ operation, each $2p\pi$-orbital is rotated by $180°$ along the C_2 axis of rotation as shown in Fig. 4.20.

$$\hat{C}_2\phi_1 = 0.372\left(\hat{C}_2\chi_1\right) + 0.602\left(\hat{C}_2\chi_2\right) + 0.602\left(\hat{C}_2\chi_3\right) + 0.372\left(\hat{C}_2\chi_4\right) \tag{4.64}$$

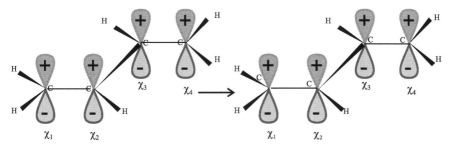

Fig. 4.19 Effect of identity operation on *trans*-butadiene

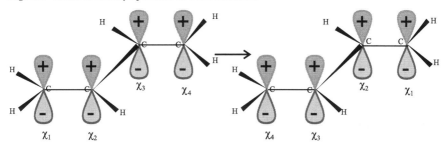

Fig. 4.20 Effect of C_2 axis of rotation on *trans*-butadiene

But, $\hat{C}_2\chi_1 = \chi_4, \hat{C}_2\chi_2 = \chi_3, \hat{C}_2\chi_3 = \chi_2$ and $\hat{C}_2\chi_4 = \chi_1$. The third symmetry operation is inversion (Fig. 4.21). The operation on the HMO can be represented as:

$$\hat{i}\phi_1 = 0.372\left(\hat{i}\chi_1\right) + 0.602\left(\hat{i}\chi_2\right) + 0.602\left(\hat{i}\chi_3\right) + 0.372\left(\hat{i}\chi_4\right) \tag{4.65}$$

But, $\hat{i}\chi_1 = -\chi_4, \hat{i}\chi_2 = -\chi_3, \hat{i}\chi_3 = \chi_2$ and $\hat{i}\chi_4 = -\chi_1$.

Now the molecule is subjected to the last element of symmetry, σ_h (Fig. 4.22):

$$\hat{\sigma}_h\phi_1 = 0.372\left(\hat{\sigma}_h\chi_1\right) + 0.602\left(\hat{\sigma}_h\chi_2\right) + 0.602\left(\hat{\sigma}_h\chi_3\right) + 0.372\left(\hat{\sigma}_h\chi_4\right) \tag{4.66}$$

It is clear that $\hat{\sigma}_h\chi_1 = -\chi_1, \hat{\sigma}_h\chi_2 = -\chi_2, \hat{\sigma}_h\chi_3 = -\chi_3$ and $\hat{\sigma}_h\chi_4 = -\chi_4$. Thus, as a result of the action of the symmetry operations $\hat{E}, \hat{C}_2, \hat{i}$ and $\hat{\sigma}h$ on ϕ_1, the orbital is

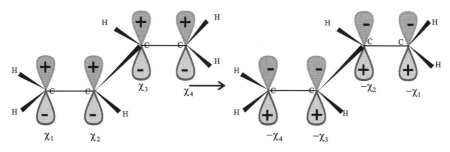

Fig. 4.21 Effect of inversion operation on butadiene

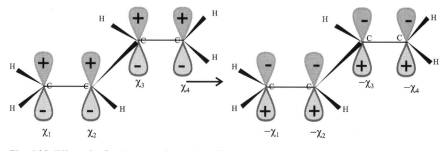

Fig. 4.22 Effect of reflection operation on butadiene

multiplied by the numbers $1, 1, -1, -1$, respectively. These numbers are the characters of the irreducible representation A_u of C_{2h}. This shows that ϕ_1 belongs to the irreducible representation A_u of C_{2h} point group (Table 4.13). Similarly, we can set the irreducible representations for ϕ_2, ϕ_3 and ϕ_4 as B_g, A_u, and B_g, respectively. Lowercase symbols for the irreducible representations are often used to denote molecular orbitals. If there is more than one orbital belonging to an irreducible representation, the symbols are preceded by numbers, starting from the lower-energy orbitals. Thus, the HMOs ϕ_1, ϕ_2, ϕ_3 and ϕ_4 can be designated as $1a_u, 1b_g, 2a_u$ and $2b_g$.

The symmetries of the orbitals can be used to decide whether the electronic transitions are allowed or forbidden. If the dipole moment vector is not zero, the transition is allowed, else it is forbidden.

$$\mu_x = -e \int \phi_{\text{final}} x \phi_{\text{initial}} \, d\tau \tag{4.67}$$

$$\mu_y = -e \int \phi_{\text{final}} y \phi_{\text{initial}} \, d\tau \tag{4.68}$$

$$\mu_z = -e \int \phi_{\text{final}} z \phi_{\text{initial}} \, d\tau \tag{4.69}$$

In general, the integral of a product of three functions over space $\int f_1 f_2 f_3 \, d\tau$ is non zero, or if the product of irreducible representations of f_1, f_2 and f_3 contains the totally symmetric irreducible representation (with all eigenvalues equal to one), then the corresponding transition is allowed. Thus for butadiene, allowed transitions are $\phi_2 \rightarrow \phi_3$ ($1b_g \rightarrow 2a_u$) and $\phi_1 \rightarrow \phi_4$ ($1a_u \rightarrow 2b_g$) and the forbidden transitions are $\phi_2 \rightarrow \phi_4$ ($1b_g \rightarrow 2b_g$) and $\phi_1 \rightarrow \phi_3$ ($1a_u \rightarrow 2a_u$). Group theory can also be used in order to simplify the HMO secular equations for symmetric molecules. This is achieved by employing symmetry-adapted linear combinations of AOs (SALCs) rather than AOs when constructing HMOs. Thus, the HMO determinantal equation is replaced by two or more equations involving smaller determinants which are easier to solve.

Table 4.13 Character table for C_{2h}

C_{2h}	E	C_2	i	σ_h	Linear functions, rotations	Quadratic functions
A_g	+1	+1	+1	+1	R_z	x^2, y^2, z^2, xy
B_g	+1	−1	+1	−1	R_x, R_y	xz, yz
A_u	+1	+1	−1	−1	z	
B_u	+1	−1	−1	+1	x, y	

4.25 Molecules Containing Heteroatoms

The HMO calculations of molecules containing heteroatoms can be done in a similar manner. In the Hamiltonian matrix, appropriate values for α and β values have to be put. This can be computed with the help of the equations. For a bond xy:

$$\beta_{xy} = k_{xy}\beta \tag{4.70}$$

For an atom x :

$$\alpha_x = \alpha + h_x\beta \tag{4.71}$$

k_{xy} and h_x values are available. Table 4.14 gives these values for common computations. It is to be noted that number of π-electrons in the molecule is no longer equal to the number of atoms. The values of k now have to be input for all bonds. For C−C bonds, $k = 1$. We have to substitute the values of h and k in Eq. 4.70 to get the corresponding α and β values.

As for example, in acrolein (CH_2=CH−CH=O), the determinant will be:

$$\begin{vmatrix} \alpha - E & \beta & 0 & 0 \\ \beta & \alpha - E & \beta & 0 \\ 0 & \beta & \alpha - E & \beta\sqrt{2} \\ 0 & 0 & \beta\sqrt{2} & \alpha + 2\beta - E \end{vmatrix} = 0 \tag{4.72}$$

Table 4.14 Values of h and k for common systems

Element		h		k
Nitrogen	N	0.5	C−N	0.8
	N	1.5	C−N	1.0
	N	2.0	N−O	0.7
Oxygen	O	1.0	C−O	0.8
	O	2.0	C=O	1.0
	O	2.5		
Chlorine	Cl	2.0	C−Cl	0.4

Table 4.15 Coefficients of the MOs of acrolein

	Atom 1	Atom 2	Atom 3	Atom 4
MO-1	0.083	0.207	0.433	0.874
MO-2	0.567	0.691	0.276	−0.354
MO-3	0.684	−0.150	−0.651	0.293
MO-4	0.452	−0.676	0.559	−0.160

Substituting appropriate values:

$$
\begin{vmatrix}
0.0 & 1.0 & 0.0 & 0.0 \\
1.0 & 0.0 & 1.0 & 0.0 \\
0.0 & 1.0 & 0.0 & 1.0 \\
0.0 & 0.0 & 1.0 & 2.0
\end{vmatrix} = 0
\tag{4.73}
$$

The coefficients of corresponding atomic orbitals can be computed, as we have seen in hydrocarbons. For acrolein, these values are tabulated in Table 4.15.

4.26 The Extended Hückel Method

The Extended Hückel Molecular Orbital Method (EHM) [6] grew out of the need to consider all valence electrons in a molecular orbital calculation. By considering all valence electrons, we could compute the molecular structure, the energy barriers for the rotation about bonds, and even determine the energies and structures of transition states for reactions.

The electronic wavefunction is taken as the product of a valence wavefunction and a core wavefunction and can be written as Eq. 4.74:

$$
\psi_{Total} = \phi_{Core} + \phi_{Valence}
\tag{4.74}
$$

The total valence electron wavefunction is described as a product of the one-electron wavefunctions:

$$
\phi_{Valence} = \psi_1(1)\psi_2(2)\psi_3(3)\ldots\psi_j(n)
\tag{4.75}
$$

where n is the number of electrons and j identifies the molecular orbital. Each molecular orbital again is given as an LCAO.

$$
\psi_j = \sum_{r=1}^{N} c_{jr}\phi_r
\tag{4.76}
$$

ϕ_r are the valance atomic orbitals chosen to include the $2s$, $2p_x$, $2p_y$, and $2p_z$ of the carbons and heteroatoms in the molecule and the $1s$ orbitals of the hydrogen atoms. The set of orbitals defined here is called a *basis set*. Since this basis set contains only the atomic-like orbitals for the valence shell of the atoms in a molecule, it is called a *minimal basis set*. We shall see more on basis sets in Chap. 5. We can deduce a matrix equation for all the molecular orbitals as in Eq. 4.77.

$$HC = SCE \tag{4.77}$$

where H is a square matrix containing the H_{rs}, the one electron energy integrals, and C is the matrix of coefficients for the atomic orbitals. Each column in C is the C' that defines one molecular orbital in terms of the basis functions. In the extended Hückel theory, the overlap is not neglected, and S is the matrix of overlap integrals. E is the diagonal matrix of orbital energies. All of these are square matrices with a size that equals the number of atomic orbitals used in the LCAO for the molecule under consideration. Similar to Hückel molecular orbital theory, Eq. 4.76 stands for an eigenvalue problem. For any extended Hückel calculation, we need to set up these matrices and then find the eigenvalues and eigenvectors. The eigenvalues are the orbital energies, and the eigenvectors are the atomic orbital coefficients that define the molecular orbital in terms of the basis functions.

The elements of the H matrix are assigned using experimental data, which makes the method a *semi-empirical* molecular orbital method. The off-diagonal Hamiltonian matrix elements are given by an approximation due to Wolfsberg and Helmholz that relates them to the diagonal elements and the overlap matrix element:

$$H_{ij} = \frac{1}{2} K \left(H_{ii} + H_{jj} \right) S_{ij} \tag{4.78}$$

The rationale for this expression is that the energy should be proportional to the energy of the atomic orbitals, and should be greater when the overlap of the atomic orbitals is greater. The contribution of these effects to the energy is scaled by the parameter K. Hoffmann assigned the value of K as 1.75 after a study of the effect of this parameter on the energies of the occupied orbitals of ethane. The H_{ii} are chosen as valence state ionization potentials with a minus sign (Table 4.16) to indicate binding.

It is common in many theoretical studies to use the extended Hückel molecular orbitals as a preliminary step to determining the molecular orbitals by a more sophisticated method, such as the CNDO/2 method and ab initio quantum chemistry methods. This leads to the determination of more accurate structures and electronic properties. A recent program for the *extended Hückel method* is YAeHMOP which stands for "yet another extended Hückel molecular orbital package". The extended Hückel method can be used for determining the molecular orbitals, but it is not very successful in determining the structural geometry of an organic molecule. It can, however, determine the relative energy of different geometrical configurations. It involves calculations of the electronic interactions in a rather simple way, where the electron-electron repulsions are not explicitly included and the total energy is just a sum of terms for each electron in the molecule. Hückel Molecular Orbital Calculator 2.0 is software which is available free, and which can compute the MO energy calculation from the following site: http://web.uccs.edu/danderso/huckel/huckel_setup.exe.

Table 4.16 H_{ii} values from the ionization potential*

Bonding site	Ionization potential (eV)	H_{ii} values (eV)
H-1s	13.60	−13.60
C-2s	21.40	−21.40
C-2p	11.40	−11.40
N-2s	25.58	−25.58
N-2p	13.90	−13.90
O-2s	32.38	−32.38
O-2p	15.85	−15.85
F-2s	40.20	−40.20
F-2p	18.66	−18.66

*(These parameters are available at http://www.op.titech.ac.jp/lab/mori/EHTB/EHTB.html)

4.27 Exercises

1. Calculate the molecular orbital energy levels (eigenvalues) and coefficients
   (eigenvectors) for the following p systems, each possessing four p orbitals
   (Fig. 4.23).

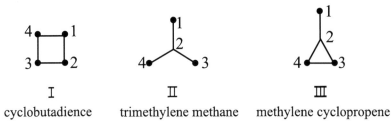

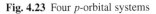

Fig. 4.23 Four p-orbital systems

2. Using the Hückel Carbon program:

   a. Compare the total p energies and the p bond orders in 1,3,5-hexatriene and
      3-methylene-1,4-pentadiene (Fig. 4.24). What can you conclude about the
      effects of branching in a conjugated pi-system?

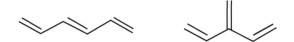

Fig. 4.24 1,3,5-hexatriene and 3-methylene-1,4-pentadiene

   b. The following bicyclic compounds (Fig. 4.25) all have ten p-electrons.
      i. Which of them exhibit aromatic stabilization?

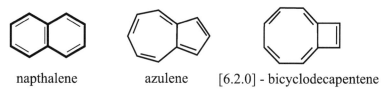

napthalene   azulene   [6.2.0] - bicyclodecapentene

Fig. 4.25 Bicyclic compounds

    ii. What unusual property of azulene is predicted by the calculations?

    iii. Why is the a position in naphthalene more reactive toward electrophilic aromatic substitution than the b position?

3. Using the Hückel Hetero program:

    a. Predict the effects of electron donating and withdrawing groups on electrophilic and nucleophilic reactions of the double bond, by comparing the appropriate HOMO and LUMO energy levels and orbitals in aminoethylene, ethylene, and acrolein (Fig. 4.26). Indicate which would be more reactive toward electrophiles, and which toward nucleophiles, and explain the regiochemistry of the reactions.

Fig. 4.26 Aminoethylene, ethylene, and acrolein

    b. Draw the molecular orbitals of formaldehyde, formamide, and urea (Fig. 4.28). Compare the delocalization energies, electron densities, and bond orders. (There is no delocalization energy for formaldehyde; rather, the p energy you calculate will serve as the localized energy for a pair of electrons in a C=O bond.) On the basis of these values, discuss the VB structures that can be written for each of these systems and show how these results are in accord with the well-known properties of the molecules (such as the fact that protonation occurs on O rather than N, and that there is limited rotation about the C−N bond).

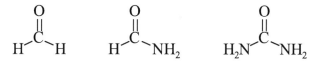

Fig. 4.27 Formaldehyde, formamide, and urea

    c. Borazole (Fig. 4.28) is an interesting analog of an aromatic system, and in fact has been called "inorganic benzene". Compare the HDE for this system with that for benzene and comment on the possible aromatic character of borazole.

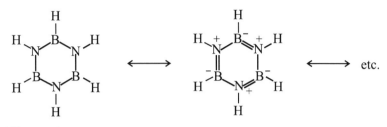

Fig. 4.28 Borazole

    d. Compare the stabilities of furan and pyrrole with that of the cyclopentadienyl anion (Fig. 4.29). Is the Hückel $4n + 2$ rule valid for heterocycles?

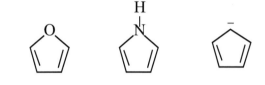

Fig. 4.29 Furan, pyrrole, and cyclopentadienyl anion

    4. Predict the aromaticities of:

      a. 16 annulene (Fig. 4.30)

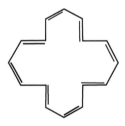

Fig. 4.30 16 annulene

      b. Cyclobutadiene
      c. Cyclopentadienylanion

    5. Which of the following reactions (Fig. 4.31) leads to a stable species?

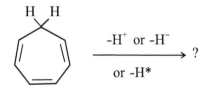

Fig. 4.31 Identifying a stable species

    6. Why is 1,3,5,7-cyclooctatetraene (C_8H_8–COT) non-planar? Why is the molecule readily reduced to the planar COT dianion ($C_8H_8^{-2}$), whereas COT has alternat-

ing carbon-carbon bonds of about 1.35 and 1.48 Å, and the dianion has a single distance of about 1.40 Å? Account for this.

7. Calculate the delocalization energies of carbocation, carbanion, and the free radical obtained from propene.
8. Find the delocalization energy of benzene.
9. Describe the structure and basis set, then generate the molecular orbitals and energy level diagram for the molecular orbital of cyclo-butadiene. How do your results vary from those of butadiene? Predict also the wavelength of its lowest energy electronic absorption.
10. Solve the Hückel problem for benzene. This time you don't have to generate the molecular orbitals, just the X vector and MOs matrix. Construct the energy level diagram for the molecular orbitals and insert electrons into your diagram.
11. Solve the Hückel problem for methylene cyclopentadiene. This time you don't have to generate the molecular orbitals, just the X vector and the MO matrix. Construct the energy level diagram for the molecular orbitals and insert electrons into your diagram. Predict also the wavelength of its lowest energy absorption.
12. For anthracene with the proper numbering, generate the MO matrix and the X vector. Predict the carbon atom(s) most likely to be the site for electrophilic aromatic substitution. Also, predict the site(s) for nucleophilic substitution. Predict the wavelength for the lowest energy absorption in the UV-visible region of the electromagnetic spectrum.
13. How is the pi-electron energy of ethyne (acetylene) calculated?
14. Derive the pi-electron wavefunctions of benzene and cyclopentadienyl anion.
15. Calculate the pi-electron energy levels and the wavefunctions of bicycle butadiene.
16. Calculate the mobile bond orders for bicycle butadiene.

References

1. Coulson CA (1947) Faraday Discussions. Chem Soc 2:9
2. Hückel E (1934) Trans Faraday Soc 30:59
3. Greenwood HH (1972) Computing Methods in Quantum Organic Chemistry. Wiley-Interscience, New York
4. Von Nagy-Felsobuki E (1989) Hückel theory and photoelectron spectroscopy. J Chem Educ 66:821
5. Hoffmann R (1963) An Extended Hückel Theory. J Chem Phys 39:1397–1412
6. Coulson CA, O'Leary B, Mallion RB (1978) Hückel Theory for Organic Chemists. Academic, London

Chapter 5
Hartree-Fock Theory

5.1 Introduction

We have seen a quantum mechanical computation with a lower level of accuracy. In a molecular orbital consisting of many electrons the wavefunctions become very complex. Since the electrons in a molecule are negatively charged, they repel each other, which clearly affects their motion. Over a period of time they may even share the same region of space providing maximum repulsive forces. Hence, at any instant, there is a strong tendency for the electrons to avoid each other, minimizing the repulsive force and thereby stabilizing the system. As a result their motions are highly *correlated*. The difficulty of finding a wavefunction for a large number of correlated electrons is one of the fundamental challenges of modern computational chemistry. The starting point of computation for most of the methods in quantum chemistry is to introduce the approximation that the motion of the particles is not correlated, and to develop a wavefunction for these *independent particles*. This approximation is known as the independent particle (4.3) approximation. These particles may still interact, but each particle experiences not an instantaneous interaction with the other particles. The interaction changes as the electrons move (which will complicate its motion). An interaction of particle resulting from a messy representation of the averaged position of all other particles can be included. When this approximation is made, the problem of finding a wavefunction for the complex systems is simplified. It is now made up of individual wavefunctions – one for each particle. Although we know that the independent particle approximation on which they are based is often a serious oversimplification, in many cases these individual wavefunctions are found to provide a great deal of insight into the chemical behavior of a molecule.

5.2 The Hartree Method

The Hartree method is a single electron approximation technique used in multi-electron systems. The molecular Hamiltonian is split up into individual single electron Hamiltonians. Consider a molecular system with N-electrons, each with degrees of freedom r_i. The wavefunction (Hartree function) $\psi_h(r_1, r_2, \ldots, r_N)$ is given by the Hartree product as shown in Eq. 5.1:

$$\psi_h(r_1, r_2, \ldots, r_N) = \phi_1(r_1).\phi_2(r_2)\ldots\phi_N(r_N) . \tag{5.1}$$

The Hamiltonian can be computed based on this concept. For the n-electron system, the Hamiltonian is given by:

$$\hat{H}_e = \hat{T}_e + \hat{V}_{ne} + \hat{V}_{ee} + \hat{V}_{nn} \tag{5.2}$$

$$\left[\text{Where, } \hat{T}_e = \sum_{i=1}^{n} \frac{-\nabla_i^2}{2}, \hat{V}_{ne} = \sum_{i}^{n} \sum_{A}^{N} \frac{-Z_A}{r_{iA}} \hat{V}_{ee} = \sum_{i}^{n} \sum_{j}^{n} \frac{1}{r_{ij}} = \hat{g}_{ij} \right.$$

$$\left. \text{and } \hat{V}_{nn} = \sum_{A}^{N} \sum_{B>A}^{N} \frac{Z_A Z_B}{R_{AB}} \right]$$

Here A and B are for representing the nuclei, i and j are for representing electrons, Z is the nuclear charge, $\hat{T}$ is the kinetic energy operator, and $\hat{V}$ the potential energy operator. Or, the Hamiltonian is written as:

$$\hat{H} = \sum_{i=1}^{n} \frac{-\nabla_i^2}{2} + \sum_{i}^{n} \sum_{A}^{N} \frac{-Z_A}{r_{iA}} + \sum_{i}^{n} \sum_{j}^{n} \frac{1}{r_{ij}} + \sum_{A}^{N} \sum_{B>A}^{N} \frac{Z_A Z_B}{R_{AB}} \tag{5.3}$$

Here, $\hat{V}_{nn}$ is independent of electronic coordinators. $\hat{T}_e$ and $\hat{V}_{ne}$ depend upon one-electron coordinators.

$$\hat{T}_e + \hat{V}_{ne} = \sum_{i=1}^{n} \frac{-\nabla_i^2}{2} + \sum_{i}^{n} \sum_{A}^{N} \frac{-Z_A}{r_{iA}} = \sum_{i=1}^{n} \left(\frac{-\nabla_i^2}{2} + \sum_{A}^{N} \frac{-Z_A}{r_{iA}} \right) = \sum_{i=1}^{n} \hat{h}_i \tag{5.4}$$

Finally there is a term $\hat{V}_{ee}$, which is the sum of $n(n-1)/2$ two-electron coordinators. Hence, the Hamiltonian becomes:

$$\hat{H} = \sum_{A}^{N} \sum_{B>A}^{N} \frac{Z_A Z_B}{R_{AB}} + \sum_{i=1}^{n} \hat{h}_i + \sum_{i}^{n} \sum_{j>i}^{n} \frac{1}{r_{ij}} \tag{5.5}$$

Substituting this Hamiltonian expression in the energy equation:

$$E = \int \psi \left[\sum_{A}^{N} \sum_{B>A}^{N} \frac{Z_A Z_B}{R_{AB}} \right] \psi \, dx + \int \psi \left[\sum_{i=1}^{n} \hat{h}_i \right] \psi \, dx + \int \psi \left[\sum_{i}^{n} \sum_{j>i}^{n} \frac{1}{r_{ij}} \right] \psi \, dx \tag{5.6}$$

The first term of the integral stands for nuclear-nuclear repulsion and is the integral over a constant (independent of coordinates). Hence:

$$V_{NN} = \int \psi \left[\sum_{A}^{N} \sum_{B>A}^{N} \frac{Z_A Z_B}{R_{AB}} \right] \psi \, dx = \left[\sum_{A}^{N} \sum_{B>A}^{N} \frac{Z_A Z_B}{R_{AB}} \right] \tag{5.7}$$

The second and third terms of Eq. 5.10 include integrals of sums, which can be written as sums of integrals. The second integral expression can be written as:

$$\sum_{i=1}^{n} \left[\int \psi \hat{h}_i \psi \, dx \right] = \sum_{i=1}^{n} \left[\int \chi_i \hat{h}_i \chi_i \, d\tau \right] = \sum_{i=1}^{n} h_{ii} \tag{5.8}$$

$$h_{ii} = \int \phi_i \hat{h}_i \phi_i \, d\tau = \int \phi_i \left[\frac{-\nabla_i^2}{2} + \sum_A^N \frac{-Z_A}{r_{iA}} \right] \phi_i \, d\tau$$

$$= \int \phi_i \left[\frac{-\nabla_i^2}{2} \right] \phi_i \, d\tau + \int \phi_i \left[\sum_A^N \frac{-Z_A}{r_{iA}} \right] \phi_i \, d\tau = \sum_{i=1}^{n} h_{ii}$$

$$\sum_{i=1}^{n} h_{ii} = \sum_{i=1}^{n} [T_{e,i} + V_{Ne,i}] = T_e + V_{Ne} \tag{5.9}$$

T_e is the electronic kinetic energy, and V_{Ne} is the potential energy due to nuclear-electronic Coulombic attraction. The third integral terms, and the two-electron terms are more complicated. In the Hartree treatment, the molecular orbital is considered as the product of single electron orbitals. Thus:

$$\psi(r_1, r_2, \ldots) = \phi(r_1)\phi(r_2)\ldots . \tag{5.10}$$

Each orbital is calculated for one electron moving in an average field of the nuclei and all other electrons.

$$\langle \Pi | \hat{g}_{ij} | \Pi \rangle = \langle [\phi_1(1)\phi_2(2)\ldots\phi_N(N)] | \hat{g}_{ij} | [\phi_1(1)\phi_2(2)\ldots\phi_N(N)] \rangle$$
$$= \langle \phi_1(1)\phi_2(2) | \hat{g}_{ij} | \phi_1(1)\phi_2(2) \rangle \langle \phi_3(3) | \phi_3(3) \rangle \ldots \langle \phi_N(N) | \phi_N(N) \rangle$$
$$= \langle \phi_1(1)\phi_2(2) | \hat{g}_{ij} | \phi_1(1)\phi_2(2) \rangle = J_{12} \tag{5.11}$$

J_{12} is known as the Coulomb integral [1]. It represents the classical repulsion between two charge distributions $\phi_1^2(1)$ and $\phi_2^2(2)$. It is to be noted that the square of the orbital function is a measure of the electronic or charge distribution. The Coulomb repulsion corresponding to a particular distance between the reference electron x_1 and another electron x_2 is weighted by a probability that the other electron is at that point in space. The results of the application of the Coulomb integral on a spin orbital depend only on the orbital function. Hence, the corresponding potential and operator are named as "local". Because $1/r$ is always positive, and ϕ^2 is a probability measure, this term contributes a positive energy, i.e., a destabilization. In general, the Coulomb integral can be written as:

$$J_{ij} = \left\langle \phi_i(1)\phi_j(2) \left| \frac{1}{r_{12}} \right| \phi_i(1)\phi_j(2) \right\rangle \tag{5.12}$$

With these generalizations, the Hamiltonian for a helium atom with one nucleus and two electrons is given by the expression:

$$\hat{H}_{He} = \sum_{i=1}^{n} \hat{h}_i + \sum_{i}^{n}\sum_{j>i}^{n} \frac{1}{r_{ij}} = \sum_{i=1}^{n} \hat{h}_i + J_{ij} = \hat{h}_1 + \hat{h}_2 + J_{12} \tag{5.13}$$

$$\left[\text{Here the nuclear repulsion } \sum_{A}^{N}\sum_{B>A}^{N} \frac{Z_A Z_B}{R_{AB}} \text{ is zero.} \right]$$

5.3 Bosons and Fermions

A quantum state of electrons can be well explained by spatial and spin coordinates. In this procedure, orbital functions are determined through the spatial position and the spin of electrons. Separating the spatial and spin functions, the orbital function can be written as:

$$\Phi_{(x,y,z,s)} = \Phi_{(x,y,z)} \times \sigma_s \qquad (5.14)$$

Here $\Phi_{(x,y,x)}$ stands for the position of electron in space and σ_s for the electronic spin. Hence, electrons present in an orbital may be symmetric with the spatial and spin functions identical or antisymmetric with the spin function alone. With this condition, all particles in nature can be classified as either *bosons* or *fermions*. Bosons are particles with integer spin. Higgs boson, pion, $^1H_1$ and $^4He_2$ in ground state are examples of bosons with spin 0. $^1H_1$ and $^4He_2$ in first excited state, ρ-meson, photon, W and Z bosons and gluons are examples of bosons with spin 1. Similarly, $^{16}O_8$ in ground state and graviton are with spin 2.

Fermions are particles with half-integer spin. Examples: spin $1/2 \rightarrow ^3He_2$ in ground state, proton, neutron, quark, electron, and neutrino, spin $3/2 \rightarrow ^5He_2$ in ground state and Δ-baryons (excitations of the proton and neutron). Excitations will change the spin only by an integer amount. The basic building blocks of atoms are all fermions; composite particles (nuclei, atoms, molecules) made of an odd number of protons, neutrons, and electrons are also fermions, whereas those made of an even number are bosons. Fermions obey the Pauli's exclusion principle. (No two fermions will occupy the same quantum state). This is the basis of atomic structure and the periodic table. There is no exclusion property for bosons, which are free to crowd into the same quantum state. This explains the spectrum of black-body radiation and the operation of lasers, the properties of liquid and superconductors.

In an orbital, electronic spin is quantized by $\pm 1/2$. The complete orbital function is generally known as *spinorbital* function. Pauli's exclusion principle further restricts electron exchange in an orbital [2]. Such an exchange should be associated with making the spin opposite without making any change in the spatial part or the part corresponding to principal, azimuthal, and magnetic quantum levels. The exchange is hence antisymmetric. The notation for orbitals is sometimes changed from the spatial orbital representation, $\phi(r)$, to the spin orbital function, $\chi(x)$. The spin orbital function is:

$$\chi(x) = \phi(r, \omega), \qquad (5.15)$$

where, ω is the spin function.

5.4 Spin Multiplicity

The spin multiplicity of a system is given by the equation, $x = (2S + 1)$, where S is the total spin of the system. A paired orbital contributes zero to the total spin as the

$+1/2$ spin given by the α-electron is nullified by the $-1/2$ spin given by the β-electron. Each unpaired electron contributes $+1/2$ to the total spin. Thus, a system with no unpaired orbital will have a spin multiplicity of 1(singlet), a system with one half filled orbital will have a multiplicity of two (doublet), a system with two unpaired orbitals will have a multiplicity of three (triplet), and so on.

5.5 The Slater Determinant

For many electron systems represented by the wavefunction

$$\left| \Psi_{(x_1,x_2,\ldots,x_i,x_j,\ldots,x_N)} \, dx_1, dx_2,\ldots, dx_i, dx_j,\ldots, dx_N \right| ,$$

the probability of finding N-electrons in a volume element

$$d\tau = x_1, x_2, \ldots, x_i, x_j, \ldots, x_N$$

is given by

$$\int \cdots \int \left| \Psi_{(x_1,x_2,\ldots,x_i,x_j,\ldots,x_N)} \right|^2 dx_1, dx_2,\ldots, dx_i, dx_j,\ldots, dx_N$$

and it should be unity. This is a consequence of the normalization condition of the avefunction. If coordinates of any two electrons are interchanged, then also the probability should remain the same. That is:

$$\int \cdots \int \left| \Psi_{(x_1,x_2,\ldots,x_i,x_j,\ldots,x_N)} \right|^2 dx_1, dx_2,\ldots, dx_i, dx_j,\ldots, dx_N =$$

$$\int \cdots \int \left| \Psi_{(x_1,x_2,\ldots,x_j,x_i,\ldots,x_N)} \right|^2 dx_1, dx_2,\ldots, dx_j, dx_i,\ldots, dx_N \qquad (5.16)$$

Bear in mind that here the coordinates of i and j are interchanged. Hence, the functional change possible during the exchange of electrons is only a spin change. Slater made a correlation between the spinorbital property and the determinant property by noting the characteristic properties of determinants. A determinant vanishes if two rows or columns are identical (the determinant is zero) and when rows or columns are interchanged, the determinant changes its sign. Hence, if spinorbitals are arrayed as a determinant, Pauli's exclusion principle can be well accommodated in it. For a two electron orbital system with electrons 1 and 2, and with spins α and β, the spinorbital function is written as:

$$\Phi_{(1,2)} = \begin{vmatrix} \phi_{(1,\alpha)} & \phi_{(2,\alpha)} \\ \phi_{(1,\beta)} & \phi_{(2,\beta)} \end{vmatrix} \qquad (5.17)$$

This allows only the antisymmetric combination:

$$\Phi_{(1,2)} = \left[\phi_{(1,\alpha)} \cdot \phi_{(2,\beta)} - \phi_{(1,\beta)} \cdot \phi_{(2,\alpha)} \right] \tag{5.18}$$

and not the symmetric combination:

$$\Phi_{(1,2)} = \left[\phi_{(1,\alpha)} \cdot \phi_{(2,\beta)} + \phi_{(1,\beta)} \cdot \phi_{(2,\alpha)} \right] \tag{5.19}$$

Such a determinant providing spinorbital property is known as the *Slater determinant*. The Hamiltonian set according to the above spinorbital function need not be close to the actual wavefunction. However, based on the variational principle, the process of approaching the spinorbital function to the real wavefunction will be associated with a decrease in energy. The Hamiltonian operator for the wavefunction can be written as:

$$\langle \hat{s} \rangle = \int \cdots \int \psi_{\text{trial}}^{*} / \hat{s} / \psi_{\text{trial}} \tag{5.20}$$

By Dirac bracket notation, it can be written as:

$$\langle \hat{s} \rangle \equiv \langle \psi_{\text{trial}}^{*} / \hat{s} / \psi_{\text{trial}} \rangle \tag{5.21}$$

In the above Eq. 5.21, ψ_{trial}^{*} is the complex conjugate of ψ_{trial}. Based on the variational principle the computed energy will be an upper bound to true energy. Hence:

$$\langle \psi_{\text{trial}} / \hat{s} / \psi_{\text{trial}} \rangle = E_{\text{trial}} \geq E_{\text{real}} = \langle \psi_{\text{real}} / \hat{H} / \psi_{\text{real}} \rangle . \tag{5.22}$$

In this expression, the complex conjugate term is avoided. Similar to

$$\Phi = \frac{1}{\sqrt{N!}} \begin{vmatrix} \phi_1(\mathbf{x}_1) & \phi_2(\mathbf{x}_1) & \cdots & \phi_N(\mathbf{x}_1) \\ \phi_1(\mathbf{x}_2) & \phi_2(\mathbf{x}_2) & \cdots & \phi_N(\mathbf{x}_2) \\ \cdots & \cdots & \cdots & \cdots \\ \phi_1(\mathbf{x}_N) & \phi_2(\mathbf{x}_N) & \cdots & \phi_N(\mathbf{x}_N) \end{vmatrix} \tag{5.23}$$

ψ_{real}, ψ_{trial} should also be finite, continuous and single-valued. If lower accepted energy results from n-functions, then the energy is said to be n-fold degenerate. Keeping in mind all the above requirements of a function, the N-electron system can be represented by the Slater determinant as given in Eq. 5.23. Here, each one-electron function $\phi_i(\mathbf{x}_i)$ stands for a spinorbital, and the pre-determinant factor is the normalization factor for the function. Generally, such a determinant can be simply represented with only the diagonal elements as given below:

$$\Phi = \hat{A} \left[\phi_1(1) \cdot \phi_2(2) \ldots \phi_N(N) \right] = \hat{A} \Pi , \tag{5.24}$$

where Π is the determinant diagonal product and $\hat{A}$ is the antisymmetrizer:

$$\hat{A} = \frac{1}{\sqrt{N!}} \sum_{p=0}^{N-1} (-1)^p \hat{P} = \frac{1}{\sqrt{N!}} \left[1 - \sum_{ij} \hat{P}_{ij} + \sum_{ijk} \hat{P}_{ijk} - \ldots \right] \quad (5.25)$$

$\hat{P}$ is a permutation operator, $\hat{P}_{ij}$ permutes two electrons, $\hat{P}_{ijk}$ permutes three electrons, and so on. It is to be noted that ψ_{real} is not a real function, while it is the function of another function. Such a function is named as functional. We shall see details of functionals in density functional theory.

5.6 Properties of the Slater Determinant

General properties of the Slater determinant with the perspective of the present context can be summarized as follows [3]:

1. It allows only antisymmetric electronic exchange within an orbital.
2. Two electrons present in an orbital should have opposite spin. If the spins were identical, then the Slater determinant would be: $\Phi_{(1,2)} = \begin{vmatrix} \phi_{(1,\alpha)} & \phi_{(2,\alpha)} \\ \phi_{(1,\beta)} & \phi_{(2,\beta)} \end{vmatrix}$, which on simplifying, we get zero. Hence, the Slater determinant wavefunction vanishes if the electrons have identical spin.
3. The wavefunction set according to Pauli's exclusion principle are said to be *antisymmetrized*.
4. Molecular orbital is obtained by the linear combination of atomic orbitals (LCAO). Hence, it is possible to have an approximation of molecular orbitals by considering them as made out of linear combination of antisymmetrized determinantal wavefunctions. *Columns are one-electron wavefunctions, molecular orbitals. Rows contain the electron coordinates.*

5.7 The Hartree-Fock Equation

The Hartree-Fock (HF) method is the most common ab initio method that is implemented in nearly every computational chemistry program. It is a modification to the Hartree treatment, which we have seen earlier. Here, we describe the many-electron wavefunction as an antisymmetrized product (the Slater determinant) of one-electron wavefunctions. Each electron moves independently in the spin orbital space and it experiences a Coulombic repulsion due to the average positions of electrons. It experiences exchange interaction due to antisymmetrization.
We have seen earlier that a one electron spinorbital integral is:

$$\langle \phi_i / \hat{o} / \phi_j \rangle = \langle i / \hat{o} / j \rangle = \int \phi_i(x_1) \hat{o}_{r_i} \phi_j(x_1) \, dx_1 \quad (5.26)$$

Similarly, a two-electron integral can be written as:

$$[\phi_i\phi_j/\phi_k\phi_l] = [ij/kl]$$
$$= \int \int \phi_i(x_1)\phi_j(x_1)\frac{1}{r_{12}}\phi_k(x_2)\,dx_1\,dx_2 \quad (5.27)$$

Here the square bracket is normally used to indicate that it is a functional and not a function. Whenever we want to determine the expectation value of a quantum operator, we multiply to the left with the conjugate complex of the wavefunction and integrate over the entire space. If the function is written as ψ_{HF} and the corresponding energy as E_{HF}, then the Schrödinger equation can be written as:

$$\left\langle \psi_{HF}/\hat{H}/\psi_{HF} \right\rangle = \left\langle \psi_{HF}/E_{HF}/\psi_{HF} \right\rangle$$
$$= \left\langle \psi_{HF}/\hat{H}/\psi_{HF} \right\rangle = E_{HF}\left\langle \psi_{HF}/\psi_{HF} \right\rangle \quad (5.28)$$

Or

$$E_{HF} = \frac{\left\langle \psi_{HF}/\hat{H}/\psi_{HF} \right\rangle}{\left\langle \psi_{HF}/\psi_{HF} \right\rangle} \quad (5.29)$$

If ψ_{HF} is known to us, E_{HF} can be easily calculated. Now, we shall see the method to calculate ψ_{HF}. The variational theorem tells us that the correct wavefunction among all possible Slater determinants is the one for which E_{HF} is minimal:

$$E_{min} = \left\langle \psi_{HF}/\hat{H}/\psi_{HF} \right\rangle < \left\langle \psi/\hat{H}_{electron}/\psi \right\rangle \quad (5.30)$$

That means that in order to find the HF wavefunction, we have to minimize the energy expression E_{HF} with respect to changes in the one electron orbitals $\phi_1 \rightarrow \phi_1 + \delta\phi_1$ from which we construct the Slater determinant Φ. The set of one-electron orbitals ϕ_i for which we obtain the lowest energy are the HF orbitals or the solutions to the HF equations. We know that the spin functions are orthonormal. That means:

$$\langle \alpha/\beta \rangle = \langle \beta/\alpha \rangle = 0 \quad (5.31)$$
$$\langle \alpha/\alpha \rangle = \langle \beta/\beta \rangle = 1 \quad (5.32)$$

Equations 5.31 and 5.32 together can be simplified as follows:

$$\left\langle \phi_i/\phi_j \right\rangle = \delta_{ij} \quad (5.33)$$

where δ_{ij} stands for the Krönecker delta which can have values 1 for $i = j$ and 0 otherwise.

Hence, the energy expression becomes:

$$E_{HF} = \left\langle \psi_{HF}/\hat{H}/\psi_{HF} \right\rangle \quad (5.34)$$

The HF function is an antisymmetrized orbital function introducing the exchange function K_{ij} in the Hamiltonian. K_{ij} can be computed as follows:

$$\langle \Pi | \hat{g}_{ij} | \hat{P}_{12} \Pi \rangle = \langle \phi_1(1)\phi_2(2) | \hat{g}_{ij} | \phi_2(2)\phi_1(1) \rangle \langle \phi_3(3) | \phi_3(3) \rangle \ldots \langle \phi_N(N) | \phi_N(N) \rangle$$
$$= \langle \phi_1(1)\phi_2(2) | \hat{g}_{ij} | \phi_2(2)\phi_1(1) \rangle = K_{12} \qquad (5.35)$$

Here K_{12} stands for the exchange integral. It does not have any classical analogue. The name *exchange integral* comes from the fact that the two electrons exchange their positions from the left to the right of the integrand. This suggests, correctly, that it has something to do with the Pauli's principle. It corresponds to the exchange of electrons in two-spin orbitals. The function depends upon all points in space as it depends upon the position of other electrons in space. Hence, the corresponding potential and operator are said to be nonlocal. The corresponding function is responsible for the formation of chemical bonds. The exchange integral (K_{ij}) is given by Eq. 5.21:

$$K_{ij} = \left\langle \phi_i(1)\phi_j(2) \left| \frac{1}{r_{12}} \right| \phi_i(2)\phi_j(1) \right\rangle \qquad (5.36)$$

However, in the derived expressions, the antisymmetrization effect should be there, somewhere. In fact, the exchange integrals "correct" the Coulomb integrals to maintain the antisymmetry of the wavefunction. We saw that the electrons (especially those of the same spin) tend to avoid each other rather more in the Slater determinant model than in the Hartree product model, so the Coulomb integrals should exaggerate the Coulomb repulsion of the electrons. The exchange integrals, which are negative, compensate for this exaggeration. In the integral term, if $i = j$, the expression leads to the potential due to the Coulomb interaction of an electron with itself. Hence, even if we compute the energy of a one-electron system, the equation gives a non-zero exchange potential. However, the HF scheme eliminates the possibility of error caused due to this self-interaction. If $i = j$, the Coulomb and exchange integrals cancel each other as they have the same value with the opposite sign. This cancels the effect of self-interaction.

For a two-electron system like helium energy, the expression becomes:

$$\hat{H}_{He} = \hat{h}_1 + \hat{h}_2 + J_{12} \pm K_{12} \qquad (5.37)$$

The exchange of electrons may be associated with an increase or decrease in energy and stability. Hence, the exchange function can be written as:

$$\psi_{\pm}(r_1, r_2) = \frac{1}{\sqrt{2}} \left(\phi_1^*(r_1)\phi_2^*(r_2) \pm \phi_1(r_1)\phi_2(r_2) \right) \qquad (5.38)$$

The HF equation may lead in to an increase or decrease in energy from the Hartree energy calculation. The spin correlation between electrons of the same spin leads to an increase in energy, while the correlation between electrons of opposite spin leads to a decrease in energy. As the decrease in energy is a stabilization condition favored by nature, electronic spins of an orbital are specified as the opposite (Pauli's exclusion principle). With this condition, the sign of K_{ij} becomes negative.

The overall contribution to the total energy of the potential energy due to the electronic–electronic Coulombic repulsion V_{ee} is therefore given as a difference of two terms:

$$V_{ee} = J_{ee} - K_{ee} = \sum_i^n \sum_j^n (J_{ee} - K_{ee}) \tag{5.39}$$

Overall, the energy of a Slater determinant is given by adding up all the terms discussed above. For the general case with matrix elements expressed as spin orbitals, one reaches the following expression:

$$E = V_{NN} + \sum_{i=1}^{n_{\text{electrons}}} h_{ii} + \sum_i^{n_{\text{electrons}}} \sum_j^{n_{\text{electrons}}} (J_{ee} - K_{ee}) \tag{5.40}$$

For a *closed-shell* system (a spin singlet where all the occupied orbitals have two electrons in them) with n-orbitals, the energy expression can be written as:

$$E = V_{NN} + 2 \sum_{i=1}^{n_{\text{orbitals}}} h_{ii} + \sum_i^{n_{\text{orbitals}}} \sum_j^{n_{\text{orbitals}}} (2J_{ee} - K_{ee}) \tag{5.41}$$

To apply the variational principle, the Coulomb and exchange integrals are written as operators:

$$E_e = \sum_{i=1}^N \langle \phi_i | \hat{h}_i | \phi_i \rangle + \frac{1}{2} \sum_i^N \sum_j^N \left(\langle \phi_j | \hat{J}_i | \phi_j \rangle - \langle \phi_j | \hat{K}_i | \phi_j \rangle \right) + V_{NN} \tag{5.42}$$

Where:

$$\hat{J}_i | \phi_j(2) \rangle = \langle \phi_i(1) | \hat{g}_{12} | \phi_i(1) \rangle \phi_j(2) \rangle \tag{5.43}$$

and:

$$\hat{K}_i | \phi_j(2) \rangle = \langle \phi_i(1) | \hat{g}_{12} | \phi_j(1) \rangle \phi_i(2) \rangle \tag{5.44}$$

The objective now is to find the best orbitals (ϕ_i, MOs) that minimize the energy (or at least remain stationary with respect to further changes in ϕ_i) maintaining the orthonormality of the orbitals. By the variational principle, the calculated energy will be always higher than the actual ground state energy. Therefore, we wish to find the set of molecular orbitals that minimizes the value of E. Since $\langle \psi / \hat{H} / \psi \rangle$ is stationary with respect to small variations in the molecular orbitals, $\delta \phi$ at the minimum, and since $\langle \psi / \psi \rangle$ must remain constant with a small $\delta \phi$, then "Lagrange's method of undetermined multipliers" may be used to derive the expression [4]. In terms of molecular orbitals, the Lagrange function can be written as:

$$L = E - \sum_{ij}^N \lambda_{ij} \left(\langle \phi_i | \phi_j \rangle - \delta_{ij} \right) \tag{5.45}$$

$$\delta L = \delta E - \sum_{ij}^N \lambda_{ij} \left(\langle \delta \phi_i | \phi_j \rangle + \langle \phi_i | \delta \phi_j \rangle \right) = 0 \tag{5.46}$$

The change in L with respect to very small changes in ϕ_i should be zero. Hence, the change in the energy with respect changes in ϕ_i becomes:

$$\delta E = \sum_{i=1}^{N} \left(\langle \delta\phi_i | \hat{h}_i | \phi_i \rangle + \langle \phi_i | \hat{h}_i | \delta\phi_i \rangle \right)$$
$$+ \sum_{ij}^{N} \left(\langle \delta\phi_i | \hat{J}_j - \hat{K}_j | \phi_i \rangle + \langle \phi_i | \hat{J}_j - \hat{K}_j | \delta\phi_i \rangle \right) \qquad (5.47)$$

Now, we introduce a new operator, $\hat{F}_i$, known as the Fock operator, $\hat{F}_i = \hat{h}_i + \sum_{j}^{N} (\hat{J}_j - \hat{K}_j)$. This operator is an effective one-electron operator, associated with the variation in the energy. Changing the energy expression in terms of the Fock operator:

$$\delta E = \sum_{i=1}^{N} \left(\langle \delta\phi_i | \hat{F}_i | \phi_i \rangle + \langle \phi_i | \hat{F}_i | \delta\phi_i \rangle \right) \qquad (5.48)$$

and:

$$\delta L = \sum_{i=1}^{N} \left(\langle \delta\phi_i | \hat{F}_i | \phi_i \rangle + \langle \phi_i | \hat{F}_i | \delta\phi_i \rangle \right) + \sum_{ij}^{N} \lambda_{ij} \left(\langle \delta\phi_i | \phi_j \rangle + \langle \phi_i | \delta\phi_j \rangle \right) = 0$$
$$(5.49)$$

According to the variational principle, the best orbitals, ϕ_i, will make $\delta L = 0$. With this substitution, and on rearrangement, we get a simple expression known as the HF equation as given below.

$$\hat{F}_i \phi_i = \sum_{ij}^{N} \lambda_{ij} \phi_j \qquad (5.50)$$

After unitary transformations, $\lambda_{ij} \rightarrow 0$ and $\lambda_{ii} \rightarrow \varepsilon_i$, HF equations in terms of canonical MOs and diagonal Lagrange multipliers can be written as:

$$\hat{F}_i \phi_i' = \varepsilon_i \phi_i' \qquad (5.51)$$

The HF equations cast in this way, form a set of pseudo-eigenvalue equations. A specific Fock orbital can only be determined once all the other occupied orbitals are known. A specific Fock orbital can only be determined if all the other occupied orbitals are known, and iterative methods must therefore be employed for determining the orbitals. A set of orbitals that is a solution to the HF equations (Eq. 5.51) are called self-consistent field (SCF) orbitals [5].

5.8 The Secular Determinant

The secular determinant equation for a multielectron system can be represented as:

$$\begin{vmatrix} (H_{11} - S_{11}E) & (H_{12} - S_{12}E) & \dots & (H_{1n} - S_{1n}E) \\ (H_{21} - S_{21}E) & (H_{22} - S_{22}E) & \dots & (H_{2n} - S_{2n}E) \\ \dots & \dots & \dots & \dots \\ (H_{n1} - S_{n1}E) & (H_{22} - S_{22}E) & \dots & (H_{nn} - S_{nn}E) \end{vmatrix} = 0 \qquad (5.52)$$

We have seen earlier that $S_{ij} = 0$ if $i \neq j$ and $S_{ij} = 1$ if $i = j$. That is, S_{ij} is orthonormal. Substituting the values of S, the secular determinant becomes:

$$\begin{vmatrix} (H_{11} - E_1) & (H_{12}) & \dots & (H_{1n}) \\ (H_{21}) & (H_{22} - E_2) & \dots & (H_{2n}) \\ \dots & \dots & \dots & \dots \\ (H_{n1}) & (H_{22}) & \dots & (H_{nn} - E_n) \end{vmatrix} = 0 \qquad (5.53)$$

For a helium atom, the secular determinant can be written as:

$$\begin{vmatrix} (H_{11} - E_1) & H_{12} \\ H_{21} & (H_{22} - E_2) \end{vmatrix} = 0 \qquad (5.54)$$

where:

$$H_{11} = H_{22} = \hat{h}_1 + \hat{h}_2 - J_{12} \qquad (5.55)$$
$$H_{12} = H_{21} = K_{12} \qquad (5.56)$$

5.9 Restricted and Unrestricted HF Models

In a closed system or a fully filled orbital system, each level is occupied by two electrons with opposite spins, whereas in an open-shell system there are partially filled levels containing only one electron. If the number of electrons present in the system is odd, it will be always an open-shell, as, for example, in an $_7N$ atomic system with the electronic configuration $1s^2, 2s^2, 2p_x^1, 2p_y^1, 2p_z^1$, three half filled orbitals are present. If the number of electrons present is even, the system needs not be always closed-shell since there may be a degenerate orbital [apart from spin-degeneracy] each containing one electron, as, for example, $_2$He with electronic configuration $1s^2$ is a closed-shell atomic system while $_8$O with electronic configuration $1s^2, 2s^2, 2p_x^2, 2p_y^1, 2p_z^1$, is an open-shell atomic system. When an electron is added into a closed-shell system, interaction of this electron with the electrons already present in the system will be different. The added electron will interact with electron of the system keeping parallel spin only. In a closed-shell system, the orbitals can be grouped in pairs with the same orbital dependence and orbital energy but with opposite spins (spin functions α and β). The setting up of the HF model by imposing the double occupancy principle is called the Restricted Hartree-Fock (RHF) model. For an open-shell system orbital, pairing does not occur in any level of computation. There are two possibilities for extending HF calculations to open-shell systems:

1. Strictly presuming that orbital pairing does not occur in any level. Each spinor-bital is allowed to have its own spatial part. This type of modeling is known as Unrestricted Hartree-Fock (UHF) modeling.
2. The RHF procedure is extended to spatial orbitals other than the orbitals which are singly occupied. Modeling of this type (Fig. 5.1) is known as restricted open-shell Hartree-Fock modeling (ROHF).

In UHF, V_{HF}^{α} and V_{HF}^{β} orbitals will have different effective potentials. UHF affords equations which are much simpler than that of ROHF. In UHF, wavefunctions are composed of single Slater determinants, while in ROHF, wavefunctions are composed of the linear combination of a few determinants, where the expansion coefficients are decided by symmetry of the state. However, the UHF Slater determinant is not an eigenfunction of the total spin operator $\hat{S}^2$. The expectation value of spin $\langle \hat{S}^2 \rangle$ may be deviated from the actual value $S(S+1)$, where S is the spin quantum number corresponding to the total spin of the system. The more the deviation, the more will be the contamination in the determinant with functions corresponding to states of higher spin multiplicity. Hence, in computational practice, the UHF approach may not be convenient. UHF and ROHF energy calculations for a nitrogen atom using GAUSSIAN 03 reports energy values, $E(\text{UHF}) = -53.9601515933$ and $E(\text{ROHF}) = -53.95988992129$ (The complete input and output details can be seen in the URL). For RHF/ROHF, α and β spins have the same spatial part. Here, the wavefunction is an eigenfunction of the $\hat{S}^2$ operator. For open-shell systems, the unpaired electron interacts differently with α and β spins. The optimum spatial orbitals are different. Restricted formalism is not suitable for spin dependent properties. For UHF, α and β spins have *different* spatial parts. The wavefunction is *not* an eigenfunction of the $\hat{S}^2$ function, and may be contaminated with states of higher multiplicity $(2S+1)$. It yields qualitatively correct spin densities. Energy

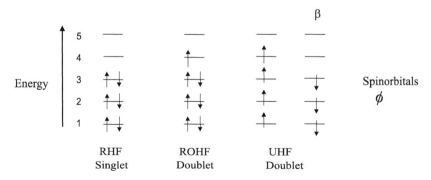

Fig. 5.1 Comparison of computed energy with different types of HF calculations

computed by UHF-method will be less than or equal to energy computed by the RHF or ROHF methods, i.e., $E_{UHF} \le E_{RHF/ROHF}$. HF methods are the starting point for more advanced calculations that include electron correlation.

5.10 The Fock Matrix

We have seen that one-electron orbitals obey the equation $\hat{F}_i \phi_i{}' = \varepsilon_i \phi_i{}'$, where $\hat{F}_i$ is the Fock operator given by the expression:

$$\hat{F}_i = \frac{1}{2}\nabla_i^2 - \frac{Z}{r_i} + \sum_j 2\hat{J}_j - \hat{K}_j \tag{5.57}$$

The term $\sum_j 2\hat{J}_j - \hat{K}_j$ is known as the Fock matrix. In the actual procedure of the computation, we make an initial guess of orbitals. From these orbitals, calculate the Fock matrix (f matrix) from which identify the new orbitals and the procedure is repeated in an iterative manner until we arrive at self consistency. The Fock matrix, in fact, is a two-dimensional array, representing the electronic structure of an atom or molecule.

5.11 Roothaan-Hall Equations

Roothaan-Hall equations are obtained by extending the concepts of the variational principle and the linear combination of atomic orbitals (LCAOs) to the HF equation, which will be obtained through certain nonorthonormal basis set functions (either Gaussian-type or Slater-type). Roothaan-Hall equations apply to closed-shell molecules or atoms where all molecular orbitals or atomic orbitals are doubly occupied. With a suitable set of basis sets (refer to Chap. 6), the function can be represented as:

$$\phi_i = \sum a_{ij} \chi_j \tag{5.58}$$

By making use of this function, the HF equation takes the form of:

$$F_i \phi_i = \varepsilon_i \sum a_{ij} \chi_j \tag{5.59}$$

where χ_j are the linear combination of the basis function. Roothaan-Hall equations are simultaneous equations of the type:

$$\sum (F_{ij} - \varepsilon_i S_{ij}) a_{ij} = 0 \tag{5.60}$$

and can be generated as given below:

$$(F_{11} - S_{11}\varepsilon_1) a_{11} + (F_{12} - S_{12}\varepsilon_1) a_{12} + \ldots + (F_{1n} - S_{1n}\varepsilon_1) a_{1n} = 0$$
$$(F_{21} - S_{21}\varepsilon_2) a_{21} + (F_{22} - S_{22}\varepsilon_2) a_{22} + \ldots + (F_{2n} - S_{2n}\varepsilon_2) a_{2n} = 0 \quad (5.61)$$
$$\ldots\ldots\ldots\ldots\ldots\ldots\ldots\ldots\ldots\ldots\ldots\ldots\ldots$$
$$(F_{n1} - S_{n1}\varepsilon_n) a_{n1} + (F_{n2} - S_{n2}\varepsilon_n) a_{n2} + \ldots + (F_{nn} - S_{nn}\varepsilon_n) a_{nn} = 0$$

In matrix notation, this can be represented as:

$$\begin{bmatrix} (F_{11} - S_{11}\varepsilon_1) & (F_{12} - S_{12}\varepsilon_1) & \ldots & (F_{1n} - S_{1n}\varepsilon_1) \\ (F_{21} - S_{21}\varepsilon_2) & (F_{22} - S_{22}\varepsilon_2) & \ldots & (F_{2n} - S_{2n}\varepsilon_2) \\ \ldots & \ldots & \ldots & \ldots \\ (F_{n1} - S_{n1}\varepsilon_n) & (F_{22} - S_{22}\varepsilon_2) & \ldots & (F_{nn} - S_{nn}\varepsilon_n) \end{bmatrix} \begin{bmatrix} a_{11} & a_{12} & \ldots & a_{1n} \\ a_{21} & a_{22} & \ldots & a_{2n} \\ \ldots & \ldots & \ldots & \ldots \\ a_{n1} & a_{n2} & \ldots & a_{nn} \end{bmatrix} = 0 \quad (5.62)$$

This can be simplified as:

$$(F - S\varepsilon)A = 0 \quad (5.63)$$
$$FA = SA\varepsilon \quad (5.64)$$

This equation is similar to those we have seen in Chap. 4, along with Hückel's MO formation. Here, the Fock matrix replaces the Hückel matrix.

5.12 Elements of the Fock Matrix

We have seen a Roothaan equation (Eq. 5.60) in Chap. 5. To solve the equation, we must express the Fock matrix elements [5] F_{rs} in terms of basis functions χ:

$$F_{rs} = \langle \chi_r(1) | \hat{F}(1) | \chi_s(1) \rangle = \langle \chi_r(1) | \hat{H} | \chi_s(1) \rangle$$
$$+ \sum_{j=1}^{n/2} \left[2 \langle \chi_r(1) | \hat{J}_j(1) \chi_s(1) \rangle - \langle \chi_r(1) | \hat{K}_j(1) \chi_s(1) \rangle \right] \quad (5.65)$$

Where the Coulomb operator,

$$\hat{J}_j(1) \chi_s(1) = \chi_s(1) \int \frac{\phi_j^*(2) \phi_j(2)}{r_{12}} dv_2$$
$$= \chi_s(1) \sum_t \sum_u c_{tj}^* c_{uj} \int \frac{\chi_t^*(2) \chi_u(2)}{r_{12}} dv_2 \quad (5.66)$$

The expansion is done by considering the Roothaan spatial orbital as a set of one-electron basis functions χ_s, $\phi_j = \sum_{s=1}^{b} c_{si} \chi_s$.

Multiplying by $\chi_r^*(1)$ and integrating over the coordinates of electron 1:

$$\langle \chi_r(1) | \hat{J}_j(1) \chi_s(1) \rangle = \sum_t \sum_u c_{tj}^* c_{uj} \int \int \frac{\chi_t^*(2) \chi_u(2)}{r_{12}} dv_1 \, dv_2$$
$$= \sum_{t=1}^{b} \sum_{u=1}^{b} c_{tj}^* c_{uj} (rs/tu) \quad (5.67)$$

The two-electron repulsion integral (rs/tu) is defined as:

$$(rs/tu) = \int \int \frac{\chi_r^*(1)\chi_s(1)\chi_t^*(2)\chi_u(2)}{r_{12}} \, dv_1 \, dv_2 \tag{5.68}$$

Similarly, the exchange operator term becomes:

$$\langle \chi_r(1) | \hat{K}_j(1)\chi_s(1) \rangle = \sum_{t=1}^{b} \sum_{u=1}^{b} c_{tj}^* c_{uj}(ru/ts) \, . \tag{5.69}$$

Substituting the integral equations in the Fock equation, we get the F_{rs} in terms of basis functions:

$$F_{rs} = H_{rs}^{core} + \sum_{t=1}^{b} \sum_{u=1}^{b} \sum_{j=1}^{n/2} c_{tj}^* c_{uj} \left[2(rs/tu) - (ru/ts) \right] \tag{5.70}$$

$$F_{rs} = H_{rs}^{core} + \sum_{t=1}^{b} \sum_{u=1}^{b} P_{tu} \left[(rs/tu) - \frac{1}{2}(ru/ts) \right] \tag{5.71}$$

$$H_{rs}^{core} = \langle \chi_r(1) | \hat{H}^{Core} | \chi_s(1) \rangle \tag{5.72}$$

$$P_{tu} = 2 \sum_{j=1}^{n/2} c_{tj}^* c_{uj} \tag{5.73}$$

(Here, t and u vary from 1 to b.)

P_{tu} are known as density matrix elements or charge bond order matrix elements. For a many-electron molecular orbital wavefunction, the probability density function of each MO is given by:

$$\rho_{(x,y,z)} = \sum_j n_j |\phi_j|^2 \tag{5.74}$$

where n_j is the number of electrons in ϕ_j.

With these generalizations, the electron probability density for closed-shell systems becomes:

$$\rho = 2 \sum_{j=1}^{n/2} \phi_j^* \phi_j = 2 \sum_{r=1}^{b} \sum_{s=1}^{b} \sum_{j=1}^{n/2} c_{rj}^* c_{si} \chi_r^* \chi_s$$

$$= 2 \sum_{r=1}^{b} \sum_{s=1}^{b} P_{rs} \chi_r^* \chi_s \tag{5.75}$$

To express the HF energy of integrals over basis functions χ, we know that:

$$\sum_{i=1}^{n/2} \varepsilon_i = \sum_{i=1}^{n/2} H_{ii}^{core} + \sum_{i=1}^{n/2} \sum_{j=1}^{n/2} (2J_{ij} - K_{ij}) \tag{5.76}$$

or

$$\sum_{i=1}^{n/2} \sum_{j=1}^{n/2} (2J_{ij} - K_{ij}) = \sum_{i=1}^{n/2} \varepsilon_i - \sum_{i=1}^{n/2} H_{ii}^{core} \tag{5.77}$$

Substituting this value in the HF equation:

$$E_{HF} = 2\sum_{i=1}^{n/2} \varepsilon_i - \sum_{i=1}^{n/2}\sum_{j=1}^{n/2} (2J_{ij} - K_{ij}) + V_{NN} \tag{5.78}$$

$$= 2\sum_{i=1}^{n/2} \varepsilon_i - \sum_{i=1}^{n/2} \varepsilon_i + \sum_{i=1}^{n/2} H_{ii}^{core} + V_{NN} =$$

$$E_{HF} = \sum_{i=1}^{n/2} \varepsilon_i + \sum_{i=1}^{n/2} H_{ii}^{core} + V_{NN} \tag{5.79}$$

$$H_{ii}^{core} = \langle \phi_i | \hat{H}^{core} | \phi_i \rangle = \sum_r \sum_s c_{ri}^* s_{si} \langle \chi_r | \hat{H}^{core} | \chi_s \rangle$$

$$= \sum_r \sum_s c_{ri}^* c_{si} H_{rs}^{core} \tag{5.80}$$

$$E_{HF} = \sum_{i=1}^{n/2} \varepsilon_i + \sum_r \sum_s \sum_{i=1}^{n/2} c_{ri}^* c_{si} H_{rs}^{core} + V_{NN} \tag{5.81}$$

$$E_{HF} = \sum_{i=1}^{n/2} \varepsilon_i + \frac{1}{2}\sum_{r=1}^{b}\sum_{s=1}^{b} P_{rs} H_{rs}^{core} + V_{NN} \tag{5.82}$$

Another important expression can be derived in the following manner.
Multiplying $\hat{F}\phi_i = \varepsilon_i \phi_i$ by ϕ_i^* and integrating:

$$\varepsilon_i = \langle \phi_i | \hat{F} | \phi_i \rangle$$

Substituting the ϕ from basis sets, $\phi_i = \sum_{s=1}^{b} c_{si}\chi_s$

$$\varepsilon_i = \sum_r \sum_s c_{ri}^* c_{si} \langle \chi_r | \hat{F} | \chi_s \rangle = \sum_r \sum_s c_{ri}^* c_{si} F_{rs} \tag{5.83}$$

we can write

$$\sum_{i=1}^{n/2} \varepsilon_i = \sum_{i=1}^{n/2}\sum_r \sum_s c_{ri}^* c_{si} F_{rs} = \frac{1}{2}\sum_r \sum_s P_{rs} F_{rs} \tag{5.84}$$

Substituting this value in E_{HF} equation:

$$E_{HF} = \frac{1}{2}\sum_{r=1}^{b}\sum_{s=1}^{b} P_{rs} (F_{rs} + H_{rs}^{core} + V_{NN}) \tag{5.85}$$

5.13 Steps for the HF Calculation

The various steps involved in an iterative HF computation are given below.

1. Giving the input data. This includes giving atomic coordinates, atomic numbers of atoms and hidden parameters such as basis sets.
2. The construction of single particle and overlap matrices.
3. Transforming the overlap matrix into the unit form.
4. Making an initial guess for the density matrix.
5. Calculating Coulomb and exchange contributions.
6. Constructing the Fock matrix.
7. Solving the eigenvalue problem.
8. Constructing a new density matrix.
9. Performing a suitable convergence test such as the convergence of acceleration after mixing or damping. Hence, $P_{pq}^{new} = \alpha P_{pq}^{last} + (1 - \alpha)P_{pq}^{previous}$ subject to $0 < \alpha < 1$
10. Getting the final output.

Construction of the Fock matrix is a crucial step in the calculation of HF energy. For higher systems, a high performance computing facility may be required.

5.14 Koopman's Theorem

Two major characteristics of an atom or molecule that are very important to computational chemists are the highest occupied molecular orbital (HOMO) and the lowest unoccupied molecular orbital or the virtual orbitals (LUMO). Together, they are called the frontier orbitals. The HOMO can be found by locating the outermost orbital containing an electron. The LUMO is the first orbital that does not contain an electron (Fig. 5.2). Koopman's theorem can be used for the computation of approximate ionization energy.

By this theorem, for closed-shell systems, the first ionization energy is equal to the energy of the HOMO. That is, the energy required to form the cation from the system provided that the ionization process is adequately represented by the removal of an electron from an orbital without change in the wavefunctions of the other electrons. Similarly, electron affinity can be found as the negative of the energy of the lowest unoccupied, i.e., virtual, orbital (the LUMO). Electron affinities calculated via Koopman's theorem are usually quite poor (Fig. 5.3).

5.15 Electron Correlation

We have seen approximation solutions to the real wavefunction. Hence, based on the variational principle, the energy computed will be higher than the ground state en-

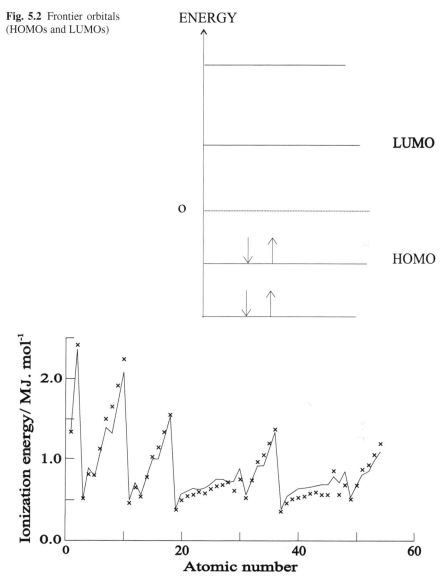

Fig. 5.2 Frontier orbitals (HOMOs and LUMOs)

Fig. 5.3 X-mark representing the calculated ionization energy according to Koopman's theorem

ergy. The difference between these two energies is known as the correlation energy. It is so named to reflect its origin in the correlated movement of electrons seeking to retain as far from each other as possible. Hence, the final step in improving MO calculations is to recover some of this correlation energy. These methods can be classified as wavefunction-based methods (configuration interaction, the Möller-Plesset perturbation theory, and coupled cluster) and electron density based methods (density functional theory). These techniques will be introduced in subsequent chapters.

5.16 Exercises

1. Compute and compare the potential energy curves for Dihydrogen with UHF and RHF calculations.
2. Write the general Hamiltonian for a condensed-matter system and explain the physical meaning of every summand. What is the main idea behind the Born-Oppenheimer approximation?
3. Consider a hydrogen atom. The Hamiltonian for a hydrogen atom (acting on the radial part of the function ϕ with the s-symmetry) reads: $\hat{H} = -\frac{1}{2}\frac{1}{r^2}\frac{\partial}{\partial r}\left(r^2\frac{\partial}{\partial r}\right) - \frac{1}{r}$. Choose a trial function in the following form: $\phi^G = \exp\left(-\beta r^2\right)$. Calculate the average energy, then, making use of the variational principle, calculate the minimum energy. Compare with the exact solution.
4. Write down the total multi-electron wavefunction for two electrons in terms of single-electron functions. Consider different spin configurations. Explain what the terms HOMO and LUMO mean.
5. Derive Hartree equations for a helium atom (with non-zero total spin).
6. Outline the HF method. Explain what closed/open-shell systems are. What is the difference between restricted and unrestricted HF methods?
7. How can we calculate the ionization potential within the HF theory? What does Koopman's theorem state?
8. Outline the Roothaan-Hall method. Which basis functions are usually used?
9. Draw the structures of H_2S and H_2O and minimize the energy using the HF method and the 3-21G* basis set. Measure the bond angles and interpret the results. Plot the electron density and electrostatic potential of each species. Comment on the results. Also, calculate the partial charges on the atoms and interpret the results.
10. Draw the structure of carbon monoxide and minimize its energy with the same basis set as above. Plot the HOMO of the species. Could CO be a ligand? If so, which end of the molecule will establish a coordinate covalent bond with a transition metal cation?
11. Calculate the structures of the isoelectronic species NH_3 and H_3O^+ and display their electron densities. Does the pattern in electron density make sense? Are there any surprises in the structural results? You will have to use the expert mode to draw the structure of the hydronium ion. In setting up the energy calculation, do not forget to set the charge of the hydronium ion.
12. Ab initio HF calculations are performed on the H, He and Li and on the isoelectronic He_2 and LiH "molecules", giving the electronic kinetic energies (T_e), and the potential energies between electrons and nuclei (V_{Ne}), electrons and other electrons (V_{ee}), and nuclei and other nuclei (V_{NN}): (all energies are in atomic units, 1 a.u. $= 2625\,kJ\,mole^{-1}$.). Calculate and discuss the bond energy of both molecules (Table 5.1).

Table 5.1 Energy values

System	T_e	V_{Ne}	V_{ee}	V_{NN}
He	2.8599	−6.7459	1.0262	0.0000
He$_2$	6.3203	−18.3003	4.1423	2.3519
H	0.4998	−0.9996	0.0000	0.0000
Li	7.4322	−17.1461	2.2819	0.0000
LiH	7.9785	−20.4234	3.4721	0.9874

References

1. Mayer I (2003) Simple Theorems, Proofs and Derivations in Quantum Chemistry Series: Mathematical and Computational Chemistry. Springer, New York
2. Ruscic B, Mayhew CA, Berkowitz J (1988) J Chem Phys 88:5580
3. Scott AP, Radom L (1996) J Phys Chem 100:5580
4. Leach AR (2001) Molecular Modelling: Priciples and Applications. Addison-Wesley Longman, Reading
5. Friesner RA (1985) Solution of Self-Consistent Field Electronic Structure Equations by a Pseudospectral Method. Chem Phys Lett 116:39

Chapter 6
Basis Sets

6.1 Introduction

A basis set is a mathematical description of orbitals of a system, which is used for approximate theoretical calculation or modeling. It is a set of basic functional building blocks that can be stacked or added to have the features that we need. By "stacking" in mathematics, we mean adding things, possibly after multiplying each of them by its own constant:

$$\psi = a_1\phi_1 + a_2\phi_2 + \ldots + a_k\phi_k \tag{6.1}$$

where k is the size of the basis set, $\phi_1, \phi_2, \ldots, \phi_k$ are the basis functions and $a_1, a_2, \ldots, a_k$ are the normalization constants. It was John C. Slater (Fig. 6.1) who first turned to orbital computation using basis sets, known as Slater Type Orbitals (STOs). The solution of the Schrödinger equation for the hydrogen atom and other one-electron ions gives atomic orbitals which are a product of a radial function that depend on the distance of the electron from the nucleus and a spherical harmonic, as is illustrated in Table 6.1.

Fig. 6.1 John C. Slater (1900–1976). Slater is recognized for calculating algorithms which describe atomic orbitals. The algorithms became known as Slater Type Orbitals (STOs). Courtesy of "Wikipedia – The Free Encyclopedia – http://en.wikipedia.org/wiki/John_C._Slater"

K. I. Ramachandran et al., *Computational Chemistry and Molecular Modeling*   115
DOI: 10.1007/978-3-540-77304-7, ©Springer 2008

Table 6.1 Radial and angular wavefunctions of orbitals. Here, z is the effective nuclear charge for that orbital of the atom, r is the radius in atomic units (1 Bohr radius = 52.9 p.m.), $e = 2.71828$ (approximately), $\pi = 3.14159$ (approximately), n = the principal quantum number, and $\rho = (2Zr)/n$

No.	Orbital	Radial wavefunction	Angular wavefunction
1	$1s$	$2 \times z^{3/2} \times e^{-\rho/2}$	$1 \times (\pi/4)^{1/2}$
2	$2s$	$(\sqrt{2}/2) \times (2 - \rho) \times z^{3/2} \times e^{-\rho/2}$	$1 \times (\pi/4)^{1/2}$
3	$2p$	$(\sqrt{6}/2) \times \rho \times z^{3/2} \times e^{-\rho/2}$	$\sqrt{3} \times (x/r) \times (\pi/4)^{1/2}$
4	$3s$	$(\sqrt{3}/9 \times (6 - 6\rho + \rho^2) \times z^{3/2} \times e^{-\rho/2})$	$1 \times (\pi/4)^{1/2}$
5	$3p$	$(\sqrt{6}/9 \times \rho\,(4 - \rho) \times z^{3/2} \times e^{-\rho/2})$	$\sqrt{3} \times (x/r) \times (\pi/4)^{1/2}$

He pointed out that we could use functions that consisted only of the spherical harmonics and the exponential term. Slater-type orbitals represent the real situation for the electron density in the valence region and beyond, but are not so good nearer to the nucleus. Strictly speaking, atomic orbitals (AOs) are the real solutions of the Hartree-Fock (HF) equations for the atom, i.e., wavefunctions for a single electron in the atom. Anything else is not really an atomic orbital function. Hence these functions are named as "basis functions" or "contractions," which are more appropriate. Earlier, the STOs were used as basis functions due to their similarity to atomic orbitals of the hydrogen atom. Many calculations over the years have been carried out with STOs, particularly for diatomic molecules. Slater fits linear least-squares to data that could be easily calculated. The general expression for a basis function is given in Eq. 6.2:

$$\text{Basis function,} \quad BF = N \times e^{(-\alpha r)} \tag{6.2}$$

where N is the normalization constant, α is the orbital exponent and r is the radius in angstroms. STOs are described by the function depending on spherical coordinates:

$$\phi_1 (\alpha, n, l, m; r, \theta, \phi) = N r^{n-1} e^{-\alpha r} Y_{l,m} (\theta, \phi) \tag{6.3}$$

The r, θ and ϕ are spherical coordinates, and $Y_{l,m}$ is the angular momentum part (the function describing the "shape"). The n, l, and m are quantum numbers: principal, angular momentum, and magnetic, respectively. Simplifying the equation for hydrogen-like systems, the STO equation takes the form of:

$$STO = \left[\frac{\alpha^3}{\pi} \right]^{0.5} e^{(-\alpha r)} \tag{6.4}$$

where α is the Slater orbital exponent. STOs are approximate solutions to the eigenvalue equation, represented by Eq. 6.1.

The orbital is resembling the atomic orbital and is shown by Fig. 6.2.

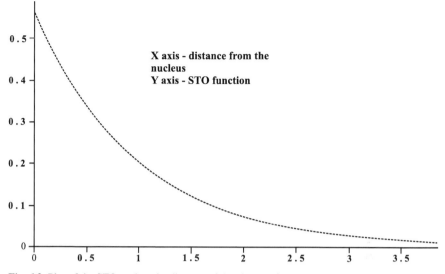

Fig. 6.2 Plot of the STO against the distance of the electron from the nucleus

6.2 The Energy Calculation from the STO Function

Thus, for hydrogen and hydrogen-like systems (containing one electron), the wave-function can be written as:

$$\psi_{1s} = \left[\frac{\alpha^3}{\pi}\right]^{0.5} e^{(-\alpha r)} \tag{6.5}$$

The Schrödinger equation in polar coordinate form can be written as:

$$-\frac{\hbar^2}{2m}\frac{d^2}{dr^2}\psi + \frac{l(l+1)\hbar^2}{2mr^2}\psi - \frac{e^2}{4\pi\varepsilon_0 r}\psi = E\psi \tag{6.6}$$

For $1s$ orbital $l = 0$. Using atomic units:

$$-\frac{\hbar^2}{2m}\frac{d^2}{dr^2}\psi - \frac{e^2}{r}\psi = E\psi \tag{6.7}$$

This can be written as:

$$-\frac{\hbar^2}{2m}\frac{1}{r^2}\frac{d}{dr}r^2\frac{d}{dr}\psi - \frac{e^2}{r}\psi = E\psi \tag{6.8}$$

Based on the variational method (see Chap. 4), the energy expression takes the form of:

$$E = \frac{\int_0^\infty \psi_{1s}/\hat{H}/\psi_{1s}d\tau}{\int_0^\infty \psi_{1s}\psi_{1s}d\tau} \tag{6.9}$$

Substituting the values of ψ_{1s} and $\hat{H}$ in the equation:

$$E = \frac{\int_0^\infty ke^{-\alpha r}\left\langle -\frac{\hbar^2}{2m}\frac{1}{r^2}\frac{d}{dr}r^2\frac{d}{dr} - \frac{e^2}{r}\right\rangle ke^{-\alpha r}4\pi r^2 dr}{\int_0^\infty \left(ke^{-\alpha r}\right)^2 \pi 4 r^2 dr} \tag{6.10}$$

where $\qquad k = \left[\frac{\alpha^3}{\pi}\right]^{0.5} \tag{6.11}$

$4\pi r^2 dr$ part accounts for the radial factor of Schrödinger equation. On simplifying, we get:

$$E = \frac{\int_0^\infty e^{-\alpha r}\left\langle -\frac{\hbar^2}{2m}\frac{1}{r^2}\frac{d}{dr}r^2\frac{d}{dr} - \frac{e^2}{r}\right\rangle e^{-\alpha r}r^2 dr}{\int_0^\infty \left(e^{-\alpha r}\right)^2 r^2 dr} \tag{6.12}$$

Separating the kinetic energy operator operating on $e^{-\alpha r}$:

$$-\frac{\hbar^2}{2m}\frac{1}{r^2}\frac{d}{dr}r^2\frac{d}{dr}e^{-\alpha r} = -\frac{\hbar^2}{2m}\left(\alpha^2 - \frac{2\alpha}{r}\right)e^{-\alpha r} \tag{6.13}$$

and applying the relation:

$$\left(e^{-\alpha r}\right)^2 = e^{-2\alpha r} \tag{6.14}$$

we get

$$E = \frac{\int_0^\infty e^{-\alpha r}\left\langle -\frac{\hbar^2}{2m}\left(\alpha^2 - \frac{2\alpha}{r}\right)e^{-\alpha r} - \frac{e^2}{r}e^{-\alpha r}\right\rangle r^2 dr}{\int_0^\infty e^{-2\alpha r}r^2 dr} \tag{6.15}$$

Simplifying the integral based on the expression:

$$\int\limits_0^\infty x^n e^{-\alpha x} dx = \frac{n!}{\alpha^{n+1}}$$ (6.16)

we get:

$$E = \frac{\frac{-\hbar^2 \alpha^2}{2m}\left(\frac{1}{4\alpha^3}\right) + \frac{-\hbar^2 \alpha}{m}\left(\frac{1}{4\alpha^2}\right) - e^2\left(\frac{1}{4\alpha^2}\right)}{\left(\frac{1}{4\alpha^3}\right)} = -\frac{\hbar^2 \alpha^2}{2m} + \frac{\hbar^2 \alpha}{m}\alpha - e^2\alpha$$

$$E = \frac{\hbar^2 \alpha^2}{2m} - e^2\alpha$$ (6.17)

The Slater coefficient α can be calculated based on the minimization condition of the energy function:

$$\frac{d}{dx}\left(\frac{\hbar^2 \alpha^2}{2m} - e^2\alpha\right) = 0$$ (6.18)

Or:

$$\frac{\hbar^2 \alpha}{m} - e^2 = 0$$ (6.19)

$$\alpha = \frac{me^2}{\hbar^2}$$ (6.20)

Substituting this value of α in the equation:

$$E = \frac{\hbar^2 \alpha^2}{2m} - e^2\alpha = \frac{\hbar^2}{2m}\left(\frac{me^2}{\hbar^2}\right)^2 - e^2\left(\frac{me^2}{\hbar^2}\right)$$

$$= \frac{1}{2}\left(\frac{me^4}{\hbar^2}\right) - \left(\frac{me^2}{\hbar^2}\right) = \frac{1}{2}\left(\frac{me^4}{\hbar^2}\right)$$ (6.21)

For the charge separation in the vacuum:

$$E = \frac{1}{2}\left(\frac{me^4}{(4\pi\varepsilon_0)^2 \hbar^2}\right) = \left(\frac{me^4}{32\pi^2 \varepsilon_0^2 \hbar^2}\right)$$ (6.22)

But $\hbar = \frac{h}{2\pi}$. Hence, the energy equation becomes:

$$E = \left(\frac{me^4}{8\varepsilon_0^2 h^2}\right)$$ (6.23)

Substituting the following values: m – mass of electron, e – the charge of the electron, ε_0, and Planck's constant:

$$E = \left(\frac{me^4}{8\varepsilon_0^2 h^2}\right)$$

$$= \frac{9.109 \times 10^{-31}\,\text{kg} \times \left(1.602 \times 10^{-19}\,\text{C}\right)^4}{8\left(8.854 \times 10^{-12}\,\text{C}^2\text{N}^{-1}\text{m}^{-2}\right)^2 \left(6.626 \times 10^{-34}\,\text{Js}\right)^2}$$

$$= -2.179 \times 10^{-18}\,\text{J}$$

6.3 The Energy Calculation of Multielectron Systems

Consider a two-electron system (a system consisting of two electrons.) such as the helium atom, making the two-electron function into two separate functions, $\psi = \psi_1 + \psi_2$.

The Hamiltonian for the electron-1:

$$\hat{H}_1 = -\frac{1}{2r_1^2}\frac{d}{dr_1}r_1^2\frac{d}{dr_1} - \frac{2}{r_1} + \int_0^\infty \psi_2\frac{1}{r_{12}}\psi_2 d\tau$$

The integral function in the above equation stands for the repulsion between electrons 1 and 2 of the atom. As we do not know the position of electron 2 due to the functional separation, we have to integrate the potential function over all possible locations of electron 2.

Similarly, the Hamiltonian for electron 2 is:

$$\hat{H}_2 = -\frac{1}{2r_2^2}\frac{d}{dr_2}r_2^2\frac{d}{dr_2} - \frac{2}{r_2} + \int_0^\infty \psi_1\frac{1}{r_{12}}\psi_1 d\tau$$

Let the functions be represented by STOs:

$$\psi_1 = \left[\frac{\alpha_1^2}{\pi}\right]^{0.5} e^{(-\alpha r_1)} \tag{6.24}$$

$$\psi_2 = \left[\frac{\alpha_2^2}{\pi}\right]^{0.5} e^{(-\alpha r_2)} \tag{6.25}$$

The repulsive function $\int_0^\infty \psi_2\frac{1}{r_{12}}\psi_2 d\tau$ becomes:

$$\int_0^\infty \psi_2\frac{1}{r_{12}}\psi_2 d\tau = \frac{1}{r_1}\left[1 - (1 + \alpha_2 r_1)e^{-2\alpha_2 r_1}\right] \tag{6.26}$$

Similarly, $\int\limits_{0}^{\infty} \psi_1 \dfrac{1}{r_{12}} \psi_1 d\tau$ becomes:

$$\int\limits_{0}^{\infty} \psi_1 \frac{1}{r_{12}} \psi_1 d\tau = \frac{1}{r_2} \left[1 - (1 + \alpha_1 r_2) e^{-2\alpha_1 r_2} \right] \tag{6.27}$$

The Hamiltonian for electrons -1 and 2 takes the forms of:

$$\hat{H}_1 = -\frac{1}{2r_1^2} \frac{d}{dr_1} r_1^2 \frac{d}{dr_1} - \frac{2}{r_1} + \frac{1}{r_1} \left[1 - (1 + \alpha_2 r_1) e^{-2\alpha_2 r_1} \right] \tag{6.28}$$

$$\hat{H}_2 = -\frac{1}{2r_2^2} \frac{d}{dr_2} r_2^2 \frac{d}{dr_2} - \frac{2}{r_2} + \frac{1}{r_2} \left[1 - (1 + \alpha_1 r_2) e^{-2\alpha_1 r_2} \right] \tag{6.29}$$

Only repulsive terms have to be added to:

$$E_1 = \frac{\alpha_1^2}{2} - Z\alpha_1 + \frac{\alpha_1 \alpha_2 \left(\alpha_1^2 + 3\alpha_1 \alpha_2 + \alpha_2^2 \right)}{(\alpha_1 + \alpha_2)^3} \tag{6.30}$$

$$E_2 = \frac{\alpha_2^2}{2} - Z\alpha_2 + \frac{\alpha_1 \alpha_2 \left(\alpha_1^2 + 3\alpha_1 \alpha_2 + \alpha_2^2 \right)}{(\alpha_1 + \alpha_2)^3} \tag{6.31}$$

It is very difficult to evaluate the necessary integrals over these STOs, especially when the orbitals in the integral are centered on three or four different atoms. The computation of integrals over STOs still remains as a difficult problem. Accurate results are just about possible, but they are very time-consuming.

6.4 Gaussian Type Orbitals

In the 1950s, Frank Boys from Cambridge University in the UK suggested a modification to the wavefunction by introducing Gaussian type functions, which contain the exponential $e^{-\beta r^2}$, rather than the $e^{-\alpha r}$ of the STOs. Such functions are very easy to evaluate. These functions neither represent the electron density of the real situation (the square of a wavefunction is a measure of electron density) nor the STOs. But we can overcome this difficulty to a large extent by using more Gaussian-type orbitals (GTOs). Some early calculations used a large number of individual GTOs. It was then suggested that the GTOs be contracted into separate functions. Each basis function in this approach consists of several GTOs combined together in a linear manner with fixed coefficients. Thus, we might define a GTO (3G) basis function as:

$$GTO(3G) = c_1 e^{-\beta_1 r^2} + c_2 e^{-\beta_2 r^2} + c_3 e^{-\beta_3 r^2} \tag{6.32}$$

where the three values of c and β are fixed, and that number is included in the designation. The values of the c and β can be found in several ways. One common way is to fit the above expression to a STO using a least squares method. Other methods involve varying them in atomic calculations to minimize the energy. Expansions of any number of GTOs are possible, but usually less than six are used due to computational reasons. Treating Gaussians as GTOs is probably a misnomer, since they are not really orbitals. They are modified and simplified forms of functions. In recent literature, they are frequently called Gaussian primitives.

A Cartesian Gaussian centered on atom a can be represented as:

$$G_{i,j,k} = N x_a^i y_a^j z_a^k e^{-\alpha r_a^2} \tag{6.33}$$

where i, j, and k are nonnegative integers, α is a positive orbital exponent, x_a, y_a, and z_a are Cartesian coordinates with the origin at a, and N is the Cartesian Gaussian normalization constant. This constant is given by the expression:

$$N = \left(\frac{2\alpha}{\pi}\right)^{3/4} \left[\frac{(8\alpha)^{i+j+k} i! j! k!}{(2i)!(2j)!(2k)!}\right]^{1/2} \tag{6.34}$$

when $i = 0$, $j = 0$, $k = 0$ and $i + j + k = 0$, then the Gaussian type function (GTF) is known as the s-type function; when $i + j + k = 1$, we have a p-type function, when $i + j + k = 2$, we have the d-type function, and so on. There are six possible d-Gaussian functions with the factors x_a^2, y_a^2, z_a^2, $x_a y_a$, $x_a z_a$ and $y_a z_a$. These d-functions can be modified into five linear combinations, as $x_a y_a$, $x_a z_a$, $y_a z_a$, $x_a^2 - y_a^2$ and $3z_a^2 - r_a^2$ to have the same angular behavior as the real $3d$ atomic orbitals. Note that the sixth possible combination $x_a^2 + y_a^2 + z_a^2 = r_a^2$ resembles the s-orbital.

6.5 Differences Between STOs and GTOs

The pre-exponential factor r^{n-1} of the STO function is dropped in the GTO function. This restricts single Gaussian primitives (corresponding to the highest energy level) for each principal quantum level. Thus, possible GTOs are: $1s$, $2p$, $3d$, $4f$, etc. This helps in fast computation [2]. Thus: possible Gaussian functions can be enlisted as: $g_{(1s)} = Ne^{-\beta r^2}$, $g_{(2p_x)} = Ne^{-\beta r^2}x$, $g_{(2p_y)} = Ne^{-\beta r^2}y$, $g_{(2p_z)} = Ne^{-\beta r^2}z$,
$g_{(3d_{xx})} = Ne^{-\beta r^2}x^2$, $g_{(3d_{yy})} = Ne^{-\beta r^2}y^2$, $g_{(3d_{zz})} = Ne^{-\beta r^2}z^2$, $g_{(3d_{xy})} = Ne^{-\beta r^2}xy$,
$g_{(3d_{xz})} = Ne^{-\beta r^2}xz$, $g_{(3d_{yz})} = Ne^{-\beta r^2}yz$, $g_{(4f_{xxx})} = Ne^{-\beta r^2}x^3$, $g_{(4f_{yyy})} = Ne^{-\beta r^2}y^3$,
$g_{(4f_{zzz})} = Ne^{-\beta r^2}z^3$, $g_{(4f_{xxy})} = Ne^{-\beta r^2}x^2 y$ etc.

The r-factor of the exponential in GTO is squared. The angular momentum factor is made into a simple function of Cartesian coordinates (Fig. 6.3 and Fig. 6.4).

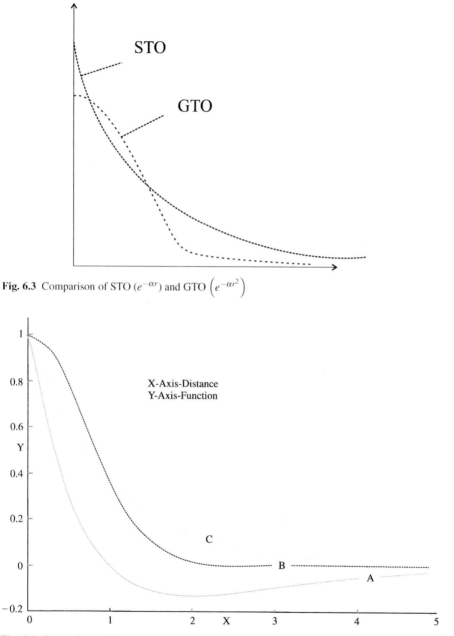

Fig. 6.3 Comparison of STO $(e^{-\alpha r})$ and GTO $\left(e^{-\alpha r^2}\right)$

Fig. 6.4 Comparison of GTO and STO with the atomic orbital. A = Atomic orbital function $[(1 - x) \times e\,(-abs(x))]$, B = GTO-function $[e\,(-abs(x^2))]$ and C = STO function $[e\,(-abs(x))]$

6.6 Classification of Basis Sets

Basis set can be broadly classified into the following types.

1. *Minimal basis sets*: STO-3G, STO-4G, STO-6G, STO-3G* – a polarized version of STO-3G.
2. *Pople basis sets*: 3-21g, 3-21g* – Polarized, 3-21+g – Diffuse, 3-21+g* – With polarization and diffuse functions, 6-31g, 6-31g*, 6-31+g*, 6-31g (3*df*, 3*pd*),6-311g, 6-311g*, 6-311+g*.
3. *Correlation consistent basis sets*: These basis sets are used for post HF calculations. They include shells of polarization (correlating) functions (*d*, *f*, *g*, etc.) that can yield convergence of the electronic energy to the complete basis set limit. Examples of these are cc-pVDZ (correlation consistent valence double zeta) cc-pVTZ (correlation consistent valence triple zeta) cc-pVQZ (correlation consistent valence quadruple zeta), cc-pV5Z (correlation consistent valence quintuple zeta), aug-cc-pVDZ (Augmented versions of cc-pVDZ), etc.
4. *Other split valence basis sets*: (They have generic names), such as SV(P), SVP, DZV, TZV, TZVPP, or valence triple-zeta plus polarization, QZVPP, valence quadruple-zeta plus polarization.
5. *Double, triple, and quadruple zeta basis sets*: Basis sets in which there are multiple basis functions corresponding to each atomic orbital, including both valence orbitals and inner orbitals, which are called zeta basis sets. The most common is the D95 basis set of Dunning.
6. *Plane wave basis sets*: In addition to localized basis sets, plane wave basis sets can also be used in quantum chemical simulations. Typically, a finite number of plane wavefunctions are used, below a specific cutoff energy which is chosen for a certain calculation. These basis sets are popular in calculations.

6.7 Minimal Basis Sets

In a minimal basis set we select one basis function for every atomic orbital that is required to describe the free atom. We take all the orbitals required for the filling up of electrons; this is also known as a single-zeta (single z, SZ) basis set. Thus, for hydrogen, the minimal basis set is just one "1*s*" orbital. For carbon, the minimal basis set consists of a "1*s*" orbital, a "2*s*" orbital and the full set of three "2*p*" orbitals. For example, the minimal basis set for the methane molecule consists of 4 "1*s*" orbitals – one per hydrogen atom – and the set of "1*s*", "2*s*", and "2*p*" as described above for carbon. The total basis set comprises nine basis functions [3].

For computing purposes, several minimal basis sets have been proposed. The most common minimal basis sets are the STO-*n*G basis sets devised by John Pople and his group. It involves a linear combination of "*n*" GTOs fitted to each STO.

The individual GTOs are called primitive orbitals, while the combined functions are called contracted functions. Thus, when a basis function contains more than one primitive Gaussian, it can be considered as "contracted."

The STO-3G basis set is a minimal basis set, where each basis function is a contraction of three primitive Gaussians. The exponents and expansion coefficients for the primitives are obtained from a least squares fit to STOs. STO-3G basis sets are available for the elements H-Xe. Their use for serious work is discouraged, but they provide a rapid way of obtaining a "quick and dirty" look at a molecule. *The STO-3G basis set for methane thus consists of a total of nine contracted functions built from 27 primitive functions.*

6.8 A Comparison of Energy Calculations of the Hydrogen Atom Based on STO-nG Basis Sets

6.8.1 STO-2G

The orbital function based on the STO-2g basis set can be written as: $STO - 2G = c_1 e^{-\beta_1 r^2} + c_2 e^{-\beta_2 r^2}$. The constants (coefficients, exponents) $c_1, c_2, \beta_1 \& \beta_2$ can be computed. Here, we give the values obtained by running the program GAUSSIAN 03W, a commercial package for Gaussian and related calculations for the hydrogen atom. The complete output is included in the book URL. Thus: $c_1 = 0.4301284983$, $c_2 = 0.6789135305$, $\beta_1 = 0.1309756377$ and $\beta_2 = 0.2331359749$. These values can be substituted in the STO-2G equation to study the functional difference on changing the distance of the electron from the nucleus.

6.8.2 STO-3G

The orbital function for STO-3G can be written as $STO - 3G = c_1 e^{-\beta_1 r^2} + c_2 e^{-\beta_2 r^2} + c_3 e^{-\beta_3 r^2}$. Similarly to STO-2G, STO-3G can be run in GAUSSIAN to generate the constants (Fig. 6.5). Refer to the Text URL to see the complete output files. The constants can be computed as: $c_1 = 0.3425250914$, $c_2 = 0.6239137298$, $c_3 =$

```
A O basis  set in the form of general basis input
10
S   2 1.00      0.000000000000
        0.1309756377D+01   0.4301284983D+00
        0.2331359749D+00   0.6789135305D+00
```

Fig. 6.5 GAUSSIAN 03 STO-2G output of the hydrogen atom

0.1688554040, $\beta_1 = 0.1543289673$, $\beta_2 = 0.5353281423$ & $\beta_3 = 0.4446345422$. Similarly, the constants of higher basis sets can also be computed. For making a graphical comparison, we find the constants with STO-6G.

6.8.3 STO-6G

Constants for STO-6G are $c_1 = 0.3552322122, c_2 = 0.6513143725,$ $c_3 = 0.1822142904, c_4 = 0.6259552659, c_5 = 0.2430767471, c_6 = 0.1001124280,$ $\beta_1 = 0.9163596281, \beta_2 = 0.4936149294, \beta_3 = 0.1685383049, \beta_4 = 0.3705627997,$ $\beta_5 = 0.4164915298$ and $\beta_6 = 0.1303340841$.

The computed energy of the hydrogen atom obtained by using different levels of basis sets is included in Table 6.2.

The energy goes on decreasing along with an increase in the number of primitives used. This leads into a slow improvement in the level of computation. It approaches the HF limit (-0.5 Hartree). (The calculation of energy up to this level is known as the HF calculation.) Note that the expansion of the wavefunction in terms of basis functions leads to a limitation of the accuracy of the ab initio HF approach only because there is a limited number of basis functions available. The greater the number of basis functions, the better the wavefunction, and the lower the energy. The limit of an infinite basis set is known as the *HF limit*. This energy is still greater than the exact energy that follows from the Hamiltonian, because of the independent particle approximation.

6.9 Contracted Gaussian Type Orbitals

Several GTOs can be grouped together to form contracted Gaussian functions. The original GTOs are referred to as *primitive Gaussian functions* or *primitives* and are centered on the same nucleus. Optimizing all the parameters with all of the primitives on each run is a huge concern of computation. Hence, as an alternative, contracted Gaussian type orbitals (CGTOs) must be considered. Grouping of Gaussian functions in CGTOs saves the computational costs to a large extent, as optimization

Table 6.2 Comparison of the energy of hydrogen computed by different basis sets

No.	Basis set	Energy
1	STO-2G	-0.454397401659
2	STO-3G	-0.466581850384
3	STO-4G	-0.469806464220
4	STO-5G	-0.470742918263
5	STO-6G	-0.471039054196

needs to be carried out once and for all with each group. CGTO functions can be represented as:

$$\chi_j = \sum_i c_{ji} g_i(\alpha, r) \qquad (6.35)$$

where $g_i(\alpha, r)$ is the Gaussian. For a larger system, such groups are combined together to get the corresponding function:

$$\psi_i = \sum_j a_{ji} \chi_j = \sum_j a_{ji} \sum_i c_{ji} g_i(\alpha, r) \qquad (6.36)$$

By making use of linear combination of Gaussian, a suitable basis function of reliable characteristics can be generated as illustrated in Fig. 6.6.

The CGTO (Table 6.3) can be written as:

$$1s = 0.0592393390 G_s(322.037)$$
$$+ 0.3514999608 G_s(48.4308)$$
$$+ 0.707657921 G_s(10.4206)$$

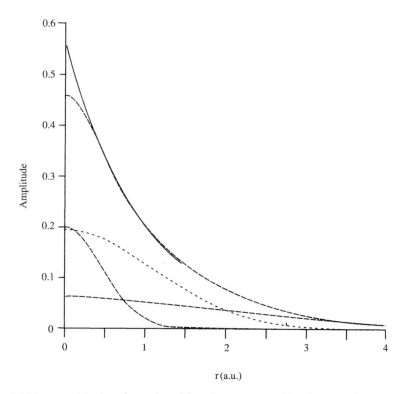

Fig. 6.6 Linear combination of a number of Gaussian groups provides a better result

Table 6.3 The 3-21G basis set for the oxygen atom generated from Gaussian 03 software

O	0		
S	3	1.00	
	322.03700000	0.0592393390	
	48.430800000	0.3514999608	
	10.420600000	0.7076579210	
SP	2	1.00	
	7.402940000	−0.4044535832	0.2445861070
	1.576200000	1.2215617610	0.8539553735
SP	1	1.00	
	0.3736840000	1.000000000	1.000000000

$G_s(322.037)$ stands for a normalized primitive s-type GTF with the orbital exponent 322.037. The large value of coefficients shows that the orbital is an inner (core) one. SP notation in the table indicate that here orbital exponents and contraction coefficients for s-type and p-type CGTFs of the valence orbitals 2s and 2p atomic orbitals are considered. The first column of the table shows the orbital exponent, and the second and third columns show the contraction coefficients. Thus, the CGTF for the valence shell becomes:

$$2s' = -0.4044535832G_s(7.40294) + 1.221561761G_s(1.5762)$$

$$2p'_x = 0.244586107G_{p_x}(7.40294) + 0.8539553735G_{p_x}(1.5762)$$

$$2p'_y = 0.244586107G_{p_y}(7.40294) + 0.8539553735G_{p_y}(1.5762)$$

$$2p'_z = 0.244586107G_{p_z}(7.40294) + 0.8539553735G_{p_z}(1.5762)$$

$$2s'' = G_s(0.373684)$$

$$2p''_x = G_{p_x}(0.373684)$$

$$2p''_y = G_{p_y}(0.373684)$$

$$2p''_z = G_{p_z}(0.373684)$$

Contraction coefficients becoming one in the valence shell proves that the primitive Gaussians are normalized. Gaussian basis sets can also be ordered from the internet (http://www.emsl.pnl. gov/forms/basisform.html).

6.10 Double- and Triple-Zeta Basis Sets and the Split-Valence Basis Sets

The split-valence basis sets were introduced by John Pople and his group in the late 1970s. They are an expansion of basis sets to make the total function more accurate and reliable. *The split-valence (SV) basis set uses one function for orbitals that are not in the valence shell and 2 functions for those in the valence shell.* The double-zeta (DZ) basis set uses two basis functions where the minimal basis set had only one function. In DZ, which is normally treated as the general split valence basis sets, each atomic orbital function is split up into two basis functions, making each basis function typically as a contraction of a small set of primitives (2–4). The D95 basis sets of Dunning and coworkers [1] are DZ-type basis sets which use 9 s-type primitive Gaussians to describe the $1s$ and $2s$ atomic orbitals and 5 p-type primitives for the atomic $2p$ orbitals. This type of basis set is available for H-Cl. The triple-zeta (TZ) basis set uses three basis functions instead of one.

The smallest split-valence basis set is denoted 3-21G (to be read as three-two-one G-available for H-Xe). It uses a three-primitive expansion for the $1s$ orbital and then splits the valence orbitals into a two basis function, the inner function being a contraction of two Gaussians and the outer function being just a single Gaussian – s, for example, for the hydrogen atom, the lone valence orbital is split up in to two Gaussian groups, ϕ_S and $\phi_{S'}$ with ϕ_S carrying two primitives and $\phi_{S'}$ carrying one primitive (Fig. 6.7).

3-21G wavefunctions for hydrogen may be represented as:

$$\phi_H = \phi_S + \phi_{S'} \tag{6.37}$$

$$\phi_S = c_1 e^{-\beta_1 r^2} + c_2 e^{-\beta_2 r^2} \tag{6.38}$$

$$\phi_{S'} = c_3 e^{-\beta_3 r^2} \tag{6.39}$$

The C-atom based on the principle of a split valence basis set, can be represented by one $1s$ inner orbital and $2(2s, 2p_x, 2p_y, 2p_z) = 2 \times 4 = 8$ valence orbitals. Similarly, the carbon atom output with a 6-31G basis set is given in Fig. 6.8.

The inner orbital of a carbon atom $(1s)$ is represented by six primitives and the four valence orbitals $(2s, 2p_x, 2p_y, 2p_z)$ are represented by two contracted orbitals. Each contracted orbital contains four primitives comprised of three contracted and one uncontracted orbitals.

Each contracted orbital contains four primitives comprising of 3 contracted and one uncontracted orbitals (Fig. 6.9). Hence, the number of primitives required to

---

Standad basis : 3 - 21G (6D. 7F)
Ao basis set in the form of genereal basis input:
10
S   2 1.00   0.000000000000
        0 . 5447178000D+01   0. 1562849787D+00
        0. 8245472400D00   0. 9046906767D+00
S   1 1. 00   0.000000000000
        0. 1831915800D+00   0.1000000000D+01

Fig. 6.7 Output of GAUSSIAN 03 run on the hydrogen atom generating the orbital functions with the 3-21G basis set

```
Standad basis : 6 - 31G  (6D,    7F)
Ao basis set in the form of genereal basis input:
1 0
S   6  1.00           0.000000000000
       0. 3047524880D+04    0. 1834737132D- 02
       0. 4573695180D+03    0. 1403722281D- 01
       0. 1039486850D+03    0. 6884262226D- 01
       0. 2921015530D+02    0. 2321844432D+00
       0. 9286662960D+01    0. 4679413484D+00
       0. 3163926960D+01    0. 3623119853D+00
 SP   3. 1 . 00           0.000000000000
       0. 7868272350D+01    -0. 1193324198D+00  0. 6899906659D- 01
       0. 1881288540D+01    -0. 1608541517D+00  0. 3164239610D+00
       0. 5442492580D+00     0. 1143456438D+ 01  0. 7443082909D+00
 SP   1  1. 00            0. 00000000000
       0. 1687144782D+00     0.1000000000D+01   0. 1000000000D+01
```

Fig. 6.8 Output of GAUSSIAN 03 run on the carbon atom generating the orbital functions with a 6-31G basis set

Fig. 6.9 Schematic representation of the 6-31G basis set
of a C-atom

| | 1s | 2Pz,inner | 2Pz,outer |
| C-atom | 6 gaussians | 3 gaussians | 1 gaussian |

represent the C-atom is 6 (inner) $+$ 4×4 (valence electrons) $= 22$ (Refer to the GAUSSIAN output given in the URL).

The 4-31G and 6-31G basis sets (the available H-Kr) are slightly larger variations on the same principle. The use of inner and outer basis functions for the valence orbitals in SV- or DZ-type basis sets allows these orbitals to expand or contract in molecular calculations, thus adjusting flexibly to the bonding requirements in the molecule. The 6-311G basis set (H-Kr) is a triply split valence basis set; it is TZ-quality in the valence part, but only minimal in the core. For the C-atom 6-311G, splitting results in 13 basis functions with 26 primitive Gaussians. Split valence basis sets allow orbitals to change size. This is the first step in providing a molecular orbital environment. In molecular orbital formulation, even the atomic orbital shape is changed. Hence, we require further refinement in the basis set.

6.11 Polarized Basis Sets

A set of Gaussian functions one unit higher in angular momentum than what are present in the ground state of the atom are added as polarization functions, again increasing the flexibility of the basis set in the valence region in the molecule. Polarization functions are p- or d-type basis functions that are added to describe the distortion of s or p orbitals, respectively. When bonds are formed in molecules, the atomic orbitals are distorted (polarized) from their original shapes to provide optimal bonding. Orbital polarization phenomenon may be well introduced by adding "polarization functions" to the basis set. This phenomenon can be introduced for computational purposes by adding basis functions representing an angular momentum higher than what is represented by the valence orbitals of the atom. The qualitative importance of polarization functions is that they permit the molecular wavefunction more flexibly to distort away from spherical symmetry in the neighborhood of each atom. The distortion of s, p, and d orbitals can be mimicked (Fig. 6.10) by the inclusion of p, d, and f functions respectively in the basis set. This leads to double-zeta plus polarization (DZP) or split-valence plus polarization (SVP) basis sets. In "Pople nomenclature," the 6-31G(d) (or 6-31G*) basis sets add a single set of (Cartesian) d-type functions to the basis sets of all non-hydrogen atoms; the 6-31G(d,p) (or 6-31G**) basis sets [4] add a set of p-type polarization functions on H as well (Fig. 6.11). In general, polarization functions significantly improve the description of molecular geometries (bond lengths and angles) as well as molecular relative energies. It is to be noted that just by using d-functions on C, we are not implying that d-orbitals are occupied in C or that d-orbitals provide a significant contribution to the bonding of C. However, the presence of the d- orbital function improves the description of the electron density around C and its bonding characteristics (Fig. 6.12) (Table 6.4).

The number of basis functions and integrals involved in computation, the dimension of matrices, etc. increase very rapidly with the increase in the number of functions in the basis set. For example, methane with a 6-31G basis set needs 17 basis functions and 38 primitive Gaussians, while with 6-31G(d), 27 basis functions and 48 primitive Gaussians, and, finally, 6-31G(d,f) requires 39 basis functions and 60 primitive Gaussians.

In practice, it has been found that good geometries can often be obtained with even simple basis sets such as 3-21G, but relative energies of systems are better de-

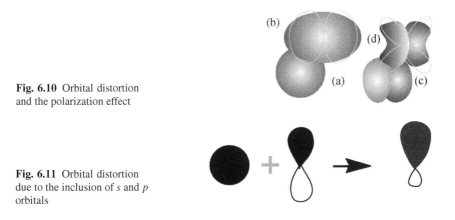

Fig. 6.10 Orbital distortion and the polarization effect

Fig. 6.11 Orbital distortion due to the inclusion of s and p orbitals

scribed using more extended basis sets. Hence, geometry optimization is generally carried out with DZ (or split-valence) plus polarization (on heavy atoms), quality basis sets, in which case bond lengths are accurate up to ±0.015 Å, and bond angles up to ±2°. Calculations including explicit electron correlation treatments via configuration interaction [QCISD (T), CCSD(T)] or perturbation theory (MPn) do not require this size of function. Correlation energy methods scale very strongly with the number of basis functions and the number of electrons. Diffuse valence functions are often added to the basis set in the computation of anions and molecules with higher bond length or in the calculations of electronically excited states. These diffuse functions (primitives with very small exponents) are used to better describe the long-range tails of the orbitals. They rarely influence geometries of covalently bonded species in any significant way, but they do improve the description of energetics associated with weak interactions and secondary bonding (van der Waals interactions, H-bonding, and electron affinities).

At the HF level of theory, most properties converge to the complete basis set limit relatively quickly with the addition of more polarization functions. However, at the correlated level of theory, the convergence is typically much slower, so that many higher functions are needed in order to reach the complete basis set limit. In particularly difficult cases, such as the computation of the dissociation energy of the N_2, the use of basis sets containing d and f polarization functions still underestimates the true value of dissociation energy by more than $10\,\text{kcal/mol}$. As the l value increases, the number of angular nodes (places where the orbital changes sign) also increases. The $l = 4$ (g) functions have 4 such nodal planes. Because of the large number of g functions and the fact that integrals over g functions are time-consuming to compute, relatively few polyatomic calculations are performed with these functions.

It is customary to perform calculations with two different forms of the higher l value Gaussians. The number of Cartesian Gaussian functions for an angular momentum quantum level l is given by:

$$N_{\text{CGF}} = l(l+1)/2 \tag{6.40}$$

The number of spherical harmonic functions is given by:

$$N_{\text{SHF}} = (2l+1) \tag{6.41}$$

Thus, for l greater than 1 (p functions) there are more Cartesian than spherical components. Certain integrals and operations are easier to code and carry out with

Fig. 6.12 Orbital distortion due to the inclusion of p and d orbitals

Table 6.4 Number of Cartesian and spherical Gaussians possible

No.	Orbital	l-value	Number of Cartesian Gaussian Functions	Number of spherical harmonic functions
1	p	1	2	3
2	d	2	6	5
3	f	3	10	7
4	g	4	15	9
5	h	5	21	11

plane wave basis functions, than with their localized counterparts. In practice, plane wave basis sets are often used in combination with core pseudopotentials, so that the plane waves are only used to describe the valence charge density. This is because core electrons tend to be concentrated very close to the atomic nuclei, resulting in large wavefunction and density gradients near the nuclei which are not easily described by a plane wave basis set unless a very high energy cutoff (and therefore small plane wavelength), is used. This combined method of a plane wave basis set with a core pseudopotential is often abbreviated as a PSPW calculation. Furthermore, as all functions in the basis are mutually orthogonal, plane wave basis sets do not exhibit basis set superposition error (Sect. 6.12). However, they are less well suited to gas-phase calculations.

6.12 Basis Set Truncation Errors

The difference between the true solution of the electronic Schrödinger equation and the experimental value corrected for non-BO effects and possibly relativistic corrections is referred to as a basis set truncation error [5]. This can further be subdivided into the basis set error associated with the limited size of the 1-electron particle basis and the n-electron error associated with the incompleteness of the n-electron basis. In the limit of a complete basis set, the basis set error vanishes and only the n-electron error remains, which is denoted by the intrinsic error of the corresponding model. One-electron basis set truncation errors may be very large, especially if basis sets chosen be inadequate to the given problem or the desired accuracy are chosen. Under the assumption that non-dynamical electron correlation plays no role, good theoretical estimates are expected.

6.13 Basis Set Superposition Error

As the atoms of interacting molecules (or of different parts of the same molecule) or two molecules approach one another, their basis functions overlap. The basic

Table 6.5 Interaction energy of helium with different basis sets

Method	r_c (pm)	BF(He) number of basis functions	Interaction energy (kJ/mol)
RHF/6-31G	323.0	2	−0.0035
RHF/cc-pVDZ	321.1	5	−0.0038
RHF/cc-p VTZ	366.2	14	−0.0023
RHF/cc-p VQZ	388.7	30	−0.0011
MP2/6-31G	321.0	2	−0.0042
MP2/cc-p VDZ	309.4	5	−0.0159
MP2/cc-p VTZ	331.8	14	−0.0211
MP2/cc-p VQZ	328.8	30	−0.0271

function of each component is influenced by functions of the nearby components, leading into an effective increase in its basis set. This improves the calculation of derived properties such as energy. If the total energy is minimized as a function of the system geometry, the short-range energies from the mixed basis sets must be compared with the long-range energies from the unmixed sets. A mismatch in this comparison leads into an error known as a basis set superposition error (BSSE). If we use finite basis sets, calculations of interaction energies are susceptible to the choice of basis set.

The interaction energy between two atoms or molecules A and B are typically calculated as the energy difference between the product complex AB and its components A and B with Eq. 6.42.

$$E_{int} = E(AB, r_c) - E(A, r_e) - E(B, r_e) \qquad (6.42)$$

Here, r_c stands for the distance between A and B in the complex AB and r_e indicates the size of the separate reactants.

The calculated interaction energies are often too large and it may lead to severe complications for systems bound through dispersion interactions or hydrogen bonds. The helium dimer is a particularly interesting example of the former situation. For helium atom, using a selection of different single-reference methods and basis sets of variable size the following results (Table 6.5) are obtained.

The best theoretical estimate is −0.091 kJ/mol for $r_c = 297$ pm. At the HF level a weakly bound minimum can be identified at interactomic distances larger than 300 pm. It is remarkable to see how the interaction energy becomes smaller and the He-He distance larger as the size of the basis set is increased. *Dispersion interaction is found to be decreased by using a basis set of small size.* Small basis sets stabilize the complex to a larger extent than the separate components, due to the basis set superposition error. While using higher basis sets, the wavefunction of the monomer is expanded in much less basis functions than the wavefunction of the complex. Each helium atom in the complex has a larger number of basis functions available than in the monomer, leading to a more flexible description of the wavefunction, and ultimately to a lower energy level.

6.14 Methods to Overcome BSSEs

There are a number of computational techniques to minimize a BSSE. Several important methods are explained below.

6.14.1 The Chemical Hamiltonian Approach

The chemical Hamiltonian approach (CHA) replaces the conventional Hamiltonian with a new one specifically designed to prevent basis set mixing *a priori* [7], by removing all the projector-containing terms which would allow basis set extension. This eliminates the terms of the Hamiltonian-making BSSE. The BSSE-free descriptions including electronic density have to be potentially determined at any level of theory.

6.14.2 The Counterpoise Method

The counterpoise method (CP) is an approximate method for estimating the size of the BSSE. While the description of the product complex is unchanged in the CP method, the separate components are provided with basis sets of identical size to those in the monomer. The CP corrected interaction energy can be computed as:

$$E_{int}(CP) = E(AB, r_c)^{AB} - E(A, r_e)^{AB} - E(B, r_e)^{AB} \qquad (6.43)$$

The superscripts AB indicate here that the complex, as well as the separate components, are calculated in the same absolute basis. In the helium example discussed before this implies that the energies of single helium atoms are calculated using the basis of the dimer complex. In this method, the BSSE is estimated as the difference between monomer energies with the regular basis functions and the energies calculated with the full set of basis functions for the whole complex. For regular basis sets, this typically stabilizes at the basis set limiting value much earlier than the uncorrected value. That need not be the case if diffuse functions are included in the basis set. The effects of increasing the basis set size are somewhat different when correlated methods are being used; it is due to the fact that the correlation energy is usually larger in the complex compared to that of the monomers. In fact, an incomplete recovery of the correlation energy weakens the complex. This effect thus compensates the BSSE effect and in such cases the final outcome of increasing the basis set size is not as obvious as in the HF level.

6.15 The Intermolecular Interaction Energy
of Ion Water Clusters

The intermolecular interaction energy of "ion-water clusters" is important in understanding a number of biochemical properties. The major problem while computing intermolecular interaction energy is the uncertainly caused by BSSEs. The BSSE [8] correction can be estimated using the counterpoise method. In the limit of a complete basis, the BSSE would be zero and it is expected that the counterpoise-uncorrected intermolecular interaction energies are equal to the counterpoise-corrected intermolecular interaction energies.

However, for enthalphy changes of $CH_3O^-(H_2O)_n$ ($n = 1, 2$) evaluated using MP2/aug-cc-pVDZ level and $CH_3S^-(H_2O)_n$ ($n = 1, 2, 3$) evaluated using MP2/6-31++G($2d,2p$) level, the counterpoise corrected values are worse than the uncorrected values. For $OH^-(H_2O)_n$ ($n = 1, 2$), the counterpoise-uncorrected intermolecular interaction energies evaluated using the MP2/aug-cc-pVDZ level is reliable. Here, the counterpoise-uncorrected intermolecular interaction energies evaluated using MP2/aug-cc-pVDZ level is close to the counterpoise-corrected intermolecular interaction energies evaluated using MP2/aug-cc-pV5Z level. Major results of the research in this field are listed below.

1. For the intermolecular interaction energies of ion water clusters $(OH^-)(H_2O)_n$ ($n = 1, 2$), $F^-(H_2O)$, $Cl^-(H_2O)$, $H_3O^+(H_2O)_n$ ($n = 1, 2$) and $NH_4^+(H_2O)_n$ ($n = 1, 2$) calculated with the correlation consistent basis sets at MP2, MP4, QCISD(T), and CCSD(T) levels, BSSE is nearly zero in the CBS limit. Here, the counterpoise-uncorrected intermolecular interaction energies are nearly equal to the counterpoise-corrected intermolecular interaction energies in the CBS limit. When the basis set is smaller, the counterpoise-uncorrected intermolecular interaction energies are more reliable than the counterpoise-corrected intermolecular interaction energies. The counterpoise uncorrected intermolecular interaction energies evaluated using the MP2/aug-cc-pVDZ level is reliable.

2. The trend for the intermolecular interaction energies of the ion water clusters calculated using B3LYP/aug-cc-pVxZ ($x = $ D,T,Q,5,6) level is extremely different from that calculated using the aug-cc-pVxZ ($x = $ D,T,Q,5,6) basis sets at MP2, MP4, QCISD(T), and CCSD(T) levels. For the intermolecular interaction energies of the ion water clusters, except for $H_3O^+(H_2O)$ calculated with correlation consistent basis sets at the B3LYP level, the CBS limit is reliable.

3. In the 6-311++G(d,p), 6-311++G($2d,2p$), 6-311++G($3d,3p$), and 6-311++G ($3df,3pd$) basis sets, the BSSE is significant for even the 6-311++G($3df,3pd$) basis set at MP2, MP4, QCISD(T), and CCSD(T) levels. In the B3LYP level, BSSE is negligible for the larger basis set; for example, the 6-311++G($3df,3pd$) basis set (except for $OH^-(H_2O)_n$ ($n = 1, 2$).

Table 6.6 Common basis sets

Basis	Options	Atoms
STO-3G	*	H–Xe
3-21G	* **	H–Cl
4-21G	* **	
4-31G	* **	H–Ne
6-21G	* **	
6-31G	+ ++ * **	H–Cl
LP-31G	* **	
LP-41G	* **	
6-311G	+ ++ * **	H–Ar
MC-311G	none	H–Ar
D95	+ ++ * **	H–Cl
D95V	+ ++ * **	H–Ne
SEC	+ ++ * **	H–Cl
		(same as SHC)
CEP-4G	+ ++ * **	H–Cl
CEP-31G	+ ++ * **	H–Cl
CEP-121G	+ ++ * **	H–Cl
LANL1MB	none	H–Bi
		(except lanthanides)
LANL1DZ	none	H–Bi
		(except lanthanides)

6.16 A List of Commonly Available Basis Sets

Note that a basis set must accompany an ab initio keyword. The "*" and "**" indicate polarization functions, i.e., 6-31G**. The "+" and "++" indicate diffuse functions.

6.17 Internet Resources for Generating Basis Sets

A number of Internet sites give information about basis sets with different software. Some of them are listed below.

1. http://www.emsl.pnl.gov/forms/basisform.htmlsss
2. http://www.molpro.net/info/molpro2006.1//molpro_basis
3. http://www.msg.ku.edu/~msg/MGM/links/bass.html
4. http://mscf.emsl.pnl.gov/software/basis_intro.shtml
5. http://www.ipc.uni-karlsruhe.de/tch/tch1/TBL/tbl.html

6.18 Exercises

1. Determine how many basis functions and Gaussian functions are used for a 6-31G calculation on HCl, and then for N,N-dimethylacetamide.
2. Prepare graphs of the errors in geometry (or %error) vs. the number of basis functions (the experimental values for H_2O are: O–H = 0.9578 A.U.; H–O–H = 104.48 deg.).
3. Compute the HF energy for the H atom converging towards the exact value, when the basis set is increased. Perform calculations with the following basis sets: STO-3G, 3-21G, 6-31G, 6-311G, cc-pVDZ, cc-pVTZ, cc-pVQZ, and cc-pV5Z.
4. Perform a series of energy computations for the H_2 molecule, at the experimental H–H distance of 0.7413 Å, using HF and the following basis sets: STO-3G, cc-pVDZ, cc-pVTZ, and cc-pVQZ. Calculate how the dissociation energy $(E_{(H_2)} \rightarrow 2E_{(H)})$ converges as a function of the basis set size. Compare this with the experimental value of 109.45 kcal/mol. Why is the HF value too low? Report the ionization potential obtained via Koopman's theorem. Compare with the experimental value of 0.585 (a.u.). Why is the agreement so good? Now perform the same calculations with the MP_2 method instead of HF. For H_2 CISD is equal to a full CI, i.e., all the electron correlation is included. Perform calculations analogous to those before but now using the method CISD. Calculate again the dissociation energy with both methods and compare the result with the experimental one. How is the rate of convergence now?
5. Compute the energy on metal-olefin complexes with the Gaussian 03W program using the LANL2DZ and 6-31G (d) basis sets and compare the results? Use $Fe(CO)_4(C_2H_4)$ complex with the route #P B3LYP/GEN 5D SCF = Tight Pop = Full IOp(3/33 = 1).
6. What changes take place in the properties of the HHeF molecule upon its complexation with N_2?
7. Why are diffuse functions important to be included into basis sets for complex calculations?

References

1. Dunning et al. (1970) J Chem Phys 53:2823
2. Thompson MA (2007) ArgusLab 4.0.1. Planaria Software LLC, Seattle, WA
3. Poirier R, Kari R, Csizmadia IG (2007) Handbook of Gaussian Basis Sets. Elsevier, New York
4. Liu B, Mclean AD (1973) J Chem Phys 59:4557
5. Boys SF, Bernardi F (1970) Mol Phys 19:553
6. Klopper W (1995) J Chem Phys 102:6168
7. Xantheas, SS (1996) J Chem Phys 104:8821
8. Weis P, Kemper PR, Bowers MT, Xantheas SS (1999) J Am Chem Soc 121:3531

Chapter 7
Semiempirical Methods

7.1 Introduction

Semiempirical methods modify Hartree-Fock (HF) calculations by introducing functions with empirical parameters. These parameters are adjusted with experimental conclusions to improve the quality of computation. The real cost of computation is due to the two-electron integrals in the Hamiltonian that has been simplified in this method. Semiempirical methods are based on three approximation schemes.

1. *The elimination of the core electrons from the calculation.*
   Inner electrons do not contribute towards chemical activity, which makes it possible to remove the core electron functions from the Hamiltonian calculation. Normally, the entire core (the nucleus and core electrons) of atoms is replaced by a parameterized function. This has the effect of drastically reducing the complexity of the calculation without a major impact on the accuracy.
2. *The use of the minimum number of basis sets.*
   In this approximation, while introducing the functions of valence electrons, only the minimum required number of basis sets will be used. This technique also reduces the complexity of computation to a large extent.
3. *The reduction of the number of two-electron integrals.*
   This approximation is introduced on the basis of experimentation rather than chemical grounds. The majority of the work in ab initio calculations is in the evaluation of the two electron integrals (Coulomb and exchange). All modern semiempirical methods are based on the modified neglect of differential overlap (MNDO) approach. In this method, parameters are assigned for different atomic types and are fitted to reproduce properties such as heats of formation, geometrical variables, dipole moments, and first ionization energies. The parameterization was carried out separately for classes of compounds like hydrocarbons, CHO systems, CHN systems, and so on. The latest versions of the MNDO method are referred to as AM1 and PM3. Another method to reduce the two-electron integral is the zero differential overlap (ZDO) approximation, which neglects all products of basis functions depending on the same electron coordinates when located on different atoms.

K. I. Ramachandran et al., *Computational Chemistry and Molecular Modeling*
DOI: 10.1007/978-3-540-77304-7, ©Springer 2008

It means that all the products of atomic orbital functions $\chi_u \chi_v$ are set to be zero, and the overlap integral $S_{uv} = \delta_{uv}$ (where δ_{uv} is the Kronecker delta, i.e., $\delta_{uv} = 0$ if $u \neq v$ and $\delta_{uv} = 1$ if $u = v$.). At the ZDO approximation, all three- and four-centered integrals vanish. This reduces the overlap matrix into a unit matrix. One-electron integrals involving three centers (two from the basis functions and one from the operator) are set to zero. All three- and four-center two-electron integrals, which are by far the most numerous of the two-electron integrals, are neglected. Parameterization is done to compensate the approximations. Hence, all the remaining integrals are replaced by proper parameters obtained by experimentation.

7.2 The Neglect of Differential Overlap Method

The neglect of differential overlap (NDO) method was first introduced by John Pople, and it is now the basis of most successful semiempirical methods. The method involves the modification of the HF equation, $FA = SA\varepsilon$, by approximating the overlap matrix S as unit matrix. This allows us to replace the HF secular equation $|H - ES| = 0$ with a simpler equation $|H - E| = 0$. We shall see some common techniques used to make the computation possible.

7.3 The Complete Neglect of Differential Overlap Method

In the complete neglect of differential overlap (CNDO) method, all integrals involving different atomic orbitals, χ_u, are ignored. Thus, the overlap matrix becomes the unit matrix, $S = 1$. The parameterization and implementation scheme of the CNDO method was also proposed by Pople.

7.4 The Modified Neglect of the Diatomic Overlap Method

The modified neglect of the diatomic overlap (MNDO) method (by Michael Dewar and Walter Thiel, 1977) is the oldest NDDO-based model that parameterizes one-center two-electron integrals based on spectroscopic data for isolated atoms, and evaluates other two-electron integrals using the idea of multipole-multipole interactions from classical electrostatics. A classical MNDO model uses only s and p orbital basis sets, while more recent MNDO/d adds d-orbitals that are especially important for the description of hypervalent sulphur species and transition metals. MNDO has a number of known deficiencies, such as the inability to describe the hydrogen bond due to a strong intermolecular repulsion. The MNDO method is characterized by a generally poor reliability in predicting heats of formation. For example, highly substituted stereoisomers are predicted to be too unstable, compared to linear isomers due to the overestimation of repulsion in sterically crowded systems.

Existing semiempirical models differ by further approximations that are made when evaluating one- and two-electron integrals and by the parameterization philosophy.

While INDO added all one-center two-electron integrals to the CNDO/2 formalism, NDDO adds all two-center integrals for repulsion between a charge distribution on one center and a charge distribution on another center. Otherwise, the zero-differential overlap approximation is used.

7.5 The Austin Model 1 Method

The Austin Model 1 (AM1) method, developed by M. J. S. Dewar and coworkers, takes a similar approach to MNDO in approximating two-electron integrals, but uses a modified expression for nuclear-nuclear core repulsion. The modified expression results in non-physical attractive forces that mimic van der Waals interactions. The modification also necessitated a re-parameterization of the model, which was carried out with a particular emphasis on dipole moments, ionization potentials, and geometries of molecules. While this allows for some description of the hydrogen bond, other deficiencies, such as systematic over-estimates of basicities, remained unsolved. Also, the lowest energy geometry for the water dimer is predicted incorrectly by the AM1 model. On the other hand, AM1 nicely improves some properties, such as heats of formation, over MNDO.

7.6 The Parametric Method 3 Model

The Parametric Method 3 (PM3) model, developed by James Stewart, uses a Hamiltonian that is very similar to the AM1 Hamiltonian, but the parameterization strategy is different. While AM1 was parameterized largely based on a small number of atomic data, PM3 is parameterized to reproduce a large number of molecular properties. In some sense, chemistry gave way to statistics with the PM3 model. A different parameterization, and a slightly different treatment of nuclear repulsion allows PM3 to treat hydrogen bonds rather well, but it amplifies non-physical hydrogen-hydrogen attractions in other cases. This results in serious problems while analyzing intermolecular interactions (methane is predicted to be a strongly-bound dimer) or conformations of flexible molecules (OH is strongly attracted to CH_3 in 1-pentanol). The accuracy of thermochemical predictions with PM3 is slightly better than that of AM1. The PM3 model has been widely used for the rapid estimation of molecular properties and has been recently extended to include many elements, including some transition metals.

7.7 The Pairwize Distance Directed Gaussian Method

The pairwise distance directed Gaussian (PDDG/PM3) method, developed by
William Jorgensen and coworkers, overcomes some of the deficiencies of the earlier
NDDO-based methods by using a functional group-specific modification of the core
repulsion function. The nPDDG/PM3 modification provides a good description of
the van der Waals attraction between atoms, and the PDDG/PM3 model appears
to be suitable for calculations of intermolecular complexes. Furthermore, careful
re-parameterization has made the PDDG/PM3 model very accurate for estimation
of heats of formation. However, some limitations common to NDDO methods remain in the PDDG/PM3 model: the conformational energies are unreliable, most
activation barriers are significantly overestimated, and the description of radicals is
erratic. So far, only C, N, O, H, S, P, Si, and halogens have been parameterized for
PDDG/PM3 [1].

7.8 The Zero Differential Overlap Approximation Method

The zero-differential overlap (ZDO) method is an approximation that is used to simplify the many electrons by ignoring two-electron repulsion integrals. If the molecular orbitals ϕ_j are expanded in terms of N basis functions, χ_s^A as:

$$\phi_j = \sum_{s=1}^{b} c_{si} \chi_s^A \tag{7.1}$$

where A is the atom the basis function is centered on, and c_{si} are the coefficients.
The two-electron repulsion integrals are then defined as:

$$\langle sv|\lambda\sigma\rangle = \int\int \chi_s^A(1)\chi_v^B(1)\frac{1}{r_{12}}\chi_\lambda^C(2)\chi_\sigma^D(2)d\tau d\tau \tag{7.2}$$

The zero-differential overlap approximation ignores integrals that contain the
product $\chi_s^A(1)\chi_v^B(1)$ where $s \neq v$. This transforms the equation to:

$$\langle sv|\lambda\sigma\rangle = \delta_{sv}\delta_{\lambda\sigma}\langle ss|\lambda\lambda\rangle \tag{7.3}$$

δ_{sv} is the Kronecker delta with $\delta_{sv} = 0$ if $s \neq v$ and $\delta_{sv} = 1$ if $s = v$. The total number of such integrals is reduced to $N(N+1)/2$ (approximately $N^2/2$) from
$[N(N+1)/2][N(N+1)/2+1]/2$ (approximately $N^4/8$) where N is the number of
orbitals.

7.9 The Hamiltonian in the Semiempirical Method

The (CNDO) method and the intermediate neglect of differential overlap (INDO) method are SCF methods solving the Roothaan equations iteratively, with approximations for the integrals in the Fock matrix. Only valence electrons are mainly considered in these methods. The Hamiltonian is:

$$\hat{H}_{\text{val}} = \sum_{i=1}^{n(\text{val})} \left[-\frac{1}{2}\nabla_i^2 + V(i) \right] + \sum_{i=1}^{n(\text{val})} \sum_{j>1} \frac{1}{r_{ij}} \qquad (7.4)$$

which can be simplified as:

$$\hat{H}_{\text{val}} = \sum_{i=1}^{n(\text{val})} \hat{H}_{\text{val}}^{\text{core}}(i) + \sum_{i=1}^{n(\text{val})} \sum_{j>1} \frac{1}{r_{ij}} \qquad (7.5)$$

where

$$\hat{H}_{\text{val}}^{\text{core}}(i) = \left[-\frac{1}{2}\nabla_i^2 + V(i) \right] \qquad (7.6)$$

Here $n(\text{val})$ stands for the number of valence electrons in the system, $V(i)$ is the potential energy of valence electron i in the field of nuclei and the core electrons, $\hat{H}_{\text{val}}^{\text{core}}(i)$ is the one-electron part of $\hat{H}_{\text{val}}$.

CNDO uses a minimal basis set of valence Slater atomic orbitals f_r with orbital exponents fixed based on the following (Slater) rules:

1. The orbital exponent ζ is given by the expression, $\zeta = (Z - s_{nl})/n$, where n is the principal quantum number, Z the atomic number, and s_{nl} is the screening constant.
2. The screening constants are determined based on the following scheme:
3. Atomic orbitals are classified into the groups $(1s)$, $(2s,2p)$, $(3s,3p)$, and $(3d)$.
4. The contribution to the screening constant is zero for electrons in groups outside the one being considered.
5. Each electrons within the group contributes a value of 0.35 excepting the $1s$ group where the value is 1.20 (in the general Slater rule scheme $1s$ is assigned 0.30).
6. For s or p orbital electrons, 0.85 from each electron whose quantum number n is one less than the orbital considered and 1.00 for each electron further in.
7. For each d orbital electron inside the group, the contribution is assigned as 1.00.
8. s_{nl} is calculated as the sum of all these contributions.

The valence orbital ϕ_i is given by:

$$\phi_i = \sum_{r=1}^{b} C_{ri} f_r \qquad (7.7)$$

The molecular electronic energy is given by:

$$E = 2 \sum_{i=1}^{n(\text{val})/2} H_{\text{val},ii}^{\text{core}} + \sum_{i=1}^{n(\text{val})/2} \sum_{j=1}^{n(\text{val})/2} (2J_{ij} - K_{ij}) + V_{cc} \qquad (7.8)$$

where V_{cc} is the core–core repulsion term, and is given by:

$$V_{cc} = \sum_{\alpha} \sum_{\beta > \alpha} \frac{C_\alpha C_\beta}{R_{\alpha\beta}} \qquad (7.9)$$

The core charge C_α on atom α equals the atomic number of atom α minus the number of core electrons on α.

The Fock matrix elements are computed by the equation:

$$F_{\text{val},rs} = H_{\text{val},rs}^{\text{core}} + \sum_{t=1}^{b} \sum_{u=1}^{b} P_{tu} \left[(rs/tu) - \frac{1}{2}(ru/ts) \right] \qquad (7.10)$$

The CNDO follows ZDO approximation.

The overlap integral $S_{rs} = \langle f_r(1) | f_s(1) \rangle = \delta_{rs}$, the Kronecker delta. By ZDO approximation, $f_r^*(1) f_s(1) dv_1 = 0$ if $r \neq s$

But,

$$(rs/tu) = \left\langle f_r(1) f_t(2) \left| \frac{1}{r_{12}} \right| f_s(1) f_u(2) \right\rangle \qquad (7.11)$$

$$(rs/tu) = \delta_{rs} \delta_{tu} (rr/tt) = \delta_{rs} \delta_{tu} \gamma_{rt} \qquad (7.12)$$

where $\gamma_{rt} = (rr/tt)$.

In the CNDO method, there are several basis valence AOs on each atom excepting the hydrogen atom. ZDO approximation neglects electron-repulsion integrals involving different AOs centered on the same atom.

The calculated values of molecular properties do not change if the coordinate axes are changed. Hence, the values are said to be rotationally invariant. Similarly, the values do not change if each basis AO on a particular atom is replaced by a linear combination of the basis AOs on that atom, or the results are hybridizationally invariant. To maintain rotational and hybridizational invariance, even after the ZDO approximation, the CNDO method introduces the following parameterization:

1. The electron repulsion integral, $\gamma_{rt} = (rr/tt)$ is considered as dependent only on the atoms where f_r and f_t are centered.
2. It does not depend on the nature of orbitals.

If the valence electrons f_r and f_t are centered on atoms A and B,
$(r_A r_A | t_B t_B) = \gamma_{r_A t_B} = \gamma_{AB}$ for all valence atomic orbitals f_r on A and all valence atomic orbitals f_t on B.

In CNDO, all one-center valence electron repulsion integrals on atom A have the value γ_{AA} and all two-center valence electron repulsion integrals involving atoms A and B have the value γ_{AB}. All three-center or four-center values are neglected by ZDO. γ_{AA} and γ_{AB} are computed using valence STOs on A and B. These values depend upon the orbital exponent, the principal quantum number of the valence electron and the distance between atoms A and B.

7.9.1 The Computation of $H^{core}_{r_A s_B}$

$H^{core}_{r_A s_B} = \beta^0_{AB} S_{r_A s_B}$ for $r \neq s$ where $S_{r_A s_B}$ is evaluated exactly, unlike the Roothaan equation. $\beta^0_{AB} = \frac{1}{2}\left(\beta^0_A + \beta^0_B\right)$ β^0_A and β^0_B are chosen to make the CNDO calculated MOs resembling the coefficients in the minimal basis ab initio MOs. When A and B are the same atoms, $S_{r_A s_B} = 0$ for $r \neq s$ by orthogonality condition of the atomic orbitals on the same atom. Then $H^{core}_{r_A s_B} = 0$.

7.9.2 The Computation of $H^{core}_{r_A r_A}$

We know that $H^{core}(1) = -\frac{1}{2}\nabla^2_1 + V(1)$, where $V(1)$ is the potential energy of valence electron 1 in the field of the core. Splitting $V(1)$ into contributions from individual atomic cores:

$$H^{core}(1) = -\frac{1}{2}\nabla^2_1 + V_A(1) + \sum_{B \neq A} V_B(1) \tag{7.13}$$

then:

$$H^{core}_{r_A r_A} = \left\langle f_{r_A}(1)\left| -\frac{1}{2}\nabla^2_1 + V_A(1) \right| f_{r_A}(1) \right\rangle + \sum_{B \neq A} \left\langle f_{r_A}(1)\left| V_B(1) \right| f_{r_A}(1) \right\rangle \tag{7.14}$$

This is simply written as:

$$H^{core}_{r_A r_A} = U_{rr} + \sum_{B \neq A} \left\langle f_{r_A}(1)\left| V_B(1) \right| f_{r_A}(1) \right\rangle \tag{7.15}$$

There are two versions of CNDO: CNDO/1 and CNDO/2. In CNDO/1, U_{rr} is computed as the negative of valence-state ionization energy from the AO f_{r_A}. The integrals $\left\langle f_{r_A}\left| V_B(1) \right| f_{r_A} \right\rangle = V_B$ are taken as equal to maintain rotational and hybridizational invariance:

$$V_{AB} = -\left\langle S_A(1)\left| \frac{C_B}{r_{1B}} \right| S_A(1) \right\rangle \tag{7.16}$$

C_B is the core charge of atom B. In CNDO/1, by using V_{AB}, two neutral molecules or atoms, separated substantially, may even experience attractive forces. This error is eliminated in CNDO/2 by taking V_{AB} as $-c_B\gamma_{AB}$. With these approximations, the Fock matrix elements are decided. Roothaan equations are solved iteratively to find the CNDO orbitals and the orbital energies.

In the INDO method, the differential overlap between AOs on the same atom is not neglected in one-center electron repulsion integrals, while two-center electron integrals are neglected. With a few more integrals added, the INDO method is an improvement to the CNDO method.

In the neglect of diatomic differential overlap (NDDO) method, the differential overlap is neglected between atoms centered on different atoms.

Hence, $f_r^*(1)f_s(1)dv_1 = 0$ when r and s are on different atoms. It satisfies the invariance conditions. Dewar and Thiel modified NDDO to make MNDO. In this method, compounds containing H, Li, Be, B, C, N, O, F, Al, Si, Ge, Sn, Pb, P, S, Cl, Br, I, Zn, and Hg have been parameterized. Valence electron Hamiltonian is given by Eq. 7.5 and the Fock matrix is given by Eq. 7.10. The MNDO Fock matrix elements can be determined as follows.

Core matrix elements (core resonance integral) $H_{\mu_A v_B}^{Core} = \langle \mu_A(1) | \hat{H}^{Core}(1) | \mu_A(1) \rangle$ with atomic orbitals centered at atoms A and B are given by:

$$H_{\mu_A v_B}^{Core} = \frac{1}{2}\left(\beta_{\mu_A} + \beta_{v_B}\right) S_{\mu_A v_B} ; A \neq B \tag{7.17}$$

where β are the parameters for each orbital. for example, carbon with valence atomic orbitals $2s$ and $2p$, centered on the same C-atom, will have parameters β_{C2s} and β_{C2p}. Core matrix elements from different atomic orbitals centered on the same atom are given by Eq. 7.13. Hence:

$$H_{\mu_A v_B}^{Core} = \left\langle \mu_A \left| -\frac{1}{2}\nabla_1^2 + V_A \right| v_A \right\rangle + \sum_{B \neq A} \langle \mu_A | V_B | v_A \rangle \tag{7.18}$$

Using group theoretical considerations $\langle \mu_A | -\frac{1}{2}\nabla_1^2 + V_A | v_A \rangle$ can be made as zero. Hence:

$$H_{\mu_A v_B}^{Core} = \sum_{B \neq A} \langle \mu_A | V_B | v_A \rangle \tag{7.19}$$

If we consider electron 1 to interact with a point core of charge C_B, then:

$$V_B = -\frac{C_B}{r_{1B}} \tag{7.20}$$

$$\langle \mu_A | V_B | v_A \rangle = -C_B \left\langle \mu_A \left| \frac{1}{r_{1B}} \right| v_A \right\rangle \tag{7.21}$$

In MNDO, $\langle \mu_A | V_B | \nu_A \rangle = -C_B \langle \mu_A \nu_A | s_B s_B \rangle$ where s_B is the valence s-orbital on atom B:

$$H_{\mu_A \nu_B}^{Core} = \sum_{B \neq A} \langle \mu_A | V_B | \nu_A \rangle = - \sum_{B \neq A} C_B (\mu_A \nu_A | s_B s_B) ; \quad \mu_A \neq \nu_A \tag{7.22}$$

Core matrix elements $H_{\mu_A \mu_A}^{Core} = \langle \mu_A(1) | \hat{H}^{Core}(1) | \mu_A(1) \rangle$ is computed using Eq. 7.14 to get:

$$H_{\mu_A \mu_A}^{Core} = \left\langle \mu_A \left| -\frac{1}{2}\nabla^2 + V_A \right| \nu_A \right\rangle + \sum_{B \neq A} \langle \mu_A | V_B | \nu_A \rangle \tag{7.23}$$

$U_{\mu_A \mu_A}^{Core} = \langle \mu_A | -\frac{1}{2}\nabla^2 + V_A | \nu_A \rangle$ is evaluated by parameterization using atomic spectra in MNDO (the parameters used for the C-atom U_{ss} and U_{pp}). Thus:

$$H_{\mu_A \mu_A}^{Core} = U_{\mu_A \mu_A}^- \sum_{B \neq A} C_B (\mu_A \nu_A | s_B s_B) \tag{7.24}$$

The evaluation of $\langle \mu_A \nu_A | s_B s_B \rangle$ *is as follows:*

1. All three-center and four-center integrals vanish with the ZDO method.
2. One-center electron repulsion integrals are either Coulomb integral $g_{uv} = \langle \mu_A \mu_A | \nu_A \nu_A \rangle$ or exchange integral $h_{uv} = \langle \mu_A \nu_A | \mu_A \nu_A \rangle$. Thus, for the C-atom, the integrals are $g_{ss}, g_{sp}, g_{pp}, g_{pp'}, h_{sp}$ and $h_{pp'}$ where p and p' are along different axes.
3. Two-center repulsion integrals are found from the values of the one-center integral and the internuclear distance using multipole expansion procedure (Dewar et al., Theor. Chim. Acta, 46, 89, 1977).
4. The core–core repulsion term is given by:

$$V_{cc} = \sum_{B > A} \sum_{A} [C_A C_B (s_A s_B | s_B s_B) + f_{AB}] \tag{7.25}$$

where:

$$f_{AB} = f_{AB}^{MNDO} = [C_A C_B (s_A s_B | s_B s_B) (e^{-\alpha_A R_{AB}} + e^{-\alpha_B R_{AB}})] \tag{7.26}$$

α_A and α_B are parameters for atoms A and B. For O–H and N–H pairs:

$$f_{AH}^{MNDO} = [C_A C_H (s_A s_H | s_H s_H) ((R_{AH}/A.U.) e^{-\alpha_A R_{AH}} + e^{-\alpha_H R_{AH}})] \alpha_A \alpha_H \tag{7.27}$$

where A is N or O.

In the MNDO method, the following parameters have to be optimized:

1. One-center one-electron integrals U_{ss} and U_{pp}.
2. The STO exponent ξ. For MNDO $\xi_s = \xi_p$.
3. β_s and β_p. MNDO assumes that $\beta_s = \beta_p$.

In the Austin model 1 (AM1), $\xi_s \neq \xi_p$. Parameterization with compounds from H, B, Al, C, Si, Ge, Sn, N, P, O, S, F, Cl, Br, I, Zn, and Hg have been made. Thus:

$$f_{AB}^{AM1} = f_{AB}^{MNDO} + \frac{C_A C_B}{R_{AB}/A.U.} \left[\sum_k a_{kA} \exp\left[-b_{kA}(R_{AB} - c_{kA})^2 \right] \right]$$

$$+ \frac{C_A C_B}{R_{AB}/A.U.} \left[\sum_k a_{kB} \exp\left[-b_{kB}(R_{AB} - c_{kB})^2 \right] \right] \tag{7.28}$$

Stewart re-parameterized the values to generate the PM series. That derived from AM1 is known as the PM3 (Parametric method 3). In the PM3, one-center electron repulsion integrals are parameterized by optimization. The core repulsion function takes only two Gaussian functions per atom. In PM3, compounds containing H, C, Si, Ge, Sn, Pb, N, P, As, Sb, Bi, O, S, Se, Te, F, Cl, Br, I, Al, Ga, In, Tl, Be, Mg, Zn, Cd, and Hg have been parameterized.

Dewar and coworkers modified AM1 to give the semi an initio model-1 (SAM-1). The differences between AM1 and SAM-1 are listed below:

1. SAM-1 evaluates two-venter electron integrals by the equation, $(\mu\nu|\lambda\sigma)_{SAM1} = g(R_{AB})(\mu\nu|\lambda\sigma)_{STO\text{-}3G}$. The integral $(\mu\nu|\lambda\sigma)_{STO\text{-}3G}$ is computed with the STO-3G basis set. $g(R_{AB})$ is a function of the internuclear distance, which reduces the size of repulsion integrals to allow electron correlation.
2. SAM-1 is slower than AM1, while it is faster than ab initio methods due to NDDO approximation.

Thiel and Voityuk extended the MNDO by introducing d-orbitals: this is called the MNDO/d method. For the elements of the first and second row of the periodic table, d-orbitals are not included, so that MNDO and MNDO/d methods are identical for them. MNDO/d method has been parameterized for a number of transition elements.

7.10 Comparisons of Semiempirical Methods

CNDO and INDO results are less accurate than minimal basis set ab initio methods. Hence these methods fail to compute accurate binding energy. Dewar's approach was to treat only valence electrons. Most of the theories such as the MINDO, MNDO, AM1, PM3, SAM1, and MINDO/d methods use a minimal basis set of valence Slater type s and p AOs to expand valence-electron MOs. A comparison of the heat of formation with MNDO, PM3, and AM1 methods has been made in Table 7.1. The CNDO method is crude, fast and can do second row elements. The INDO method is better for first row elements, while the MINDO3 and MNDO methods are more reliable. The AM1 method is better for estimating H bonds. The PM3 method, developed from AM1, includes more main-group elements. For ordinary molecules, AM1 or PM3 are probably the best to try. Semiempirical methods are highly useful for better geometry optimization than force fields, especially geome-

Table 7.1 Heat of formation of some MNDO, PM3, and AM1 compounds

| Compound | Heat of formation | | | |
	MINDO/3	MNDO	AM1	PM3
CH_4	−6.3	−11.9	−8.8	−13.0
LiH	—	+23.2	—	—
BeO		+38.6		+53.0
NH_3	−9.1	−6.4	−7.3	−3.1
CO2	−95.7	−75.1	−79.8	−85.0
SiH	+82.9	+90.2	+89.8	+94.6
H_2S	−2.6	+3.8	+1.2	−0.9
HCl	−21.1	−15.3	−24.6	−20.5
HBr	—	+3.6	−10.5	+5.3
$HgCl_2$	—	−36.9	−44.8	−32.7
ICl	—	−6.7	−4.6	+10.8
TlCl	—	—	—	−13.4
PbF	—	−22.6	—	−21.0

tries for molecules including atoms which are not parameterized in a force field. It makes the qualitative prediction of IR frequencies and the total electron density surface for graphical display. They are not really good enough for reaction energies and equilibrium constants; even quite low level ab initio methods are better for energetics.

Semiempirical methods cannot do anything with core electrons, e.g., NMR shielding. The ZINDO method can deal with excited states, which are more difficult to do than ground states in ab initio methods. Hence, predictions of UV/visible spectrum absorption wavelengths are possible. A comparative study of different semiempirical methods on the basis of theory has been made in Table 7.2.

It is well known that the strength of H-bonds in charged systems is proportional to the difference in proton affinities (PAs) of their components. The evaluation of PAs is very important in predicting the strength of H-bonds in biomolecules such as enzymes, on their models. Bliznyuk and Voityuk used the MNDO method to estimate PAs of DNA base pairs and in their complexes and found that the MNDO method was in good agreement with theory.

A highly symmetric zinc(II) complex with $\{[Zn(tren)]_4(\mu_4\text{-}ClO_4)\}^{7+}$ structure unit (tren=tris(2-aminoethyl)amine) was characterized by single-crystal X-ray diffraction and was compared by the calculation from the MNDO method by Fu et al. [2].

The *syn,syn* configurational preference of compounds of the type R–NSN–R, where the substituent R is $SiMe_3$, is rationalized in terms of anti-periplanar hyperconjugation between the in plane nitrogen lone pairs on the NSN fragment and the electropositive silicon-H/Me σ bonds. MNDO and ab initio calculated energies and geometries were reported for a range of electropositive and electronegative substituents R and discussed in terms of stereoelectronic interactions by Rzepa and Woollins [3].

Table 7.2 A comparative study of different semiempirical methods

Acronym	Full name	Underlying approximation	Parameters	Fitted parameters
CNDO	Complete neglect of differential overlap	CNDO	—	—
INDO	Intermediate neglect of differential overlap	INDO	—	—
MINDO/3	Modified intermediate neglect of differential overlap, version 3	INDO	10	2
MNDO	Modified neglect of differential overlap	NDDO	10	5
AM1	Austin model 1	NDDO	13	8
PM3	Parametric model number 3	NDDO	13	13

In the search for new beta-lactam antibiotics (penicillins fall in this class of compounds), it was found that sulphur-based drugs (thiamazins) displayed no activity, while the traditional oxygen-based drugs (oxamazins) were useful. The explanation of this surprising behavior was partially done by semiempirical calculations, which indicated that the structure of the inactive drugs results in a poor fit with the "active site." Boyd et al. conducted a series of studies in this regard (Boyd, Eigenbrot, Indelicato, Miller, Pasini, Woulfe, J. Med. Chem. 1987, 30, 528.) These calculations (which utilized the AM1, MNDO, and MINDO/3 methods) were also able to identify several other factors (which may not be important), which lowers the "likelihood" that potentially useful drugs will be eliminated without consideration.

Myclobutanil is a broad spectrum, agrochemical fungicide. After narrowing the possible types of compounds that appeared useful by field testing, differences between the activity of these molecules were correlated by Boyd with a number of molecular properties, including an analysis of molecular charges calculated using the semiempirical MNDO method. The eventual development of myclobutanil was credited as a direct result of this analysis.

It is estimated that over 400,000 tons of zeolites are used annually, primarily in petroleum refining processes. Since these are solid state materials, both experimental and theoretical investigations are quite difficult. However, it has been shown that the results of quantum mechanical calculations on isolated molecules can be successfully applied to enhance the understanding of some of the properties of these solid-state materials. The research conducted by Earley (C. W. Inorg. Chem. 1992, 31, 1250) concluded that AM1 calculations on molecules containing as few as two or three silicon centers can be used to explain one of the basic structural features of these molecules. Semiempirical calculations on larger molecules have been used to determine the most acidic sites.

The antipsoriatic drug anthralin has been in use for over 60 years. The AM1 study conducted by Holder and Upadrashta (J. Pharm. Sci. 1992, 81, 1074) explains some of the properties that make the drug active and suggests further directions for research.

Clinical trials of an aldose reductase inhibitor conducted by Kador and Sharpless (Molec. Pharm. 1983, 24, 521) suggests that these types of compounds can prevent certain eye problems (cataract formation and corneal re-epithelialization) in diabetic patients. Clinical studies indicate that no "universally potent" inhibitor exists, emphasizing the need to find new drugs of this type. A comparison of the activities of several of these drugs with results of quantum mechanical calculations (energies of lowest unoccupied molecular orbitals and atomic charges) showed strong correlations, which aided in the prediction of the minimal requirements for an active drug.

GABA (gamma-aminobutyric acid) is a mediator of the central nervous system and has been implicated as a contributor in chemically-induced depression. A theoretical study using the AM1 method on GABA and two derivatives of this compound conducted by Kehl and Holder (J. Pharm. Sci. 1991, 80, 139) was able to show that one of these derivatives is more closely related to the parent system than the second. This result is in agreement with the actual experimental results.

The phospholiphase A2 enzyme is thought to be involved in the breakdown of phospholipids, important components in living systems. This study was undertaken by Ripka, Sipio, and Blaney (Lect. Heterocyc. Chem. 1987, IX, S95) to show that theoretical methods can be successfully applied to drug design. The analysis of the geometries of a number of proteins suggested one key structural component. Quantum mechanical calculations not only supported these findings, but were also able to offer a simple explanation for this phenomenon.

Quantum mechanical calculations on a number of simple sugars conducted by Szarek, Smith, and Woods (J. Am. Chem. Soc. 1990, 112, 4732) provided an explanation of the relative sweetness of these compounds. An analysis of the structural features observed in the calculated geometries of these compounds suggests that a previously neglected feature of these molecules may be important in determining "sweetness."

Carotenoids are light-gathering agents in the pigments of eyes. In order to understand the efficiency of these compounds in transferring light energy, a theoretical study using the AM1 method was performed by Wasielewski, Johnson, Bradford, and Kispert (J. Chem. Phys. 1989, 91, 6691) The explanation for the high efficiency of this process obtained from these calculations was in agreement with the results of experimental studies.

Applications of these MNDO type methods usually involve exploration of multidimensional potential surfaces which is greatly facilitated if the gradient of the energy with respect to the nuclear coordinates can be evaluated efficiently. Once a stationary point on a potential energy surface is found, the second derivatives of the energy with respect to the nuclear coordinates provide the harmonic force constants and the harmonic vibrational frequencies. They may also be used for characterizing stationary points and for locating transition states on potential surfaces. Other molecular properties such as infrared vibrational intensities, polarizabilities magnetic susceptibilities, magnetic shielding tensors, and spin-spin coupling constants at equilibrium geometries may also be of interest in a theoretical investigation. These physical quantities can be conveniently expressed as partial derivatives of the energy, and thus share a significant portion of the underlying mathematical formalism.

Semiempirical calculations are much faster than their ab initio counterparts. Their results, however, can be very wrong if the molecule being computed is not similar enough to the molecules in the database used to parameterize the method. Semiempirical calculations have been most successful in the description of organic chemistry, where only a few elements are used extensively and molecules are of moderate size.

Despite their limitations, semiempirical methods are often used in computational chemistry because they allow the study of systems that are out of reach of more accurate methods. For example, modern semiempirical programs allow the study of molecules consisting of thousands of atoms while ab initio calculations that produce similar thermochemical accuracy are feasible on molecules consisting of less than 50–70 atoms. Semiempirical calculations can be useful in many situations, such as the following:

1. The computational modeling of structure-activity relationships to gain insight about reactivity or property trends for a group of similar compounds.
2. The design of chemical synthesis or process scale-up, especially in industrial settings where getting a qualitatively correct answer today is more important than getting a highly accurate answer after some time.
3. The development and testing of new methodologies and algorithms, for example, the development of hybrid quantum mechanics/molecular mechanics (QM/MM) methods for the modeling of biochemical processes.
4. Checking for gross errors in experimental thermochemical (e.g., heat of formation) data.
5. The preliminary optimization of geometries of unusual molecules and transition states that cannot be optimized with molecular mechanics methods.
6. In many applications, where qualitative insight about electronic structure and properties is sufficient.

For large systems, either molecular mechanics or semiempirical quantum mechanics could be used for the optimization and calculation of conformational energies. The molecular mechanics approach is faster and in most cases it produces more accurate conformational energies and geometries. Some molecular mechanics methods, such as MM3 and MM4, can also predict the thermochemistry of stable species reasonably well. On the other hand, if there is no suitable force field for the system (e.g., in case of reactive intermediates or transition states), semiempirical methods may be the only choice. For a small system, the compromise must be made between the semiempirical approach and the more reliable but much more time-consuming ab initio calculations. In general, semiempirical results can be trusted only in situations when they are known to work well (e.g., systems similar to molecules in the parameterization set). Finally, it is not correct to assume that for modeling all larger systems, semiempirical methods can be used. No computational insight may be better than a wrong computational insight.

7.11 Software Used for Semiempirical Calculations

AMPAC, GAMESS, GAUSSIAN, MOLCAS, MOPAC, POS, VASP, Spartan, and Hyperchem are some of the common types of software used for semiempirical calculations. Most of the software can use all the methods mentioned in this chapter. Some typical semiempirical computational input and output files of molecules with different software have been included in the URL.

7.12 Exercises

1. Create acetonitrile (CH_3CN) in the SPARTAN builder or Gaussian and set up an AM1 or PM3 semiempirical calculation. Include molecular orbitals, frequencies, and the Mulliken populations in the output file. Add any surfaces you would like to look at, such as the electron density and the HOMO, and optimize the structure. Examine the output file, the vibrational animations, and the orbital pictures and answer the following questions:

   a. What are the energies of the HOMO and the LUMO?
   b. In which MOs are the two C–N p bonds mostly localized?
   c. Which MO and which AOs appear to be the locus of the unshared pair on nitrogen?
   d. What is the calculated stretching frequency of the CN triple bond?
   e. What is the calculated enthalpy of formation?

2. The semiempirical module can compute solvation energies using the SM5.4 solvation model. Select a simple amino acid. Create both the neutral and the zwitter ion. Optimize each geometry using the PM3 Hamiltonian (when you set up the calculations, select the "E. Solvation" button).

   a. Obviously, the zwitterion should have the greater solvation energy.
   b. How do the HOMO and LUMO energies change from the neutral to the zwitterion?
   c. Is there any significant difference in the optimized geometry between the two structures?

3. Model the Wittig reaction using gas-phase semiempirical AM1 calculations.
4. Calculate the geometry of NH_3 (C_{3v} symmetry) with MOPAC. Compare these values with the experimental values $r_{N-H} = 1.012$ Å and $\theta_{HNH} = 106.7°$.
5. Calculate the geometry and energy of planar NH_3 (D_{3h} symmetry). The difference in energy between this planar structure and pyramidal ammonia represents the barrier to the "umbrella" inversion in ammonia. Compare the computed value with the experimental barrier of 24.3 kJ/mol.
6. In this exercise you will calculate the rearrangement barrier for the reaction:

$$HNC \rightarrow HCN$$

(Hint: First, calculate the structure and energy of HNC and HCN using MOPAC and the PM3 parameter set. Compare your geometries with the experimental values. (For HNC, C–N = 1.169 Å and N–H = 0.994 Å; for HCN, C–N = 1.153 Å and C–H = 1.065 Å). Repeat the calculations with the MNDO and AM1 parameter sets. How do the results change? Which method is found to be the best? Justify your answer. Refer to the MOPAC manual for details.

References

1. Bliznyuk AA, Voityuk (1989) Proton affinities of nucleic bases and their complexes. Zh Phys Khim 63:1227-1230
2. Fu H et al. (2004) A novel perchlorate-bridged tetranuclear zinc(II) structure with tris(2-aminoethyl)amine ligand. Inorg Chem Comm 7:7 pp 906–908
3. Rzepa HS, Woollins JD (1988) Stereoelectronic effects in R–NSN–R systems. An MNDO and ab initio SCFMO study. J Chem Soc Dalt Trans pp 3051–3053

Chapter 8
The Ab Initio Method

8.1 Introduction

Electrons present in a system will be influenced by the remaining electrons present in the same system. In the single electron approximation techniques, which we have considered so far, this interaction is neglected. The interaction between electrons in a quantum system is known as *electronic correlation*. Within the Hartree-Fock (HF) limit of computation, the antisymmetric wavefunction is approximated by

a single Slater determinant, which does not include the Coulomb correlation leading to the total calculated electronic energy different from the exact solution of the non-relativistic Schrödinger equation within the Born-Oppenheimer approximation. The difference in energy between the HF limit and the actual (theoretical) one is known as the *correlation energy* (given by Löwdin).

It is to be noted that a certain level of electron correlation is already considered within the HF approximation, found in the electron exchange term describing the correlation between electrons with parallel spin. The effect of the correlation can be explained through electron density. In the immediate vicinity of an electron, there is a reduced probability of finding another electron. For electrons of opposite spin, this is often referred to as the Coulomb hole; the corresponding phenomenon for electrons of the same spin is the Fermi hole. We shall discuss correlation through electron density in the next chapter.

There is also a correlation related to the overall symmetry or total spin of the considered system. The solution to the Schrödinger equation through a single electron Slater determinant (SD) comes in the vicinity of the HF method. An additional approximation to the HF limits leads to semiempirical methods, while the introduction of additional determinants to the computation makes the solution exact. Electron correlation techniques will come under that category. The above concept is schematically represented in Fig. 8.1.

sSDs, taking account of the Pauli's exclusion principle (orbital asymmetry) are most suitable for describing many-electron basis functions. Automatically, the first step in correlation technique will be to set up a multi-determinant trial wave function ψ_{trial}, describing the total wave function in a "coordinate" system of an SD equation

K. I. Ramachandran et al., *Computational Chemistry and Molecular Modeling*   155
DOI: 10.1007/978-3-540-77304-7, ©Springer 2008

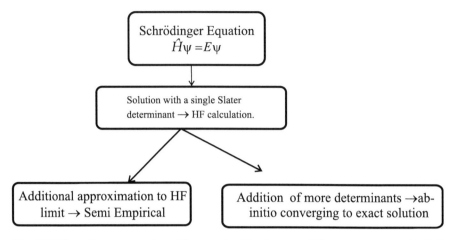

Fig. 8.1 Schematic representation giving relationships between different quantum mechanical methods

as given in Eq. 8.1. The procedure involves an expansion of the N-electron wave function as a linear combination of SD (in which each element is a one-electron function of the molecular orbital):

$$\psi_{\text{trial}} = a_0 \phi_{\text{HF}} + \sum_{i=1} a_i \phi_i \tag{8.1}$$

We have seen earlier that the basis set determines the size of the one-electron basis and thus limits the description of the one-electron functions (the MOs). Similarly, the number of determinants included decides the size of the many-electron basis and the extent of electron correlation.

8.2 The Computation of the Correlation Energy

The correlation energy can be expressed as given in Eq. 8.2:

$$E_{\text{HF}}^{C} = E_0 - E_{\text{HF}} \tag{8.2}$$

Where E_0 is the energy calculated by the Born-Oppenheimer approximation and E_{HF} is the energy computed by the HF approximation. It is a measure of the error introduced through the HF scheme. The development of methods to determine the correlation contributions accurately and efficiently is still a highly active research area in conventional quantum chemistry. Electron correlation is mainly caused by the instantaneous repulsion of the electrons, which is not covered by the effective HF potential. Pictorially speaking, the electrons often get too close to each other in the HF scheme, because the electrostatic interaction is treated in only an average manner. As a consequence, the electron-electron repulsion term is too large, resulting in E_{HF} being above E_0.

8.3 The Computation of the SD of the Excited States

The computation of the restricted Hartree-Fock (RHF) energy of a system containing N-electrons and M-basis function generates $(N/2)$ occupied molecular orbitals and $(M - N/2)$ unoccupied molecular orbitals. For example, the computation of dioxygen with a 3-21G basis set, keeping 16 electrons and 18 basis functions will carry 8 occupied molecular orbitals and 10 virtual molecular orbitals (refer to the book URL to see the output). An SD is determined by $N/2$ spatial MOs multiplied by two spin functions $(\alpha \& \beta)$ to yield N spinorbitals. By replacing MOs which are occupied in the HF determinant by MOs which are unoccupied, a whole series of determinants may be generated. These orbitals can be designated on the basis of the number of occupied HF MOs which have been replaced by unoccupied MOs, i.e., SDs which are singly, doubly, triply, quadruply, etc. excited "relative to the HF determinant", may reach up to a maximum of N excited electrons. These determinants are often referred to as Singles (S), Doubles (D), Triples (T), Quadruples (Q) etc. (Fig. 8.2).

The total number of determinants that can be generated from a given basis set depends on the size of the basis set. The larger the basis, the higher will be the number of virtual MOs generated, and the higher will be the possibility of generating excited determinants. If all the possible determinants in a given basis set are included, all the electron correlation can be recovered from the function. Automatically, the Schrödinger equation can be fully solved if we choose a basis set of infinite size. Methods which include electron correlation are thus two-dimensional; the larger the one-electron expansion (basis set size) and the larger the many-electron expansion (number of determinants), the better are the results.

Energy

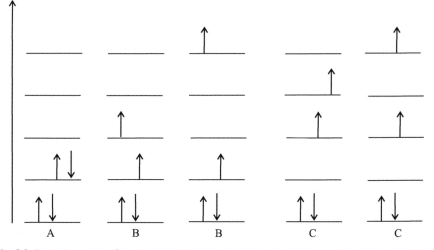

Fig. 8.2 Excited states configuration. A: HF ground state, B: singly excited (Singles or S) and C: doubly excited (Doubles or D)

8.4 Configuration Interaction

This method is based on the variational method similar to the HF formulation. Just as the lowest eigenvalue has been shown to be an upper bound to the exact ground-state energy, more generally, any eigenvalue calculated will be an upper bound to the exact excitation energy. We start with proposing a trial wavefunction, which is written as a linear combination of determinants with the expansion coefficients determined based on the variational principle. The wavefunction with the configuration interaction (ψ_{CI}) can be written as Eq. 8.2:

$$\psi_{CI} = a_0 \phi_{SCF} + \underbrace{\sum a_S \phi_S}_{\text{Singles(S)}} + \underbrace{\sum a_D \phi_D}_{\text{Doubles(D)}} + \ldots = \sum_{i=0} a_i \phi_i \tag{8.3}$$

Based on the linear variation method, the linear expansion $|\psi\rangle = \sum_i c_i |\Phi_i\rangle$ is repeated by varying coefficients c_i so as to minimize energy, $E = \left(\langle \psi | \hat{H} | \psi \rangle / \langle \psi | \psi \rangle \right)$. But, due to the additional normalization condition, the computation is turned into a constraint optimization problem. In this constraint optimization problem, we apply Lagrange's method of undetermined multipliers, and we minimize the Lagrange functional L (Eq. 8.3), which has the same minimum energy as E when the function is normalized:

$$L = \langle \psi_{CI} | H | \psi_{CI} \rangle - \lambda \left[\langle \psi_{CI} | \psi_{CI} \rangle - 1 \right] \tag{8.4}$$

where $\langle \psi_{CI} | H | \psi_{CI} \rangle$ is the energy of the ψ_{CI} wave function, $\langle \psi_{CI} | \psi_{CI} \rangle$ is the norm of the wave function, and λ the Lagrange multiplier. Substituting the values of energy function and the norm in the Lagrange functional:

$$L = \sum_{ij} a_i a_j \langle \Phi_i | \hat{H} | \Phi_j \rangle - \lambda \left(\sum_{ij} a_i a_j \langle \Phi_i | \Phi_j \rangle - 1 \right) \tag{8.5}$$

$$\sum_{ij} a_i a_j \langle \Phi_i | \hat{H} | \Phi_j \rangle = \sum_{i=0} a_i^2 E_i + \sum_{i=0} \sum_{j \neq 0} a_i a_j \langle \Phi_i | \hat{H} | \Phi_j \rangle$$

$$\sum_{ij} a_i a_j \langle \Phi_i | \Phi_j \rangle = \sum_{i=0} \sum_{j=0} a_i a_j \langle \Phi_i | \Phi_j \rangle = \sum_{i=0} a_i^2 \langle \Phi_i | \Phi_i \rangle = \sum_{i=0} a_i^2$$

$$L = \left(\sum_{i=0} a_i^2 E_i + \sum_{i=0} \sum_{j \neq 0} a_i a_j \langle \Phi_i | \hat{H} | \Phi_j \rangle \right) - \lambda \left(\sum_{i=0} a_i^2 - 1 \right) \tag{8.6}$$

$$\frac{\delta L}{\delta a_i} = 2 \sum_j a_j \langle \Phi_i | \hat{H} | \Phi_j \rangle - 2\lambda a_i = 0$$

$$= a_i \left(\langle \Phi_i | \hat{H} | \Phi_i \rangle - \lambda \right) + \sum_{j \neq i} a_j \langle \Phi_i | \hat{H} | \Phi_j \rangle = 0 \tag{8.7}$$

$$= a_i (E_i - \lambda) + \sum_{j \neq i} a_j \langle \Phi_i | \hat{H} | \Phi_j \rangle = 0 \tag{8.8}$$

If only a single determinant is there, then $a_i = 1$, and CI energy is the Lagrange multiplier ($\lambda = E$):

$$\frac{\delta L}{\delta a_i} = a_i (H_{ii} - \lambda) + \sum_{j \neq i} a_j H_{ij} = 0 \tag{8.9}$$

Where $H_{ij} = \langle \Phi_i | \hat{H} | \Phi_j \rangle$, $E_i = H_{ii} = \langle \Phi_i | \hat{H} | \Phi_i \rangle$

Eq. 8.9 is the variational requirement for energy minimization.

8.5 Secular Equations

The variational problem setup can be converted into a problem of solving secular equations. Equation 8.9 can be expanded to get secular equations for each element corresponding to each i:

$$a_0 (H_{00} - E) + a_1 H_{01} + \ldots + a_j H_{0j} = 0$$
$$a_0 H_{10} + a_1 (H_{11} - E) + \ldots + a_j H_{1j} = 0 \tag{8.10}$$
$$\ldots\ldots\ldots\ldots\ldots\ldots\ldots\ldots\ldots\ldots\ldots\ldots$$
$$a_0 H_{j0} + a_1 H_{j1} + \ldots + a_j (H_{jj} - E) = 0$$

The matrix equation corresponding to Eq. 8.10 can be represented as Eq. 8.11:

$$\begin{bmatrix} (H_{00} - E) & H_{01} & \ldots & H_{0j} \\ H_{10} & (H_{11} - E) & \ldots & H_{1j} \\ \ldots & \ldots & \ldots & \ldots \\ H_{j0} & H_{j1} & \ldots & (H_{jj} - E) \end{bmatrix} \begin{bmatrix} a_0 \\ a_1 \\ \ldots \\ a_j \end{bmatrix} = \begin{bmatrix} 0 \\ 0 \\ \ldots \\ 0 \end{bmatrix} \tag{8.11}$$

The matrix obtained in the above equation is known as the configuration interaction (CI) matrix. Solving the secular equations is equivalent to diagonalizing the CI matrix. The configuration interaction energy is obtained as the lowest eigenvalue of the CI matrix, and the corresponding eigenvector contains the a_i coefficients. The second lowest eigenvalue corresponds to the first excited state, the third lowest eigenvalue corresponds to the second excited state, and so on.

8.6 Many-Body Perturbation Theory

Many-body perturbation theory (MBPT) is a method to explain electron correlation by treating it as a perturbation to the HF wavefunction. Here, we start with a simple system and gradually turn on an additional "perturbing" Hamiltonian, representing a weak disturbance to the system. If the disturbance is not too large, various physical quantities associated with the perturbed system (e.g., its energy levels and eigenstates) will be continuously generated from those of the simple system. We can, therefore, study the former based on our knowledge of the latter. The solution to the present problem will be closely related to the previous one, though not identical. The starting point in our development of MBPT is the eigenvalue equation for the exact system:

$$H_0 \psi_n = \varepsilon_n \psi_n \qquad (8.12)$$

Once the solution to this problem is known, we will switch over to finding the eigenvalues (E_n) and eigenfunctions (ψ_n) of the perturbed system:

$$H \psi_n = E_n \psi_n \qquad (8.13)$$

The basic idea of perturbation theory is to expand the energy and wavefunctions of the perturbed system in powers of the small potential V:

$$H = H_0 + \lambda V \qquad (8.14)$$

where H_0 is the Hamiltonian of the previous computation, which is solved exactly or approximately, λ is a perturbation parameter, which measures the extent (power) of perturbation made to the initial Hamiltonian, and V-is the perturbation operator. It is assumed that the correction factor is small compared to the initial Hamiltonian so that the perturbed wave function and energy can be expressed in the form of Taylor expansion in powers of the perturbation parameter. Next, we write the eigenvalues and eigenfunctions of the perturbed system as:

$$E^n = E_0^n + \lambda E_1^n + \lambda^2 E_2^n + \dots \qquad (8.15)$$
$$\psi^n = \psi_0^n + \lambda \psi_1^n + \lambda^2 \psi_2^n + \dots \qquad (8.16)$$

Terms with the suffix 0 stand for zero-order terms, terms with the suffix 1 stand for first-order correction terms, terms with suffix 2 stand for second-order correction terms, and so on. In the computation procedure our aim is to use the minimum number of terms in this expansion that are necessary to achieve satisfactory approximations for E^n and ψ^n.

It is customary to consider the perturbed wavefunctions to be intermediately normalized. Hence:

$$\langle \psi | \phi_0 \rangle = 1 . \qquad (8.17)$$

Substituting ψ:

$$\langle \lambda^0 \psi_0 + \lambda^1 \psi_1 + \lambda^2 \psi_2 + \dots | \phi_0 \rangle = 1 . \qquad (8.18)$$

Rearranging:

$$\langle \psi_0 | \phi_0 \rangle + \lambda \langle \psi_1 | \phi_0 \rangle + \lambda^2 \langle \psi_2 | \phi_0 \rangle + \dots = 1 . \qquad (8.19)$$

This confirms that:

$$\langle \psi_{i \neq 0} | \phi_0 \rangle = 0 \tag{8.20}$$

Similarly, the total wavefunction is also treated as normalized.

The perturbed Schrödinger equation can be written as:

$$(H_0 + \lambda V)(\lambda^0 \psi_0 + \lambda^1 \psi_1 + \lambda^2 \psi_2 + \ldots) =$$
$$(\lambda^0 \psi_0 + \lambda^1 \psi_1 + \lambda^2 \psi_2 + \ldots)(\lambda^0 \psi_0 + \lambda^1 \psi_1 + \lambda^2 \psi_2 + \ldots) \tag{8.21}$$

If $\lambda = 0$, then Eq. 8.21 becomes:

$$H\psi = E_0 \psi_0 \tag{8.22}$$

It is known as the zero-order perturbation equation.

If $\lambda = 1$, the first-order perturbation equation takes the form of:

$$(H_0 \psi_1 + V \psi_0) = (E_0 \psi_1 + E_1 \psi_0) \tag{8.23}$$

In general, the n-th-order perturbation equation takes the form of:

$$(H_0 \psi_n + V \psi_{n-1}) = \sum_{i=0}^{n} E_i \psi_{n-1} \tag{8.24}$$

The computation of the n-th-order energy correction can be calculated from Eq. 8.23 by multiplying from the left by ϕ_0, and integrating, and using the "turnover rule":

$$\langle \phi_0 | H_0 | \psi_i \rangle = \langle \psi_i | H_0 | \phi_0 \rangle^*$$

$$\langle \phi_0 | H_0 | \psi_n \rangle + \langle \phi_0 | V | \psi_{n-1} \rangle = \sum_{i=0}^{n-1} E_i \langle \phi_0 | \psi_{n-1} \rangle + E_n \langle \phi_0 | \psi_0 \rangle \tag{8.25}$$

$$E_0 \langle \phi_0 | \psi_n \rangle + \langle \phi_0 | V | \psi_{n-1} \rangle = E_n \langle \phi_0 | \psi_0 \rangle \tag{8.26}$$

$$E_n = \langle \phi_0 | V | \psi_{n-1} \rangle \tag{8.27}$$

Hence, so as to find the energy of the n-th order, the wavefunction of $(n-1)$ order is required.

8.7 The Möller-Plesset Perturbation

The unperturbed HF function is subjected to MBPT to deliver the Möller-Plesset perturbation theory. The MP unperturbed Hamiltonian is taken as the sum of the one-electron Fock operator:

$$\hat{H}^0 = \sum_{m=1}^{n} \hat{f}(m) \tag{8.28}$$

Where
$$\hat{f}(m) = -\frac{1}{2}\nabla_m^2 - \sum_a \frac{Z_a}{r_{ma}} + \sum_{j=1}^{n} [\hat{J}_j(m) - \hat{k}_j(m)] \tag{8.29}$$

The ground state HF wavefunction Φ_0 is the SD $|u_1, u_2, \ldots, u_n|$ of spin orbitals, which is an antisymmetrized product of the spin orbitals. Each term in the expansion of Φ_0 is an eigenfunction of Möller-Plesset $\hat{H}^0$. For the spin-orbital (the spin orbital is represented by u and spatial orbital by ϕ), the HF equation for electron m in an n-electron species is given by:

$$\hat{f}(m)u_i(m) = \varepsilon_i u_i(m) \ . \tag{8.30}$$

For a four-electron system, the equation becomes:

$$\left[\hat{f}(1) + \hat{f}(2) + \hat{f}(3) + \hat{f}(4)\right] u_1(3)u_2(2)u_3(4)u_4(1) =$$
$$(\varepsilon_4 + \varepsilon_2 + \varepsilon_1 + \varepsilon_3) u_1(3)u_2(2)u_3(4)u_4(1) \tag{8.31}$$

Each other term is an eigenfunction of $\hat{H}^0$ with the same eigenvalue:

$$\hat{H}^0 \Phi_0 = \left(\sum_{m=1}^{n} \varepsilon_m \right) \Phi_0 \tag{8.32}$$

Eigenfunctions of $\hat{H}^0$ are an unperturbed (zeroth order) wavefunction. Hence, the HF ground state function Φ_0 is one of the zeroth order wave functions. The Hermitian operator $\hat{f}(m)$ has a complete set of eigenfunctions (all the possible spin-orbital functions). The molecule has n-occupied spin-orbitals and infinite virtual spin-orbitals. The eigenfunction of $\hat{H}^0$ are all possible products of any n of the spin orbital. We must antisymmetrize these zeroth order wavefunctions through the SD [1].

The perturbation $\hat{H}'$ is the difference between the true molecular electronic Hamiltonian and $\hat{H}^0$.

Hence, the perturbation:

$$\hat{H}' = \left(\hat{H} - \hat{H}^0\right) = \sum_l \sum_{m>l} \frac{1}{r_{lm}} - \sum_{m=1}^{n} \sum_{j=1}^{n} \left[\hat{J}_j(m) - \hat{k}_j(m)\right] \tag{8.33}$$

It is the difference in energy between true interelectronic repulsion and the HF interelectronic potential. The Möller-Plesset first order correlation to the ground state energy is:

$$E_0^{(1)} = \left\langle \psi_0^{(0)} \left| \hat{H}' \right| \psi_0^{(0)} \right\rangle = \int \psi_0^{(0)*} \hat{H}' \psi_0^{(0)} \, d\tau = \left\langle \Phi_0 \left| \hat{H}' \right| \Phi_0 \right\rangle \tag{8.34}$$

(The subscript 0 stands for the ground state).

$$E_0^{(0)} + E_0^{(1)} = \left\langle \psi_0^{(0)} \left| \hat{H}^0 \right| \psi_0^{(0)} \right\rangle + \left\langle \Phi_0 \left| \hat{H}' \right| \Phi_0 \right\rangle$$
$$= \left\langle \Phi_0 \left| \hat{H}^0 + \hat{H}' \right| \Phi_0 \right\rangle = \left\langle \Phi_0 \left| \hat{H} \right| \Phi_0 \right\rangle \tag{8.35}$$

But, $\left\langle \Phi_0 \left| \hat{H} \right| \Phi_0 \right\rangle$ is the variational HF integral, E_{HF}.

Hence:

$$E_0^{(0)} + E_0^{(1)} = E_{HF} \tag{8.36}$$

Usually, one computes corrections to the energy using second-order perturbation theory, which is abbreviated MBPT(2). This is usually also called second-order Möller-Plesset perturbation theory, or MP2. For some problems, MP2 is more reliable than DFT. It is virtually always an improvement on HF. From Eq. 8.35, the zeroth order eigenfunction Φ_0 of $\hat{H}^0$ has the eigenvalues

$$\sum_{m=1}^{n} \varepsilon_m$$

and

$$E_0^{(0)} = \sum_{m=1}^{n} \varepsilon_m$$

Second order energy correction $E_n^{(2)}$:

$$E_0^{(2)} = \sum_{s \neq 0} \frac{\left| \left\langle \psi_s^{(0)} \left| \hat{H}' \right| \Phi_0 \right\rangle \right|^2}{E_0^{(0)} - E_s^{(0)}} \tag{8.37}$$

Let the occupied spin-orbitals be represented by $i, j, k, \ldots$ and virtual spin-orbitals by $a, b, c, \ldots$ for the HF function Φ_0. Depending upon the number of virtual spin orbitals the unperturbed wavefunction contains, it can be classified. This number is often known as the "excitation level." For example, Φ_i^a denotes the singly excited (excitation level = 1) determinant, which differs from Φ_0 by replacing the occupied orbital u_i by the virtual orbital u_a. Similarly, Φ_{ij}^{ab} denotes the doubly excited determinant, and so on.

In the matrix elements of the $\left\langle \psi_m^{(0)} \left| \hat{H}' \right| \Phi_0 \right\rangle$ of Eq. 8.36, it can be seen that for all singly excited states, the integral disappears:

$$\left\langle \psi_m^{(0)} \left| \hat{H}' \right| \Phi_0 \right\rangle = 0$$

Similarly, if the excitation level is equal to or higher than three, then the integral also vanishes (Condon-Slater rules). Hence, only the doubly excited states need to be considered.

The doubly excited function Φ_{ij}^{ab} is an eigenfunction of $\hat{H}^0 = \sum_m \hat{f}(m)$ with an eigenvalue which varies from the eigenvalue of Φ_0 by the following:

1. ε_i is replaced by ε_a.
2. ε_j is replaced by ε_b.

Hence, for the doubly excited function:

$$E_0^{(0)} - E_s^{(0)} = \varepsilon_i + \varepsilon_j - \varepsilon_a - \varepsilon_b$$

Substituting these values in the $E_0^{(2)}$ equation:

$$E_0^{(2)} = \sum_{b=a+1}^{\infty} \sum_{a=n+1}^{\infty} \sum_{i=j+1}^{n} \sum_{j=1}^{n-1} \frac{\left| \left\langle ab \left| \frac{1}{r_{12}} \right| ij \right\rangle - \left\langle ab \left| \frac{1}{r_{12}} \right| ji \right\rangle \right|^2}{(\varepsilon_i + \varepsilon_j - \varepsilon_a - \varepsilon_b)} \qquad (8.38)$$

where n is the number of electrons and

$$\left\langle ab \left| \frac{1}{r_{12}} \right| ij \right\rangle = \int \int u_a^*(1) u_b^*(2) \frac{1}{r_{12}} u_i(1) u_j(2) \, d\tau_1 \, d\tau_2 \qquad (8.39)$$

In MP2 (MBPT(2)) the molecular energy is computed as:
$E^{(0)} - E^{(1)} + E^{(2)} = E_{HF} + E^{(2)}$. Similarly, with higher correction factors, higher MPs can also be computed. An MP with a correction through $E^{(2)}$ is called MP2, a correction through $E^{(3)}$ is called MP3, and so on.

The general procedure for MPn calculation can be listed as follows:

1. Choose a basis set.
2. Compute Φ_0, E_{HF}, and the virtual orbitals.
3. Evaluate $E^{(n)}$ correction evaluating integrals over the basis set.
4. Expand the basis function to use the entire basis set.
5. Perform SCF calculation to calculate the exact E_{HF} and the entire virtual orbitals.

MP calculations are not variational, and the computed energy may be less than the true energy. MP calculations with lower basis sets are of no practical use. The normal basis set used is 6-31G*. For a DZP basis set, MP2 yields up to about 95% basis set correction energy. Moreover, with this basis set, highly dependable equilibrium geometries and vibrational energies are obtained.

Experiments indicate that in most electron-correlation calculations, the basis set truncation error is larger than correlation truncation error. Hence, an increase in the basis set from 6-31G* to TZ2P, the error in a MP2 predicted equilibrium single bond length, are reduced by a factor of 2 or 3 while moving up from MP2/TZ2P to MP3/TZ2P; no improvement in geometry accuracy is obtained.

There are two types of MP2 computations: direct MP2 and conventional MP2. In direct MP2, no external storage is used, while in conventional MP2 all the integrals are stored.

Localized MP2 (LMP2) is a modification to MP2 to speed up the computation [2]. Here, instead of using canonical SCF MOs in the HF reference Φ_0, one takes the localized MOs. Similarly, instead of taking virtual orbitals, we use orthogonal localized occupied MOs. It can be further modified by adding pseudospectral data. For species involving open-shell ground states (O_2, NO_2, and OH) unrestricted MPn can be computed. Mp calculations do not work well far away from equilibrium geometries.

MP calculations are not applicable to excited states. For excited states, CI calculations are widely used. Instead of starting with an SCF wavefunction as the zeroth-order wavefunction, we can start with MCSCF. CASSCF is the most common type among them.

8.8 The Coupled Cluster Method

The coupled cluster method was introduced by Coester and Kümmel in 1958. It is a numerical technique used for describing many electron systems [3].

The wavefunction of the coupled-cluster theory is written as an exponential:

$$\psi = e^{\hat{T}} \Phi_0 \tag{8.40}$$

where Φ_0 is an SD usually constructed from HF molecular orbitals. $\hat{T}$ is an excitation operator which, when acting on Φ_0, produces a linear combination of excited SDs.

The cluster excitation operator is written in the form:

$$\hat{T} = \hat{T}_1 + \hat{T}_2 + \hat{T}_3 + \ldots + \ldots, \tag{8.41}$$

where $\hat{T}_1$ is the operator of all single excitations, $\hat{T}_2$ is the operator of all double excitations, and so on. In the formalism of second quantization, these excitation operators are conveniently expressed as:

$$\hat{T}_1 \Phi_0 = \sum_{a=n+1}^{\infty} \sum_{i=1}^{n} t_i^a \Phi_i^a \tag{8.42}$$

$$\hat{T}_2 \Phi_0 = \sum_{b=a+1}^{\infty} \sum_{a=n+1}^{\infty} \sum_{j=i+1}^{n} \sum_{1=1}^{n-1} t_{ij}^{ab} \Phi_{ij}^{ab} \tag{8.43}$$

where Φ_i^a is a singly excited SD, and $\hat{T}_1$ converts SD $|u_1, u_2, \ldots u_n| = \Phi_0$ into a linear combination of all possible singly excited SDs. Similarly, $\hat{T}_2$ is the doubly excited SD. Since for an "n-electron system", not more than n-electrons can be excited, no operator beyond $\hat{T}_n$ appears in the cluster operator. By definition, when $\hat{T}_n$ operates on a determinant containing occupied and virtual spin orbitals, the resulting sum contains a determinant with excitations from those spin orbitals that are occupied in Φ_0 and not from virtual spin orbitals [4].

Thus, $T_1^2 \Phi_0 = \hat{T}_1(\hat{T}\Phi_0)$ contains only doubly excited determinants and $\hat{T}_2^2 \Phi_0$ contains only quadruply excited determinants. When T_1 operates on a determinant containing only virtual orbitals, the result will be zero. The $e^{\hat{T}}$ operator converts ψ into a linear combination with all excited states. A full CI calculation with a complete basis set gives the exact ψ. In CC, we work with an individual SD. The main computation of the CC method involves calculating the amplitude coefficients $t_i^a, t_{ij}^{ab}, t_{ijk}^{abc}, \ldots$ and so on. From these coefficients, ψ is determined. The following approximations are made for the computations:

1. Instead of using a complete basis set, a finite basis set is used. This leads to a basis set truncation error.
2. Instead of using all the operators $\hat{T} = \hat{T}_1 + \hat{T}_2 + \hat{T}_3 + \ldots + \ldots$ only a few operators are used, especially $\hat{T}_2$.

Thus:

$$\psi_{CCD} = e^{\hat{T}_2} \Phi_0 \tag{8.44}$$

This method is referred to as the coupled-cluster doublet (CCD) method.

But, by the Taylor expansion:

$$e^{\hat{T}_2} = 1 + \hat{T}_2 + \frac{\hat{T}_2^2}{2!} + \frac{\hat{T}_2^3}{3!} + \dots \tag{8.45}$$

Hence, the wavefunction contains determinants with multiple substitution. The CCD quadruple excitations are produced from $\frac{\hat{T}_2^2}{2!}$. Hence, the coefficients of the quadruply substituted determinant are determined as products of doubly substituted coefficients [5].

The Hamiltonian takes the form of:

$$\hat{H} e^{\hat{T}} \Phi_0 = E e^{\hat{T}} \Phi_0 \tag{8.46}$$

Or, multiplying with $\Phi*_0$ and integrating:

$$\left\langle \Phi_0 \left| \hat{H} \right| e^{\hat{T}} \Phi_0 \right\rangle = E \left\langle \Phi_0 \left| e^{\hat{T}} \Phi_0 \right\rangle \right. \tag{8.47}$$

Because of the orthogonality of orbitals, $\left\langle \Phi_0 \left| e^{\hat{T}} \Phi_0 \right\rangle = 1 \right.$

$$\left\langle \Phi_0 \left| \hat{H} \right| e^{\hat{T}} \Phi_0 \right\rangle = E \tag{8.48}$$

Similarly, multiplying with Φ_{ij}^{ab*} and integrating:

$$\left\langle \Phi_{ij}^{ab} \left| \hat{H} \right| e^{\hat{T}} \Phi_0 \right\rangle = E \left\langle \Phi_{ij}^{ab} \left| e^{\hat{T}} \Phi_0 \right\rangle \right. \tag{8.49}$$

Substituting the value of E from the above equation:

$$\left\langle \Phi_{ij}^{ab} \left| \hat{H} \right| e^{\hat{T}} \Phi_0 \right\rangle = \left\langle \Phi_0 \left| \hat{H} \right| e^{\hat{T}} \Phi_0 \right\rangle \left\langle \Phi_{ij}^{ab} \left| e^{\hat{T}} \Phi_0 \right\rangle \right. \tag{8.50}$$

Now $\hat{T} \approx \hat{T}_2$

$$\left\langle \Phi_{ij}^{ab} \left| \hat{H} \right| e^{\hat{T}_2} \Phi_0 \right\rangle = \left\langle \Phi_0 \left| \hat{H} \right| e^{\hat{T}_2} \Phi_0 \right\rangle \left\langle \Phi_{ij}^{ab} \left| e^{\hat{T}_2} \Phi_0 \right\rangle \right.$$

$$\left\langle \Phi_{ij}^{ab} \left| \hat{H} \right| e^{\hat{T}_2} \Phi_0 \right\rangle = \left\langle \Phi_0 \left| \hat{H} \right| \left(1 + \hat{T}_2 + \frac{\hat{T}_2^2}{2!} + \frac{\hat{T}_2^3}{3!} + \dots \right) \Phi_0 \right\rangle$$

$$\left\langle \Phi_0 \left| \hat{H} \right| \Phi_0 \right\rangle + \left\langle \Phi_0 \left| \hat{H} \right| \hat{T}_2 \Phi_0 \right\rangle + 0$$

$$= E_{HF} + \left\langle \Phi_0 \left| \hat{H} \right| \hat{T}_2 \Phi_0 \right\rangle \tag{8.51}$$

Thus, $\hat{T}_2 \Phi_0$ differs from Φ_0 by four spin orbitals. By the Condon-Slater rule, the matrix elements of $\hat{H}$ between the SDs differing by four spin-orbitals are zero.

$$\left\langle \Phi_{ij}^{ab} \left| \hat{H} \right| e^{\hat{T}_2} \Phi_0 \right\rangle = \left\langle \Phi_{ij}^{ab} \left| \hat{H} \right| \left(1 + \hat{T}_2 + \frac{\hat{T}_2^2}{2!} \right) \Phi_0 \right\rangle \tag{8.52}$$

With orthogonality conditions:

$$\left\langle \Phi_{ij}^{ab} \left| e^{\hat{T}_2} \Phi_0 \right. \right\rangle = \left\langle \Phi_{ij}^{ab} \left| \hat{T}_2 \Phi_0 \right. \right\rangle \tag{8.53}$$

From Eqs. 8.50, 8.51, and 8.53:

$$\left\langle \Phi_{ij}^{ab} \left| \hat{H} \right| \left(1 + \hat{T}_2 + \frac{\hat{T}_2^2}{2!} \right) \Phi_0 \right\rangle = \left(E_{\mathrm{HF}} + \left\langle \Phi_0 \left| \hat{H} \right| \hat{T}_2 \Phi_0 \right\rangle \right) \left\langle \Phi_{ij}^{ab} \left| \hat{T}_2 \Phi_0 \right. \right\rangle \tag{8.54}$$

Here i varies from 1 to $(n-1)$, j varies from $(i+1)$ to n, a varies from $(n+1)$ to infinity, and b varies from $(a+1)$ to infinity.

$\hat{T}_2$ can be replaced by amplitude coefficients. The net result is a set of simultaneous nonlinear equations for the unknown amplitudes t_{ij}^{ab} in the form of:

$$\sum_{s=1}^{m} a_{rs} \chi_s + \sum_{t=2}^{m} \sum_{s=1}^{t-1} b_{rst} \chi_s \chi_t + c_r = 0 \tag{8.55}$$

where r varies from 1 to m, $\chi_1, \chi_2, \ldots, \chi_m$ are the unknown t_{ij}^{ab}; a_{rs}, b_{rst} and c_r are constants involving orbital energies and repulsion integrals over the basis functions, and m is the number of unknown amplitudes t_{ij}^{ab}. This set of equations is solved iteratively [6].

Depending upon the highest number of excitations allowed in the definition of $\hat{T}$, CC is further classified.

1. S for single excitations (shortened to *singles* in coupled-cluster terminology)
2. D for double excitations (*doubles*)
3. T for triple excitations (*triples*)
4. Q for quadruple excitations (*quadruples*)

Thus, the CCD can be further modified by introducing $\hat{T}_1$ in $e^{\hat{T}}$ to give the CC singles and doubles method (CCSD). Similarly, by introducing $\hat{T}_3$ in addition to $\hat{T}_2 (\hat{T} = \hat{T}_1 + \hat{T}_2 + \hat{T}_3)$, CC singles, doubles and triples (CCSDTs) has been designed. Several approximate forms of CCSDT are available: CCSD(T), CCSDT-1, CCSD+T(CCSD), and so on. Pople and co-workers developed the nonvariational quadratic configuration interaction method (QCI), which is intermediate between CC and CI methods.

Terms in round brackets indicate that these terms are calculated based on perturbation theory. For example, a CCSD(T) approach simply means:

1. A coupled-cluster method.
2. It includes singles and doubles fully.

3. Triples are calculated with perturbation theory. The complexity of equations and the corresponding computer codes, as well as the cost of the computation, increases sharply with the highest level of excitation. For many applications, the sufficient accuracy may be obtained with CCSD, and the more accurate (and more expensive) CCSD (T) is often called "the gold standard of quantum chemistry" for its excellent compromise between the accuracy and the cost for the molecules near-equilibrium geometries [7]. More complicated coupled-cluster methods such as CCSDT and CCSDTQ are used only for high-accuracy calculations of small molecules. The inclusion of all n levels of excitation for the n-electron system gives the exact solution of the Schrödinger equation within the given basis set.

8.9 Research Topics

Major research areas in ab initio technique can be summarized as follows:

1. Basis set convergence and extrapolation to the 1-particle basis set limit.
2. Correction for higher-order correlation effects.
3. The effect of inner-shell correlation.
4. The study of scalar relativistic effects.
5. The study of rotational-vibrational anharmonicity.
6. Structural and functional studies of biologically important proteins, systems, and problems.
7. Work on therapeutic (inhibitor) discovery and nanobiotechnology.
8. Simulations with empirical interatomic potentials, such as core-shell models, are very important in mineralogy and will continue to be for a long time because of the large unit cells (super lattice cells) needed both in static and molecular dynamics simulations. Therefore, an important role of ab initio calculations is to monitor and fine-tune these empirical potentials.
9. Ab initio calculations of the electronic excited states of molecules, the electronic structure, and the circular dichroism of proteins, protein folding and evolution, bioinformatics, computer-aided drug design, drug resistance and so on.
10. Ab initio polymer quantum theory: structural and vibrational properties [8].

8.10 Exercises

1. Ethanol and dimethyl ether are isomers of C_2H_6O. Evaluate the energy difference between the two isomers at the HF/STO-3G, HF/6-31G**, and MP2/6-31G**//HF/6-31G** levels of theory.
2. Make a computational analysis of the nonlinear optical properties of the linear complexes $[M(I)(PH_3)_2]+(M=Cu, Ag, Au)$.

3. Find the conformational minima for the following molecules using the MMFF force field (Figs. 8.3 and 8.4).

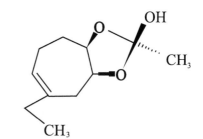

Fig. 8.3 Molecule example 1

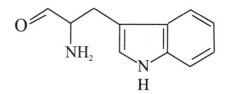

Fig. 8.4 Molecule example 2

4. Find the rotation barrier for the aryl-aryl bond in the following compound (Fig. 8.5): Build the molecule and minimize it (MM/MMFF). In Spartan, you can go to "Build", then "Define Profile". Select "Dihedral", then select the four atoms that define the dihedral angle. You will want to drive the dihedral from approximately $+90°$ to $-90°$ or from $+90°$ to $+270°$ (depending on the direction of rotation). Save the molecule, then set up calculations for an Energy Profile, using MM/MMFF as the method/force field.

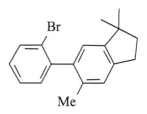

Fig. 8.5 Molecule example 3

5. Make an ab initio level study of "annulation effects" on the valence isomerization of paracyclophanes.
6. Calculate the energy of ionization for tert-butyl chloride and benzyl choride at the AM1 level by computing the heats of formation of the reactants, the carbocations, and chloride ion. For each optimized species, calculate the CI stabilization. In Spartan, use the default 6-level CI calculation by inserting the CI keyword and performing a single point calculation.

7. Using VSEPR, predict the bond angles in NO_2, NO_2^+, and NO_2^-. What do you find for the angles from AM1 and PM3 calculations? Are the bond lengths consistent with your expectations? Explain. (Note that at least one of these molecules has an odd number of electrons. When you choose the semiempirical method, you must go into the options box, and be certain that the total charge is set to the charge on the species $(0, +1, \text{ or } -1)$ and the spin multiplicity is set to the appropriate value (remember that the spin multiplicity is always one more than the number of unpaired electrons)).

8. Use AM1 semi-empirical calculations and 3-21G(*) and 6-31G*ab initio calculations to compare the relative stabilities and the major geometrical parameters within the isomeric series: 1,1-dichloroethylene, cis-1,2-dichloroethylene, and trans-1,2-dichloroethylene.

9. Perform a CASSCF calculation for CH_2. The active space consists of four electrons in four orbitals (CAS(4,4)). (a) How many determinants will you get for this configuration space? (b) Which of the configuration state functions would you expect to contribute to the energy of a CIS calculation? Identify the functions to CID and MP2 calculation. Carry out geometrical optimization of an ozone molecule with MP2, QCISD, and QCISD(T) to generate the $O-O$ bond length and the $O-O-O$ bond angle. Compare the results with the experimental values (Bond length $= 1.272$ A.U., Bond angle $= 116.8°$).

References

1. Häser M, Ahlrichs R (2004) Improvements on the direct SCF method. J Comp Chem 10:1 pp 104–111
2. Levine I (1991) Quantum Chemistry. Prentice Hall, Englewood Cliffs, NJ
3. Cramer CJ (2002) Essentials of Computational Chemistry. John Wiley & Sons, Chichester
4. Jensen F (2007) Introduction to Computational Chemistry. John Wiley & Sons, Chichester
5. Colegrove BT, Schaefer HF III (1990) Disilyne (Si_2H_2) revisited. J Phys Chem 94:5593
6. Grev RS, Schaefer HF III (1992) The remarkable monobridged structure of Si_2H_2. J Chem Phys 97:7990
7. Palágyi Z, Schaefer HF III, Kapuy E (1993) Ge_2H_2: A molecule with a low-lying monobridged equilibrium geometry. J Amer Chem Soc 115 pp 6901–6903
8. Stephens JC, Bolton EE, Schaefer HF III, Andrews L (1997) Quantum mechanical frequencies and matrix assignments to Al_2H_2. J Chem Phys 107 pp 119–223

Chapter 9
Density Functional Theory

9.1 Introduction

Electrons are, in fact, quantum mechanical spin particles. Density functional theory (DFT) allows us to compute all properties of systems by the electron density $\rho(r)$ which is a function of three variables: $\rho(r) = f(x,y,z)$. As density is the function of the wavefunction, it is referred to as functional. It is an elegant formulation of N-particle quantum mechanics with conceptual simplicity and computational efficiency. The major development in this field are as follows:

1. The introduction of the Thomas-Fermi model (1920)
2. Hohenberg-Kohn proving the existence of DFT (1964)
3. The introduction of the Kohn-Sham (KS) scheme (1965)
4. DFT in molecular dynamics (Car-Parrinello, 1985)
5. Becke and LYP functionals (1988)
6. Walter Kohn receives the Nobel prize for developing a complete DFT (1998)

9.2 Electron Density

The square of a wavefunction, in fact, is a direct measure of electron density. Total electron density due to N electrons can be defined as N-times the integral of square of wavefunctions over the spin coordinates of all electrons and over all but one of the spatial variables:

$$\rho(r) = N \int \ldots \int |\psi(x_1, x_2, \ldots, \ldots, x_N)|^2 \, ds_1 \, dx_2 \ldots, \ldots, dx_N \qquad (9.1)$$

Here $\rho(r)$ is the probability of finding any of the N-electrons within a volume element $d(r)$ with arbitrary spin. Other $(N-1)$ electrons will be having arbitrary positions and spin as is given by the wavefunction. The probability density is known as electronic probability density or electronic density. However, since electrons are in-

K. I. Ramachandran et al., *Computational Chemistry and Molecular Modeling*   171
DOI: 10.1007/978-3-540-77304-7, ©Springer 2008

distinguishable, the probability of finding any electron at this position is just N times the probability for one particular electron. Unlike the wavefunction, the electron density is observable and can be measured experimentally, e.g., by X-ray diffraction.

9.3 Pair Density

The probability of finding a pair of electrons is known as pair density. If two electrons, 1 and 2, with spins σ_1 and σ_2 are present in two volume elements dr_1 and dr_2, respectively, then the pair density is given by Eq. 9.2:

$$\rho_2(x_1,x_2) = N(N-1) \int \cdots \int |\psi(x_1,x_2,\ldots,\ldots,x_N)|^2 \, dx_3,\ldots dx_N \qquad (9.2)$$

All other electrons (other than the electrons specified) will have arbitrary positions and spins. Pair density contains all information about electron correlation. Electron density and pair density are nonnegative. Pair density is symmetric in the coordinates and normalized to the total number of $N(N-1)$ non-distinct pairs. This is a measure of finding both the electrons simultaneously in the same volume element.

9.4 The Development of DFT

Electron density is more attractive and effective in explaining properties as it is measurable. It depends only on the Cartesian axes, x, y, and z. For a system with N electrons, the electron density depends on $3N$ variables (or $4N$ if you count in spin). There are two types of electron densities for spin polarized systems, one for spin up electrons $\rho \uparrow (r)$ and the other for spin down electrons $\rho \downarrow (r)$. The fact that the ground state properties are functionals of the electron density $\rho(r)$ was introduced by Hohenberg and Kohn (1964) and it is the basic framework for modern Density functional (DF) methods [1].

The total ground state energy of an electron system can be written as a functional of the electronic density. This energy is at a minimum if the density corresponds to the exact density for the ground state. The theorem of Hohenberg and Kohn is a proof of such a functional, but there is no method for constructing it. Once this functional is fully characterized, quantum chemistry would be able to help us in establishing the properties. Unfortunately we do not know the exact form of the energy functional. It is necessary to use approximations regarding parts of the functional dealing with kinetic energy and exchange and correlation energies of the system of electrons.

The simplest approximation is the local density approximation (LDA) which leads to a Thomas-Fermi (Fermi, 1928; Thomas, 1927) term for kinetic energy and the Dirac (1930) term for the exchange energy. The corresponding functional is called the Thomas-Fermi-Dirac energy. These functionals can be further improved but the results are not that encouraging for molecular systems. On the other hand, improvements on the Thomas-Fermi-Dirac method lead into the true DF method, where all components of energy are expressed through density alone rather than using many particle wavefunctions. However, for the time being, it seems that there is no way to avoid wavefunctions in molecular calculations and for accurate calculations they have to be used as a mapping step between the energy and density. While pure DFTs are very useful in studying a solid phase (e.g., conductivity), they fail to provide meaningful results for molecular systems. For example, the Thomas-Fermi theory could not predict chemical bonds. The real predecessor of the modern chemical approaches to the DFT was the Slater method formulated in 1951. It was developed as an approximate solution to the Hartree Fock (HF) equations. In this method, the HF exchange was approximated by:

$$E_{Xa[\rho\uparrow,\rho\downarrow]} = -\frac{9}{4}\alpha\left(\frac{3}{4\pi}\right)^{1/3}\int\left[\rho_\uparrow^{4/3}(r) + \rho_\downarrow^{4/3}(r)\right]dr \tag{9.3}$$

The exchange energy E_{Xa} given here are the functional of densities for spin up ($\uparrow$) and spin down ($\downarrow$) electrons and it contains an adjustable parameter α. This parameter was empirically optimized for each atom of the periodic table and its value was between $0.7-0.8$ for most atoms. For a special case of homogenous electron gas, its value is exactly $2/3$.

9.5 The Functional

The functional is a function of another function. It takes a function and provides a number. It is usually written with the function in square brackets as $F[f] = a$. For example, consider a function subjected to integration. It is represented as Eq. 9.4:

$$F[f] = \int_{-\infty}^{+\infty} f(x)\,dx \tag{9.4}$$

Functionals can also have derivatives, which behave similarly to traditional derivatives for functions. The differential of the functional is defined as:

$$\delta F[f] = F[f + \delta f] - F[f] = \int \frac{\delta F}{\delta f(x)}\delta f(x)\,dx \tag{9.5}$$

The functional derivatives have properties similar to traditional function derivatives, e.g.:

$$\frac{\delta}{\delta f(x)}(C_1 F_1 + C_2 F_2) = C_1\frac{\delta F_1}{\delta f(x)} + C_2\frac{\delta F_2}{\delta f(x)} \tag{9.6}$$

$$\frac{\delta}{\delta f(x)}(F_1 F_2) = \frac{\delta F_1}{\delta f(x)}F_2 + \frac{\delta F_2}{\delta f(x)}F_1 \tag{9.7}$$

9.6 The Hohenberg and Kohn Theorem

Hohenberg and Kohn (HK) in their theorem propose the following:

1. Every observable of a stationary quantum mechanical system (including energy), can be calculated, in principle exactly, from the *ground-state density* alone, i.e., every observable can be written as a functional of the ground-state density.
2. The ground state density can be calculated, in principle exactly, using the variational method involving only density. (The original theorem refers to the time independent-stationary-ground state, but are being extended to excited states and time-dependent potentials) [2].

Within a Born-Oppenheimer approximation, the ground state of the system of electrons is a result of the positions of the nuclei. In the Hamiltonian, the kinetic energy of electrons and the electron-electron interaction adjust themselves to the external (i.e., coming from the nuclei) potential $\hat{V}_{ext}$. Actually, once the external potential starts functioning on a system, everything else, including electron density, adjusts themselves to give the lowest possible total energy of the system. Hence, the external potential is the only variable term required in the equation.

HK posed three interesting question in this regard. Is $\hat{V}_{ext}$ uniquely determined from the knowledge of electron density $\rho(r)$? Can we characterize the nucleus (find out where and what the nuclei are), from the density $\rho(r)$ of the system in the ground state? Is there a precise mapping from $\rho(r)$ to $\hat{V}_{ext}$?

Mapping from $\rho(r)$ to $\hat{V}_{ext}$ is expected to be accurate within a constant, since Schrödinger equations with $\hat{H}_{ele}$ and $\hat{H}_{ele}$ + constant yield exactly the same eigenfunctions and the energies will be simply elevated by the value of this constant. Note that all energy measurements are within some constant, which establishes the framework of reference. If this is true, the knowledge of density may provide total information about the system. Since $\rho(r)$ determines number of electrons, N:

$$N = \int \rho(r)\, dr \qquad (9.8)$$

and ρ determines the $\hat{V}_{ext}$, the knowledge of the total density is as good as that of ψ, the wavefunction describing the state of the system. They proved it through a contradiction:

1. Consider an exact ground state density $\rho(r)$, which is nondegenerate (i.e., there is only one wave function ψ for this ground state, though HK theorems can be easily extended for degenerate ground states.)
2. Assume that for the density $\rho(r)$ there are two possible external potentials: $\hat{V}_{ext}$ and $\hat{V}'_{ext}$, which obviously produce two different Hamiltonians: $\hat{H}_{ele}$ and $\hat{H}'_{ele}$, respectively with two different wavefunctions for the ground state, ψ and ψ'. They correspond to energies:

$$E_0 = \langle \psi |H| \psi \rangle \qquad (9.9)$$

$$E_0' = \left\langle \psi' \left| H' \right| \psi' \right\rangle \qquad (9.10)$$

respectively.

3. Now, let us calculate the expectation value of energy for the ψ' with the Hamiltonian $\hat{H}$ and using the variational theorem:

$$E_0 < \left\langle \psi' \left| H \right| \psi' \right\rangle = \left\langle \psi' \left| H' \right| \psi' \right\rangle + \left\langle \psi' \left| H - H' \right| \psi' \right\rangle \qquad (9.11)$$

But:

$$\left\langle \psi' \left| H' \right| \psi' \right\rangle = E_0' \qquad (9.12)$$

$$\left\langle \psi' \left| H - H' \right| \psi' \right\rangle = \int \rho(r) \left[\hat{V}_{ext} - \hat{V}_{ext}' \right] dr \qquad (9.13)$$

Hence:

$$E_0 < E_0' + \int \rho(r) \left[\hat{V}_{ext} - \hat{V}_{ext}' \right] dr \qquad (9.14)$$

4. Similarly, let us calculate the expectation value of energy for the ψ with the Hamiltonian $\hat{H}'$:

$$E_0' < \left\langle \psi \left| H' \right| \psi \right\rangle = \left\langle \psi \left| H \right| \psi \right\rangle + \left\langle \psi \left| H' - H \right| \psi \right\rangle \qquad (9.15)$$

But:

$$\left\langle \psi \left| H \right| \psi \right\rangle = E_0 \qquad (9.16)$$

$$\left\langle \psi \left| H' - H \right| \psi \right\rangle = \int \rho(r) \left[\hat{V}_{ext}' - \hat{V}_{ext} \right] dr \qquad (9.17)$$

$$E_0' < E_0 - \int \rho(r) \left[\hat{V}_{ext} - \hat{V}_{ext}' \right] dr \qquad (9.18)$$

5. From Eqs. 9.14 and 9.18, we obtain:

$$E_0 + E_0' < E_0' + E_0 \qquad (9.19)$$

and it leads to a contradiction.

Since $\rho(r)$ determines N and $\hat{V}_{ext}$, it should also determine all properties of the ground state, including the kinetic energy of electrons T_e and the energy of interaction among electrons U_{ee}, i.e., the total ground state energy is a functional of density with the following components:

$$E[\rho] = T_e[\rho] + U_{ee}[\rho] + V_{ext}[\rho] \qquad (9.20)$$

(V_{ext} is the energy corresponding to external potential).

Additionally, HK grouped together all functionals which are secondary (i.e., which are responses) to the $V_{ext}[\rho]$:

$$E[\rho] = V_{ext}[\rho] + F_{HF}[\rho] = \int \rho(r) \hat{V}_{ext}(r) \, dr + F_{HF}[\rho] \qquad (9.21)$$

The $F_H K$ functional operates only on density and is universal, i.e., its form does not depend on the particular system under consideration (note that N-representable densities integrate to N, and the information about the number of electrons can be easily obtained from the density itself). The second HK theorem provides variational extension to electron density representation $\rho(r)^3$.

For a trial density $\tilde{\rho}(r)$ such that $\tilde{\rho}(r) \geq 0$ and for which $\int \tilde{\rho}(r) \, dr = N$:

$$E_0 \leq E[\tilde{\rho}] \qquad (9.22)$$

where $E[\tilde{\rho}]$ is the energy functional. In other words, if some density represents the correct number of electrons N, the total energy calculated from this density cannot be lower than the true energy of the ground state. By the N-representability (Chap. 10), the trial density $\tilde{\rho}$ has to sum up to N electrons by simple rescaling. It is automatically insured if by nature $\rho(r)$ is mapped to some wave function. Assuring that the trial density has V_{ext}-representability also (usually denoted in the literature as v-representability) is not that easy. Levy (1982) and Lieb (1983) proposed some reasonable trial densities, which are not the ground state densities for any possible V_{ext} potential. These densities do not map to any external potential. Such trial densities will not correspond to any ground state. Or, optimization of the system with this trial density will not lead to a ground state. Moreover, during energy minimization, we may take a wrong turn, and get stuck into some non v-representable density and never be able to converge to a physically relevant ground state density. Assuming that we restrict ourselves only to trial densities which are both N and v representable, the variational principle for density is easily established, since each trial density $\tilde{\rho}$ defines a Hamiltonian $\hat{H}_{el}$. From the Hamiltonian we can derive the corresponding wavefunction $\tilde{\psi}$ for the ground state represented by this Hamiltonian. Furthermore, according to the traditional variational principle, this wavefunction $\tilde{\psi}$ will not be a ground state for the Hamiltonian of the real system $\hat{H}_{el}$:

$$\langle \tilde{\psi} | H | \tilde{\psi} \rangle = E[\tilde{\rho}] \geq E[\rho_0] \equiv E_0 \qquad (9.23)$$

where $\rho_0(r)$ is the true ground state density of the real system.

The condition of minimum for the energy functional:

$$\delta E[\rho(r)] = 0 \qquad (9.24)$$

It needs to be constrained by the N-representability of the density which is optimized. The Lagrange method of undetermined multipliers is a very convenient approach for the constrained minimization problems. In this method, we represent constraints in such a way that their value is exactly zero to make the optimization easier. In our case, the N representability constraint can be represented as:

$$\text{Constraint} = \int \rho(r)\,dr - N = 0 \tag{9.25}$$

These constraints are then multiplied by an undetermined constant and added to a minimized function or functional to get Eq. 9.26.

$$E[\rho(r)] - \mu \left[\int \rho(r)\,dr - N \right] \tag{9.26}$$

where μ is yet undetermined Lagrange multiplier. Minimizing this condition by making the first derivative zero:

$$\delta \left\{ E[\rho(r)] - \mu \left[\int \rho(r)\,dr - N \right] \right\} = 0 \tag{9.27}$$

Solving this differential equation will provide us with a prescription of finding a minimum which satisfies the constraint. In our case it leads to:

$$\delta E[\rho(r)] - \mu \delta \left\{ \int \rho(r)\,dr \right\} = 0 \tag{9.28}$$

since μ and N are constants. Using the definition of the differential of the functional:

$$F[f + \delta f] - F[f] = \delta F = \int \frac{\delta F}{\delta f(x)} \delta f(x)\,dx \tag{9.29}$$

and the fact that differential and integral signs may be interchanged, we obtain:

$$\int \frac{\delta E[\rho(r)]}{\delta \rho(r)} \delta \rho(r)\,dr - \mu \int \delta \rho(r)\,dr = 0 \tag{9.30}$$

Since integration runs over the same variable and has the same limits, we can write both expressions under the same integral:

$$\int \left\{ \frac{\delta E[\rho(r)]}{\delta \rho(r)} - \mu \right\} - \delta \rho(r)\,dr = 0 \tag{9.31}$$

which provides the condition for constrained minimization and defines the value of the Lagrange multiplier at the minimum. It is expressed here through external potential from Eq. 9.21.

$$\mu = \frac{\delta E[\rho(r)]}{\delta \rho(r)} = \hat{V}_{\text{ext}}(r) + \frac{\delta F_{\text{HK}} \rho(r)}{\delta \rho(r)} \tag{9.32}$$

DFT gives a firm definition of the chemical potential, and leads to several important general conclusions.

9.7 The Kohn and Sham Method

The above equations provide a method of minimizing energy by changing corresponding density. Unfortunately, the expression relating kinetic energy to density is not known with a satisfactory level of accuracy. The current expressions, which are improved upon from the original Thomas-Fermi theory, are quite crude and unsatisfactory for atoms and molecules in particular. On the other hand, the kinetic energy is easily calculated from the wave function. For that reason, Kohn and Sham proposed an ingenious method, the KS method, of combining wavefunctions and the density approach. They repartitioned the total energy functional into the following parts:

$$E[\rho] = T_0[\rho] + \int \left[\hat{V}_{ext}(r) + \hat{U}_{el}(r) \right] \rho(r) \, dr + E_{xc}[\rho] \tag{9.33}$$

where $T_0[\rho]$ is the kinetic energy of electrons in a system which has the same density ρ as the real system, but in which there is no electron-electron interactions. This is frequently considered as a system with noninteracting electrons. However, the term *noninteracting* is not fully correct as the electrons interact with nuclei [3].

$$\hat{U}_{el}(r) = \int \frac{\rho(r')}{|r' - r|} \, dr' \tag{9.34}$$

is a pure Coulomb (classical) interaction between electrons. It includes electron self-interaction explicitly, since the corresponding energy is:

$$E_{el}[\rho] = \int \int \frac{\rho(r')\rho(r)}{|r' - r|} \, dr \, dr' \tag{9.35}$$

and it represents interaction of ρ with itself. $\hat{V}_{ext}(r)$ is the external potential, i.e., the potential effected from nuclei:

$$\hat{V}_{ext} = \sum_a \frac{-Z_a}{|R_a - r|} \tag{9.36}$$

The last functional, $E_{xc}[\rho]$, is called the exchange-correlation energy. $E_{xc}[\rho]$ includes all the energy contributions which were not accounted for by the previous terms, i.e.:

1. Electron exchange.
2. Electron correlation, since non-interacting electrons need to correlate their movements. Please note, however, that this correlation component is not the same as defined by Lowdin for ab initio methods.
3. A portion of the kinetic energy which is needed to correct $T_0[\rho]$ to obtain the true kinetic energy of a real system $T_e[\rho]$.
4. Correction for self-interaction introduced by the classical coulomb potential.

In fact, all the difficult things were "swept under the carpet" in this functional to make the computation easier. However, better approximations for this functional are being published. To conclude the derivation of KS equations, let us assume that we know the energy functional reasonably well. In a similar fashion, as was done for the equations defining chemical potential (Eqs. 9.31 and 9.32) we may apply the variational principle and obtain:

$$\mu = \frac{\delta E[\rho(r)]}{\delta \rho(r)} = \frac{\delta T_0[\rho(r)]}{\delta \rho(r)} + \hat{V}_{\text{ext}}(r) + \hat{U}_{\text{el}}(r) + \frac{\delta E_{\text{xc}}[\rho(r)]}{\delta \rho(r)} \tag{9.37}$$

This can be simply written as:

$$\mu = \frac{\delta E[\rho(r)]}{\delta \rho(r)} = \frac{\delta T_0[\rho(r)]}{\delta \rho(r)} + \hat{V}_{\text{eff}}(r) \tag{9.38}$$

Here we combined together all terms, excepting noninteracting electron kinetic energy, into an effective potential $\hat{V}_{\text{eff}}(r)$ depending upon r:

$$\hat{V}_{\text{eff}}(r) = \hat{V}_{\text{ext}}(r) + \hat{U}_{\text{el}}(r) + \hat{V}_{\text{xc}}(r) \tag{9.39}$$

where the exchange correlation potential is defined as a functional derivative of the exchange correlation energy:

$$\hat{V}_{\text{xc}}(r) = \frac{\delta E_{\text{xc}}[\rho(r)]}{\delta \rho(r)} \tag{9.40}$$

The form of Eq. 9.40 asks for a solution to the Schrödinger equation for noninteracting particles as seen in Eq. 9.41:

$$\left[-\frac{1}{2} \nabla_i^2 + \hat{V}_{\text{eff}}(r) \right] \phi_i^{\text{KS}}(r) = \epsilon_i \, \phi_i(r)^{\text{KS}} \tag{9.41}$$

Equation 9.41 is very similar to the eigenequation of the HF method and is much simpler. The Fock operator in the above equation contains the potential which is non local, i.e., it will be different for different electrons.

The KS operator depends only on r, and not upon the index (nature) of the electron. It is the same for all electrons. The KS orbitals, $\phi_i(r)^{\text{KS}}$, which are quite easily derived from this equation, can be used immediately to compute the total density:

$$\rho(r) = \sum_{i=1}^{N} \left| \phi_i^{\text{KS}}(r) \right|^2 \tag{9.42}$$

which can be used to calculate an improved potential $\hat{V}_{\text{eff}}(r)$, leading to a new cycle of self-consistent field. Density can also be used to calculate the total energy from Eq. 9.33, in which the kinetic energy $T_0[\rho]$ is calculated from the corresponding orbitals, rather than the density itself:

$$T_0[\rho] = \frac{1}{2} \sum_{i=1}^{N} \left\langle \phi_i^{\text{KS}} \left| \nabla_i^2 \right| \phi_i^{\text{KS}} \right\rangle \tag{9.43}$$

and the rest of the total energy as:

$$\hat{V}_{\text{eff}}(r) = \int \hat{V}_{\text{eff}}(r)\rho(r)\,dr \tag{9.44}$$

In practice, the total energy is calculated economically using orbital energies $\in_i$, according to Eq. 9.45:

$$E_{\text{el}}[\rho] = \sum_{i=1}^{N} \in_i - \frac{1}{2} \int \int \frac{\rho(r)\rho(r')}{|r-r'|}\,dr\,dr' - \int \hat{V}_{\text{xc}}(r)\rho(r)\,dr + E_{\text{xc}}[\rho] \tag{9.45}$$

It is a popular misconception to look at this method as describing noninteracting electrons moving in a potential given by nuclei. In fact, they move in an effective potential $\hat{V}_{\text{eff}}(r)$ which includes electron interaction, though in an artificial or indirect manner. This appears to be philosophical rather than physical. In KS equations, the electron-electron interaction is replaced by the interaction of electrons with some medium which mimics the electron-electron interaction. This medium actually exaggerates the interaction between electrons. The correction which needs be added to T_0 ($\Delta T = T_e - T_0$ is embedded in E_{xc}) is positive, i.e., the "noninteracting electrons" move slower than the real, interacting ones.

It has to be stressed that KS orbitals (given by $\phi_i(r)^{\text{KS}}$) are not the real orbitals, and they do not correspond to any real physical system. Their only role in the theory and computation is to provide a proper mapping between kinetic energy and density. The total KS wavefunction is a single determinant and is unable to model situations where more determinants are needed such as molecules dissociating to atoms. An interesting discussion on symmetry of this wavefunction is given by Dunlap (1991, 1994) [4].

9.8 Implementations of the KS Method

In the original presentation of the KS method, a non-polarized electron density was used, and occupation numbers for Ks orbitals were assumed as one. However, extensions exist both for polarized spin densities (i.e., different orbitals for spin up and spin down electrons), and for nonintegral occupation numbers in the range (0; 1).

KS orbitals are artifacts with no real physical significance. However, they are quite close to the HF orbitals. The KS formalism can be extended to the fractional occupation numbers $0 \leq n_i \leq 1$. The orbital energies $\in_i$ can be written as:

$$\in_i = \frac{\partial E}{\partial n_i} \tag{9.46}$$

One immediate application of the KS formalism (Eq. 9.46) is to integrate energy from $(N-1)$ to N electrons, and to calculate the ionization potential. The derivatives of energy versus occupation numbers provide other response functions such as the chemical potential, electro negativity, softness, hardness and so on.

The first implementations of the KS method used the local approximations to the exchange correlation energy. The appropriate functionals were taken from data on homogenous electron gas. There were two variants of the method, spin unpolarized local density functional/approximation (LDF/LDA) and spin polarized local spin density (LSD) where arguments require both α and β electron densities, rather than a total density.

The exchange correlation energy was partitioned into 2 parts: the exchange energy, and the correlation energy, as given in Eq. 9.46:

$$E_{xc}[\rho] = E_x[\rho] + E_c[\rho] \tag{9.47}$$

This partition is quite arbitrary, since the exchange and the correlation have slightly different meanings than in ab initio approaches. The exchange energy in LDF/LSD was approximated with the homogenous gas exchange result given by Eq. 9.3 with $\alpha = 2/3$. The correlation energy can be expressed as:

$$E_c[\rho] = \int \rho(r) \in_c [\rho \uparrow (r)\rho \downarrow (r)] \, dr \tag{9.48}$$

where $\in_c [\rho \uparrow (r)\rho \downarrow (r)]$ is the correlation energy per one electron in a gas with spin densities $\rho \uparrow (r)$ and $\rho \downarrow (r)$. This function is not known analytically, but is constantly improved on the basis of quantum Monte Carlo simulations, and fitted to analytical expansion. The local functionals derived from electron gas data worked surprisingly well, taking into account that they substantially underestimate the exchange energy (by as much as 15%) and grossly overestimate the correlation energy, sometimes by 100%. The error in exchange is, however, larger than the correlation error in absolute values. LSD/LDF is known to overbind normal atomic bonds. On the other hand, it produces too weak hydrogen bonds.

Early attempts to improve functionals by the gradient expansion approximation (GEA), in which $E_{xc}[\rho]$ was expanded in the Taylor series versus ρ and truncated at a linear term, did not improve results very much. Only the generalized gradient approximation (GGA) provided notable improvements by expanding $E_{xc}[\rho]$. The expansion here is not a simple Taylor expansion, but tries to find the right asymptotic behavior and scaling for the usually nonlinear expansion. These enhanced functionals are frequently called nonlocal or gradient corrections, since they depend upon the density and magnitude of the gradient of the density at a given point. Most of the nonlocal functionals are quite complicated functions in which the value of density and its gradient are integral parts of the formula.

9.9 Density Functionals

In the following, ρ^{α} and ρ^{β} are the α, β spin densities; the total and spin densities are:

$$\rho = \rho^{\alpha} + \rho^{\beta} \tag{9.49}$$

and:

$$\hat{\rho} = \rho^\alpha - \rho^\beta \tag{9.50}$$

The gradients of the density enter through:

$$\sigma = \nabla\rho.\nabla\rho, \quad \hat{\sigma} = \nabla\rho.\nabla\hat{\rho}, \quad \hat{\hat{\sigma}} = \nabla\hat{\rho}.\nabla\hat{\rho}, \quad \upsilon = \nabla^2\rho, \quad \hat{\upsilon} = \nabla^2\hat{\rho},$$

Additionally, the kinetic energy density for a set of (KS) orbitals generating the density can be introduced through:

$$\tau = \left(\sum_i^\alpha + \sum_i^\beta \right) |\nabla\phi_i|^2 \tag{9.51}$$

$$\hat{\tau} = \left(\sum_i^\alpha - \sum_i^\beta \right) |\nabla\phi_i|^2 \tag{9.52}$$

All of the available functionals are of the general form:

$$F = \left[\rho, \hat{\rho}, \sigma, \hat{\sigma}, \hat{\hat{\sigma}}, \tau, \hat{\tau}, \upsilon, \hat{\upsilon} \right] \tag{9.53}$$

$$= \int d^3 r K \left(\rho, \hat{\rho}, \sigma, \hat{\sigma}, \hat{\hat{\sigma}}, \tau, \hat{\tau}, \upsilon, \hat{\upsilon} \right) \tag{9.54}$$

Now, let us see some common exchange energy, functional, and potential terms used in DFT.

9.10 The Dirac-Slater Exchange Energy Functional and the Potential

The Dirac-Slater exchange energy functional and the potential are given by the following equations:

$$
\begin{aligned}
E_X^{LSD}[\rho_\alpha, \rho_\beta] &= \int d\rho\, \varepsilon_x(\rho, \zeta) \\
\varepsilon_x(\rho, \lambda) &= \varepsilon_x^0(\rho) + \left[\varepsilon_x^1(\rho) - \varepsilon_x^0(\rho) \right] f(\zeta) \\
\varepsilon_x^0(\rho) &= \varepsilon_x(\rho, 0) = C_x \rho^{1/3} \; ; \varepsilon_x^1(\rho) = \varepsilon_x(\rho, 1) = 2^{1/3} C_x \rho^{1/3} \\
C_x &= \frac{3}{4} \left(\frac{3}{\pi} \right)^{1/3} \; ; f(\zeta) = \frac{(1+\zeta)^{4/3} + (1-\zeta)^{4/3} - 2}{2\left(2^{1/3} - 1\right)} \\
\zeta &= \frac{\rho_\alpha - \rho_\beta}{\rho_\alpha + \rho_\beta} \; ; \upsilon_{x\sigma}^{LSD} = \frac{\delta E_x^{LSD}}{\delta\rho_\sigma} = \left(\frac{6}{\pi}\rho_\sigma \right)^{1/3}
\end{aligned}
\tag{9.55}
$$

9.11 The von Barth-Hedin Exchange Energy Functional and the Potential

The von Barth-Hedin exchange energy functional and the potential are given by the following equations:

$$E_X^{VBH}\left[\rho_\alpha,\rho_\beta\right] = \int dr \rho \varepsilon_x^{VBH}(\rho,x)$$

$$\varepsilon_x^{VBH} = \varepsilon_x^P + \gamma^{-1}\mu_x^P f(x) \;; \varepsilon_x^P(r_s) = -\frac{\varepsilon_x^0}{r_s} \;; \mu_x^P = \frac{4}{3}\varepsilon_x^P(r_s)$$

$$f(x) = \frac{x^{4/3} + (1-x)^{4/3} - \alpha}{1-\alpha}$$

$$x = \frac{\rho_\alpha}{\rho} \;; \gamma = \frac{4}{3}\left(\frac{a}{1-a}\right) \;; \alpha = 2^{-1/3} \;; \varepsilon_x^0 = \frac{3}{4\pi a_0} \approx 0.45815 \;;$$

$$a_0 = \left(\frac{4}{9\pi}\right)^{1/3} \approx 0.52106 \;; v_{x\alpha}^{VBH} = \mu_x^P(2x)^{1/3} \;; v_{x\beta}^{VBH} = \mu_x^P[2(1-x)]^{1/3} \quad (9.56)$$

9.12 The Becke Exchange Energy Functional and the Potential

The Becke exchange energy functional and the potential are given by the following equations:

$$E_X^{BEC}\left[\rho_\alpha,\rho_\beta\right] = E_X^{LSD}\left[\rho_\alpha,\rho_\beta\right] - \sum_\sigma^{\alpha,\beta} \int dr \rho_\sigma \varepsilon_x^{NL}$$

$$= E_X^{LSD}\left[\rho_\alpha,\rho_\beta\right] - \sum_\sigma \int dr \rho_\sigma^{4/3} \frac{bX_\sigma^2}{1+6bX_\sigma \sinh^{-1} X_\sigma} \;;$$

$$X_\sigma = \frac{|\nabla\rho_\sigma|}{\rho_\sigma^{4/3}} \;; b = 0.0042 \;;$$

$$v_{X\sigma}^{BEC} = v_{X\sigma}^{LSD} + \frac{\partial\left(\varepsilon_x^{NL}\rho\right)}{\partial\rho_\sigma} - \sum_i \frac{\partial}{\partial x_i}\frac{\partial\left(\varepsilon_x^{NL}\rho\right)}{\partial\rho_{\sigma,x_i}} \;;$$

$$v_{X\sigma}^{NL} = -bF\rho_\sigma^{-4/3}\frac{4}{3}\rho_\sigma^{5/3}X_\sigma^2 - \nabla^2\rho_\sigma\left(1+F\left(1-\frac{6bX_\sigma^2}{\sqrt{1+X_\sigma^2}}\right)\right) +$$

$$6bF\nabla\rho_\sigma.\nabla X_\sigma\left\{(1+2F)\sinh^{-1}X_\sigma\right\} +$$

$$\frac{X_\sigma}{\sqrt{1+X_\sigma^2}}\left[\frac{1}{1+X_\sigma^2}+2F\left[2-\frac{6bX_\sigma^2}{\sqrt{1+X_\sigma^2}}\right]\right]$$

$$F = \frac{1}{1+6bX_\sigma \sinh^{-1} X_\sigma} \qquad (9.57)$$

9.13 The Perdew-Wang 91 Exchange Energy Functional and the Potential

The Perdew-Wang 91 exchange energy functional and the potential are given by the following equations:

$$E_x^{PW91}\left[\rho_\alpha,\rho_\beta\right] = \frac{1}{2}E_x^{PW91}\left[2\rho_\alpha\right] + \frac{1}{2}E_x^{PW91}\left[2\rho_\beta\right]$$

$$E_x^{PW91}[\rho] = \int dr\rho\varepsilon_x\left(r_s,0\right)F(s)$$

$$\varepsilon_x\left(r_s,0\right) = -\frac{3k_F}{4\pi} \; ; k_F = (3\pi^2\rho)^{1/3} = \frac{1.91916}{r_s}$$

$$s = \frac{|\rho|}{2k_F\rho} \; ;$$

$$F(s) = \frac{1 + 0.19645s\sinh^{-1}(7.7956s) + (0.2743 - 0.1508\exp(-100s^2)s^2)}{1 + 0.19645s\sinh^{-1}(7.7956s) + 0.004s^4}$$

$$\upsilon_{xc} = \frac{\delta E_x\left[\rho_\alpha,\rho_\beta\right]}{\delta\rho_\sigma} = \frac{1}{2}\frac{\delta E_x[2\rho_\sigma]}{\delta\rho_\sigma} = \frac{1}{2}\upsilon_x\left(2\rho_\sigma,\frac{s_\sigma}{2^{1/3}}\right)$$

$$\upsilon_x = \frac{1}{2}\upsilon_x^{LDA}\left(\frac{4}{3}F(s) - \left(\frac{\nabla\rho.\nabla|\nabla\rho|}{\rho^2(2k_F)^3} - \frac{4}{3}s^3\right)F_{ss} - \frac{\nabla^2\rho_\sigma}{\rho(2k_F)^2}F_s\right)$$

$$F_s = P_3^2P_5P_6 + P_3P_7 \; ; F_{ss} = P_2^3\left(P_5P_9 - P_6P_8\right) + 2P_3P_5P_6P_{11} + P_3P_{10} + P_7P_{11}$$

$$P_0 = (1 + (7.7956s)^2)^{-1/2} \; ; P_1 = \sinh^{-1}(7.7956s) \; ; P_2 = \exp(-100s^2) \; ;$$

$$P_3 = \frac{1}{1 + 0.19645sP_1 + 0.004s^4} \; ;$$

$$P_4 = 1 + 0.19645sP_1 = (0.2743 - 0.15084P_2)s^2 \; ;$$

$$P_5 = 0.004s^2 - 0.15084P_2 - 0.2743 \; ; P_6 = 0.19645s(P_1 + 7.7956sP_0) \; ;$$

$$P_7 = 0.5486 - 0.30168P_2 + 2015.084s^2P_2 - 0.016s^2F(s)$$

$$P_8 = 2s(0.004 - 15.084P_2)$$

$$P_9 = 0.19645P_1 + 7.7956 \times 0.19645sP_0\left(3 - (7.7956sP_0)^2\right)$$

$$P_{10} = 60.336sP_2\left(2 - 100s^2\right) - 0.032sF\left(s_0 - 0.016s^3F_5\right)$$

$$P_{11} = -P_3^2\left(0.19645P_1 + 7.7956 \times 0.19645sP_0 + 0.016s^3\right) \quad\quad (9.58)$$

9.14 The Perdew-Zunger LSD Correlation Energy Functional and the Potential

The Perdew-Zunger LSD correlation energy functional and the potential are given by the following equations:

$$E_c^{LSD}\left[\rho_\alpha,\rho_\beta\right] = \int dr\rho \varepsilon_c^{LSD}(r_s,\zeta)$$

$$\varepsilon_c^{LSD}(r_s,\zeta) = \varepsilon_c^0 + \left[\varepsilon_c^1(r_s) - \varepsilon_c^0(r_s)\right]f(\zeta)$$

$$v_c^\sigma(r_s,\zeta) = v_c^0(r_s) + \left[v_c^1(r_s) - v_c^0(r_s)\right]f(\zeta)$$

$$+ \left[\varepsilon_c^1(r_s) - \varepsilon_c^0(r_s)\right](\text{sgn}(\sigma) - \zeta)\frac{df}{d\zeta}$$

$$f(\zeta) = \frac{(1+\zeta)^{4/3} + (1-\zeta)^{4/3} - 2}{2(2^{1/3} - 1)}$$

where $\text{sgn}(\sigma)$ is 1 for $\sigma = \alpha$ and -1 for $\sigma = \beta$, and the low density limit $r_s \geq 1$:

$$\varepsilon_c^i = \frac{\gamma_i}{1 + \beta_1^i\sqrt{r_s} + \beta_2^i r_s}$$

$$v_c^i = \left[1 - \frac{r_s d}{3 dr_s}\right]\varepsilon_c^i = \varepsilon_c^i \frac{1 + \frac{7}{6}\beta_1^i\sqrt{r_s} + \frac{4}{3}\beta_2^i r_s}{1 + \beta_1^i\sqrt{r_s} + \beta_2^i r_s}$$

and the high density limit $0 \leq r_s \leq 1$:

$$\varepsilon_c^i = A_i \ln r_5 + B_i + C_i r_5 \ln r_5 + D_i r_5$$

$$v_c^i = A_i \ln r_5 + \left(B_i - \frac{1}{3}A_i\right) + \frac{2}{3}C_i r_5 \ln r_5 + \frac{1}{3}(2D_i - C_i)r_5 \qquad (9.59)$$

Constants in these equations are included in Table 9.1.

Table 9.1 Constants used in the Perdew-Zunger parametrization of the Ceperley-Alder quantum Monte-Carlo results for a homogeneous electron gas

Parameter	$i = 0$	$i = 1$
γ	−0.1423	−0.0843
β_1	1.0529	1.3981
β_2	0.3334	0.2611
A	0.0311	0.0155
B	−0.0480	−0.0269
C	0.0020	0.0007
D	−0.0116	−0.0048

9.15 The Vosko-Wilk-Nusair Correlation Energy Functional

The Vosko-Wilk-Nusair correlation energy functional is given by the following equations:

$$E_c^{VWN}\left[\rho_\alpha,\rho_\beta\right] = \int dr \rho \varepsilon_c^{VWN}\left(\rho_\alpha,\rho_\beta\right)$$

$$\varepsilon_c^{VWN}\left(\rho_\alpha,\rho_\beta\right) = \varepsilon_i(\rho_\alpha,\rho_\beta) + \Delta\varepsilon_c(r_s,\zeta)$$

$$\varepsilon_i(\rho_\alpha,\rho_\beta) = A_i\left[\ln\frac{x^2}{X(x)} + \frac{2b}{Q}\tan^{-1}\left(\frac{Q}{2x+b}\right)\right.$$

$$\left. - \frac{bx_0}{X(x_0)}\left(\ln\frac{(x-x_0)^2}{X(x)} + \frac{2(b+2x_0)}{Q}\tan^{-1}\left(\frac{Q}{2x+b}\right)\right)\right]$$

$$x = r_s^{1/2}; Q = \left(4c_i - b_i^2\right)^{1/2}; X(x) = x^2 + b_i x + c_i : (i = I, II);$$

$$\Delta\varepsilon_c(r_s,\zeta) = \varepsilon_{III}\left(\rho_\alpha,\rho_\beta\right)\left[\frac{f(\zeta)}{f''(0)}\right]\left[1 + \beta_i(r_s)\zeta^4\right]$$

$$\beta_i(r_s) = \left[\frac{f''(0)}{\varepsilon_{III}\left(\rho_\alpha,\rho_\beta\right)}\right]\Delta\varepsilon(r_s,1) - 1$$

$$\Delta\varepsilon_c(r_s,1) = \varepsilon_I\left(\rho_\alpha,\rho_\beta\right) - \varepsilon_{II}\left(\rho_\alpha,\rho_\beta\right) \tag{9.60}$$

Constants for the Vosko-Wilk-Nusair parametrization are included in Table 9.2.

Table 9.2 Constants for the Vosko-Wilk-Nusair parametrization

Parameter	I	II	III
A_i	0.0621841	0.0310907	−0.033774
b_i	3.72744	7.06042	1.131071
c_i	12.9352	18.0578	13.0045
x_{0i}	−0.10498	−0.32500	−0.0047584

9.16 The von Barth-Hedin Correlation Energy Functional and the Potential

The von Barth-Hedin correlation energy functional and the potential are given by the following equations:

$$E_c^{VBH}\left[\rho_\alpha,\rho_\beta\right] = \int dr \rho \varepsilon_c^{VBH}(\rho,x)$$

$$\varepsilon_c^{VBH} = \varepsilon_c^P + \gamma^{-1}\upsilon_c^P f(x) ; \upsilon_c = \gamma\left(\varepsilon_c^F - \varepsilon_c^P\right)$$

$$\varepsilon_c^P = -c^P F\left(\frac{r_s}{r^P}\right) ; \varepsilon_c^F = -c^F F\left(\frac{r_s}{r^F}\right) ;$$

$$F(z) = \left(1 + z^3\right)\ln\left(1 + \frac{1}{z}\right) + \frac{z}{2} - z^2 - \frac{1}{3}$$

$$C^P = 0.0252 \; ; r^P = 30 \; ; c^F = 0.0127 \; ; r^F = 75$$
$$v_{c\alpha}^{\text{VBH}} = v_c(2x)^{1/3} + \mu_c^P - v_c + \tau_c f(x) \; ;$$
$$v_{c\beta}^{\text{VBH}} = v_c(2(1-x))^{1/3} + \mu_c^P - v_c + \tau_c f(1-x) \; ;$$
$$\mu_c^P(r_s) = -c^P \ln\left(1 + \frac{r^P}{r_s}\right) \; ; \mu_c^F(r_s) = -c^F \ln\left(1 + \frac{r^F}{r_s}\right)$$
$$\tau_c = \mu_c^F - \mu_c^P - \frac{4}{3}\left(\varepsilon_c^F - \varepsilon_c^P\right) \tag{9.61}$$

9.17 The Perdew 86 Correlation Energy Functional and the Potential

The Perdew 86 correlation energy functional and the potential are given by the following equations:

$$E_c^{\text{P86}}\left[\rho_\alpha, \rho_\beta\right] = E_c^{\text{LSD}}\left[\rho_\alpha, \rho_\beta\right] + \int drd^{-1} \exp(-\Phi)C(\rho)\frac{|\nabla\rho|^2}{\rho^{4/3}}$$

$$\Phi = 1.745\tilde{f}\left[\frac{C(\infty)}{C(\rho)}\right]\frac{|\nabla\rho|}{\rho^{7/6}}$$

$$C(\rho) = 0.001667 + \frac{0.002568 + \alpha r_s + \beta r_s^2}{1 + rs + \delta r_s^2 + 10^4 \beta r_s^3}$$

$$d = 2^{1/3}\left[\left(\frac{1+g}{2}\right)^{5/3} + \left(\frac{1-g}{2}\right)^{5/3}\right]^{1/2}$$

$$\alpha = 0.023266 \; ; \beta = 7.389 \times 10^{-6} \; ; \gamma = 8.723 \; ; \delta = 0.472 \; ; \tilde{f} = 0.11$$
$$v_{c\alpha}^{\text{P86}} = v_{c\sigma}^{\text{LSD}} - d^{-1}\exp(-\Phi)C(\rho)\rho^{-1/3}$$

$$\times \left[\frac{(2-\Phi)\nabla^2\rho}{\rho} - \left(\frac{4}{3} - \frac{11\Phi}{3} + \frac{7\Phi^2}{6}\right)\frac{|\nabla\rho|^2}{\rho^2} + \frac{\Phi(\Phi-3)\nabla\rho.\nabla|\nabla\rho|}{\rho|\nabla\rho|}\right.$$
$$\left. - \frac{5\rho^{1/3}n_{-\sigma}^{2/3}}{6d^2\rho^4}\left[2^{2/3}(1-\Phi)\rho_{-\sigma}|\nabla\rho|^2 - 2^{2/3}(2-\Phi)\rho\nabla\rho_{-\sigma}\nabla\rho\right]\right]$$
$$+ d - 1\exp(-\Phi)\frac{|\nabla\rho|^2}{\rho^{4/3}}\left(\Phi^2 - \Phi - 1\right)\frac{dC}{d\rho} \tag{9.62}$$

9.18 The Perdew 91 Correlation Energy Functional and the Potential

The Perdew 91 correlation energy functional and the potential are given by the following equations:

$$E_c^{\text{P91}}\left[\rho_\alpha, \rho_\beta\right] = \int dr\rho\left[\varepsilon_c^{\text{LSD}}(r_s, \zeta) + H(t, r_s, \zeta)\right]$$

$$H = H_0 + H_1$$

$$H_0 = g^3 \frac{\beta^2}{2\alpha} \ln\left[1 + \frac{2\alpha}{\beta}\left(\frac{t^2 + At^4}{1 + At^2 + A^2t^4}\right)\right]$$

$$A = \frac{2\alpha}{\beta} \frac{1}{\exp(-2\alpha\varepsilon_c^{LDA}(r_s - \zeta)/(g^3\beta^2)) - 1}$$

$$H_1 = 15.7559(C_c(r_s) - 0.003521)g^3t^2 \exp\left[-100g4\left[\frac{k_s}{k_F}\right]^2 t^2\right]$$

$$t = \frac{|\nabla\rho|}{2gk_s\rho}$$

$$k_s = \left(\frac{4k_F}{\pi}\right)^{1/2}; k_F = (3\pi^2\rho)^{1/3}; g = \frac{\left[(1+\zeta)^{2/3} + (1-\zeta)^{2/3}\right]}{2}$$

$$\alpha = 0.09; \beta = \gamma C_c; C_c(0) = 0.004235; C = -0.001667; \gamma = \left(\frac{16}{\pi}\right)(3\pi^2)^{1/3}$$

$$v_c^\sigma = \varepsilon_c^{LSD} - \frac{r_s}{3}\frac{\partial\varepsilon_c^{LSD}}{\partial\zeta} - (\zeta - \text{sgn}(\sigma))\frac{\partial\varepsilon_c^{LSD}}{\partial\zeta} + H - \frac{r_s}{3}\frac{\partial H}{\partial r_s}$$

$$- (\zeta - \text{sgn}(\sigma))\left[\frac{\partial H}{\partial\zeta} - \frac{g'}{g}t^2\left(t^{-1}\frac{\partial H}{\partial}\right)\right]$$

$$+ \frac{1}{6}t^2\left(t^{-1}\frac{\partial H}{\partial}\right) + \frac{7}{6}t^3\frac{\partial}{\partial}\left(t^{-1}\frac{\partial H}{\partial}\right)$$

$$- \frac{\nabla\rho\nabla|\nabla\rho|}{(2gk_s)^3\rho^2}\frac{\partial}{\partial}\left(t^{-1}\frac{\partial H}{\partial}\right) - \frac{\nabla^2}{(2gk_s)^2\rho}\left(t^{-1}\frac{\partial H}{\partial}\right)$$

$$- \frac{\nabla\rho\nabla\zeta}{(2gk_s)^2\rho}\left[\left[t^{-1}\frac{\partial^2 H}{\partial\partial\zeta}\right] - \frac{g'}{g}\left\{2\left(t^{-1}\frac{\partial H}{\partial}\right) + t\frac{\partial}{\partial}\left(t^{-1}\frac{\partial H}{\partial}\right)\right\}\right] \quad (9.63)$$

9.19 The Lee, Yang, and Parr Correlation Energy Functional and the Potential

The Lee, Yang and Parr correlation energy functional and the potential are given by the following equations:

$$E_c^{LYP}[\rho_\alpha, \rho_\beta] = -a\int dr \frac{\gamma(r)}{1 + d\rho^{-1/3}}\left\{\rho + 2b\rho^{-5/3}\left[2^{2/3}C_F\rho_\beta^{8/3}\right.\right.$$

$$\left.\left. - \rho t_w + \frac{1}{9}\left(\rho_\alpha t_w^\alpha + \rho_\beta t_w^\beta\right) + \frac{1}{18}(\rho_\alpha\nabla^2\rho_\alpha)\right]\exp\left(-c\rho^{-1/3}\right)\right\}$$

$$\gamma(r) = 2\left(1 - \frac{\rho_\alpha^2(r) + \rho_\beta^2(r)}{\rho^2(r)}\right)$$

$$t_w(r) = \frac{1}{8}\frac{|\nabla\rho(r)|^2}{\rho(r)} - \frac{1}{8}\nabla^2\rho$$

$$C_F = \frac{3}{10}\left(3\pi^2\right)^{2/3} \; ; \alpha = 0.04918 \; ; b = 0.132 \; ; c = 0.2533 \; ; d = 0.349$$

$$\upsilon_{c\sigma}^{LYP} = -a(F_2'\rho + F_2) - 2^{5/3}abC_F\left[G_2'\left(\rho_\alpha^{8/3} + \rho_\beta^{8/3}\right) + \frac{8}{3}G_2\rho_\beta^{8/3}\right]$$

$$-\frac{ab}{4}\left[\rho\nabla^2 G_2 + 4\nabla G_2\nabla\rho + 4G_2\nabla^2\rho + G_2'\left(\rho\nabla^2\rho - |\nabla\rho|^2\right)\right]$$

$$-\frac{ab}{36}\left[3\rho_\alpha\nabla^2 G_2 + 4\nabla\rho_\alpha\nabla G_2 + 4G_2\nabla^2\rho_\alpha + 3G_2'\left(\rho_\alpha\nabla^2\rho_\alpha + \rho_\beta\nabla^2\rho_\beta\right)\right]$$

$$+ G_2'\left(|\nabla\rho_\alpha|^2 + |\nabla\rho_\beta|^2\right)$$

$$F_2 = \frac{\gamma(r)}{1+d\rho^{-1/3}} \; ; G_2 = F_2(\rho)\rho^{-5/3}\exp(-c\rho^{-1/3})$$

$$F_2' = \frac{\partial F_2}{\partial\rho_\sigma} \; ; G_2' = \frac{\partial G_2}{\partial\rho_\sigma} \tag{9.64}$$

9.20 DFT Methods

DFT would yield the exact ground state energy and electron density if the exchange-correlation functional was known. In practice, the exact functional is unknown but one may try some approximate form. This has led to an extensive search for functionals with new variations being published on a regular basis. Because the quality of the results depends critically on the functional, selecting a suitable form will be a vital factor in using the module. DFT methods are broadly classified into two methods: pure DFT and hybrid DFT. They are designated on the basis of type of correlation energy functional, the exchange energy functional, and the potential.

The pure DFT method consists of:

1. SVWN5 (also known as LDA)
2. BLYP
3. PW91
4. HCTH-93
5. HCTH-120
6. HCTH-147
7. HCTH-402
8. Becke97GGA-1

Similarly, the hybrid DFT method consists of:

1. BH&HLYP
2. B3PW91
3. mPW1PW91
4. PBE0

Table 9.3 Basis set dependence on SVWN

Type of bond	6-31 G(d,p)	6-311++G(d,p)	Basis set free data	Experiment
H – H	-/0.765	-/0.765	-/0.765	-/0.741
C – C	1.513/1.105	1.510/1.101	1.508/1.100	1.526/1.088
C = C	1.330/1.098	1.325/1.094	1.323/1.093	1.339/1.085
C ≡ C	1.212/1.078	1.203/1.073	1.203/1.074	1.203/1.061

5. Becke97
6. Becke97-1
7. Becke98
8. mPW1k

For example, in SVWN5 (which is also known as LDA) keeps the Slater exchange with the Vosko-Wilk-Nusair expression 5 for the correlation energy.

LDA geometries depend up on the choice of basis set. SVWN-optimized bond length for hydrocarbons are included in Table 9.3.

9.21 Applications of DFT

Applications of modern DFT calculations have been extended from small molecules for testing the accuracy to transition metal complexes. For complex molecules, DFT appears to be the method of choice at present. In the last few years, people have begun to apply DFT methods to a variety of systems such as biomolecules, polymers, macromolecules, and so on. Recently, researchers started examining spin densities in bio-inorganic complexes. These are very challenging calculations. involving up to hundreds of electrons. In about 1985, Car and Parrinello introduced a new method whereby one can solve for the electron density for a configuration of nuclei, and then move the nuclei based on the resulting forces, resolve the electronic structure problem, and so on. This means one can do real-time simulations without using any "made up" force fields. This technique has been applied to many problems in chemistry and materials science. Examples are water and ions in water, the proton in water, silicon surfaces, chemical reactions, etc. In the last few years, a lot of work has been done in developing methods which scale linearly with system size (RDM is one example, which is discussed in the next chapter).

The single geometry SCF cycle or geometry optimization involves the following steps:

1. Start with a density (for the 1st iteration, a superposition of atomic densities is typically used).
2. Establish a grid for the charge density and the exchange correlation potential.
3. Compute the KS matrix (equivalent to the F matrix in the HF method) elements and overlap the integrals matrix.
4. Solve the equations for expansion coefficients to obtain the KS orbitals.

5. Calculate a new density $\rho = \sum_{i=occ} |\phi_i(r)|^2$.
6. If the density or energy changed substantially, go to step 1.
7. If the SCF cycle converged and geometry optimization is not requested, go to step 10.
8. Calculate the derivatives of the energy vs. the atom coordinates, and update the atom coordinates. This may require denser integration grids and the recomputing of the Coulomb and the exchange correlation potential.
9. If the gradients are still large, or the positions of the nuclei moved appreciably, go back to step 1.
10. Calculate the properties and print the results.

It is quite popular to limit expense of numerical integration during the SCF cycle. This is frequently done by fitting auxiliary functions to the charge density and the exchange correlation potential. This allows for a much faster integral evaluation. These auxiliary fitting functions are usually uncontracted Gaussians (though quite different from the atomic basis sets) for which the integrals required for the KS matrix can be calculated analytically. Different auxiliary sets are used for fitting the charge density and the exchange correlation potential.

9.22 The Performance of DFT

We have a short list of DFT applications. The G1 database of Pople and coworkers is a remarkable proof of accuracy of the traditional ab initio methods. The database contains 55 molecules for which experimental values of atomization energies are within the limit of permitted error (± 1 kcal/mol). With the G2 procedure, Curtiss et al. (1991) achieved the 1.2 kcal/mol mean absolute error for these 55 atomization energies, which is a quite involved prescription incorporating higher order correlated methods. Becke (1992) was able to reproduce values in this database with a mean absolute error of 3.7 kcal/mol using his NUMOL program with gradient corrected functionals. This result was additionally improved by Becke (1993) to 2.4 kcal/mol by calculating the exchange correlation energy with the KS orbitals While the error in DFT is considered still too big, these results were obtained with a method which is substantially less computationally demanding than original correlated ab initio procedures used by Pople and coworkers. Rather than the absolute atomization energy the differences are usually computed much better with DFT methods. We will be concerned with only the difference in energy associated with a change. Hence, the method is highly appreciated.

Even without gradient corrections, DFT results for bond dissociation energies are usually much better than the HF results, though they have an overbinding tendency. The LDA results are found to be approximately of MP2 quality. The inclusion of gradient corrections to DFT provides a better computation of bond dissociation energies with the level of MP4 and CC computations. Molecular geometries even with LSD are much better than corresponding HF results and are of the MP2 quality.

However, LSD fails to explain hydrogen bonding. This defect is overcome by using gradient corrections. DFT methods are supportive to molecules such as OOF, FON, and metal organic or inorganic moieties, where traditional ab initio methods fail to be supportive. In most cases, if ab initio methods are not working properly, we have the possibility to at least try with DFT. In most cases, this method gives promising results. Transition states of organic molecules are frequently not reproduced well with pure DFT methods. However, it seems that hybrid methods give improved results.

Vibrational frequencies are well reproduced even by LSD, though gradient corrections improve agreement with the experiment even further. Ionization potentials, electron affinities, and proton affinities are reproduced fairly well within gradient corrected DFT. Using DFT methods for high spin species gives promising results.

The scope of applications for DFT grows rapidly with the calculations of new molecular properties being added to actively developed software. Recent extensions include parameters for NMR and ESR spectroscopy, diamagnetic properties, polarizabilities, relativistic calculations, and others.

9.23 Advantages of DFT in Biological Chemistry

Computational demands with DFT methods are much less than with ab initio methods of similar quality. Hence, DFT methods are widely used in computing larger molecules such as biomolecules. Metals are frequently present in active centers of enzymes. Traditional ab initio methods have severe problems with transition metals. In fact, the HF equation cannot be solved for the true metallic state. It is related to the fact that there is a difficulty to converge HF when the highest occupied orbitals are very close in energy (the situation very popular for transition metals). The DFT, similar to ab initio methods, is nonparametric, i.e., it is applicable to any molecule. We may think that basis sets which are used as parameters for ab initio and DFT methods are parametric. It is not completely true, as basis sets can be easily derived from atomic calculations. Moreover, basis sets were derived a long time ago for most of the elements of the periodic table with proper experimental and theoretical proofs. The restriction of DFT being applicable to the ground state only is not usually a problem, unless you study the interaction of radiation with biological molecules (e.g., UV-induced mutations).

9.24 Exercises

1. Optimize the geometry of a water molecule using Molecular Mechanics (MM3), two semiempirical methods (AM1 and PM3) and DFT, an ab initio method (DFT-B88LYP). Measure the bond length and bond angle and compute the heat

of formation. Compare the computed results with the computed experimental values.

2. Optimize the carbon dioxide molecule by the following methods: HF, SVWN, SVWN5, BLYP, B3LYP, and MP2. Compute zero point energy by all these methods. Compute single point energies of carbon and oxygen using tight SCF convergence. Calculate the total atomization energy.

3. Perform optimization for F_2O_2 using B3LYP/6-31+G(d) and B3LYP/6-31G($2d$) and compare the O–O, and O–F bond lengths, the bond angle and the dihedral angle.

4. Find the spin polarization in the $CH_2=CH-XH_n$ species where n is R=O, R=Be, R=Mg, and R=S using B3LYP.

5. Compute the effect of ozone depletion by chlorine. Use B3LYP/6-31+G(d).

6. Compute the atomization energy of carbon monoxide and dinitrogen by a suitable DFT method.

7. Find the atomization energy of water molecule by the DFT method. Compare the result with HF and ab initio methods.

8. Compute the proton affinity of phosphene in the G2 level (G2 key word of Gaussian 03).

For answers to these questions see the URL.

References

1. Parr RG, Yang W (1989) Density Functional Theory of Atoms and Molecules. Oxford, New York
2. Dreizler RM, Gross EKU (1990) Density Functional Theory. Springer, New York
3. Springborg M (1997) Density Functional Methods in Chemistry and Materials Science. Wiley, New York
4. Szabo A, Ostlund NS (1989) Modern Quantum Chemistry. McGraw-Hill, New York
5. Foresman JB, Frisch A (1996) Exploring Chemistry with Electronic Structure Methods. Gaussian, Pittsburgh, PA

Chapter 10
Reduced Density Matrix

10.1 Introduction

The solution of an N-body Schrödinger equation through the ground state properties of a fermion system (5.4) in an applied external potential for the analysis of a boundless variety of physical situations remains a focus of research. It was J. E. Mayer in 1955 who first identified that for non-relativistic electrons (which interact via pair forces alone), the system energy depends only upon the two-body reduced density matrix (2-RDM). In fact, only two combinations are possible in this regard; the pair density (2-RDM) and the one-body reduced density matrix (1-RDM). The former one keeps four-particle degrees of freedom while the latter one keeps only two particle degrees of freedom. Mayer suggested the possibility of computing the ground state energy and density matrix information by simply carrying out a Rayleigh-Ritz minimization with respect to the pair density and 1-RDM. However, the initial computations resulted in horrible results due to the ignoring of a number of necessary restrictions or constraints. Progress in this very promising approach could be possible, if and only if we include all the necessary restrictions.

10.2 Reduced Density Matrices

The N-fermion problem can be treated as a discrete orthonormal basis of single particle wavefunctions. Let ψ be the ground state normalized wavefunction for an N-fermion system. Hence:

$$\langle \psi | \psi \rangle = 1 \qquad (10.1)$$

K. I. Ramachandran et al., *Computational Chemistry and Molecular Modeling*
DOI: 10.1007/978-3-540-77304-7, ©Springer 2008

1-RDM (γ) can be defined as:

$$\gamma(i,i') = \left\langle \psi \left| a_{i'}^+ a_i \right| \psi \right\rangle \tag{10.2}$$

Here, a_i and $a_{i'}^+$ are the annihilation and creation operators for the single particle state i for the chosen basis set. An annihilation operator is an operator that lowers the number of particles in a given state by one. A creation operator is an operator that increases the number of particles in a given state by one, and it is the adjoint of the annihilation operator. Similarly, 2-RDM (Γ) is given by:

$$\Gamma(i,j;i',j') = \left\langle \psi \left| a_{i'}^+ a_{j'}^+ a_j a_i \right| \psi \right\rangle \tag{10.3}$$

(a_j and $a_{j'}^+$ are the annihilation and creation operators single particle state j for the chosen basis set.).

$\Gamma(i,j;i',j')$ is antisymmetric under the interchange of i and j and also under the interchange of i' and j'; γ and Γ are hermitian.

The Hamiltonian of the N-fermion system involving only one-body and two-body interactions can be written as Eq. 10.4:

$$\hat{H} = \sum_{i,i'} h_1(i,i') a_i^+ a_{i'} + \frac{1}{2} \sum_{i,j,i',j'} h_2(i,j,i',j') a_i^+ a_j^+ a_{j'} a_{i'} \tag{10.4}$$

(h_1 and h_2 are single particle Hamiltonians).

The ground state energy E can be expressed exactly in terms of the 1-RDM and 2-RDM:

$$E = Tr(h_1 \gamma) + \frac{1}{2} Tr(h_2 \Gamma) \tag{10.5}$$

Tr stands for trace of the operator.

$$Tr(h_1 \gamma) = \sum_{i,i'} h_1(i,i') \gamma(i',i) \tag{10.6}$$

$$Tr(h_2 \Gamma) = \sum_{i,j,i',j'} h_2(i,j,i',j') \Gamma(i',j',i,j) \tag{10.7}$$

The pair (γ, Γ) is used as a trial function in the space of functions satisfying the stated antisymmetry and hermiticity conditions. In the computation it seeks to minimize the right-hand side of Eq. 10.5 (the variational principle).

For an N-fermion system from the definition of 1-RDM and 2-RDM we come across the following conditions.

The linear equality condition:

$$\sum_k \Gamma(i,k,i'k) = (N-1) \gamma(i,i') \tag{10.8}$$

and trace conditions:

$$\sum_i \gamma(i,i) = N \qquad (10.9)$$

$$\sum_{i,j} \Gamma(i,j;i,j) = N(N-1) \qquad (10.10)$$

Linear equality and convex inequality conditions are imposed on (γ, Γ) that are necessary to ensure that the trial pair lies in the convex hull of density matrices that are actually derived from N-fermion wavefunctions. These additional conditions were introduced by Coleman, Garrod, and Percus.

10.3 *N*-Representability Conditions

Besides the conditions mentioned above (Eqs. 10.9 and 10.10) convex inequality conditions that do not explicitly involve the particle number N have to be included. For the 1-RDM, a complete set of representability conditions was given by Coleman. Basically, the γ matrix should be positive semidefinite. Hence, $\gamma \succeq 0$ or $(I - \gamma) \succeq 0$, where I stands for the identity matrix. That is, all its eigenvalues of the matrix are nonnegative. He also made two more conditions known as P and Q conditions.

The P condition states that $\Gamma \succeq 0$. Here, Γ is identified as a hermitian operator on the space of antisymmetric two-body wavefunctions. Hence, for any antisymmetric function $g(i,j)$, based on this condition:

$$\sum_{i,j,i',j'} g*(i,j)\Gamma(i,j;i',j')g(i',j') \geq 0 \qquad (10.11)$$

The Q condition follows from the positive semidefinite property of the operator A^+A where:

$$A = \sum_{i,j} g(i,j)a_i^+ a_j^+ \qquad (10.12)$$

Hence, $\langle \psi | A^+A | \psi \rangle \geq 0$ Or:

$$\sum_{i,j,i',j'} g*(i,j) \left\langle \psi \left| a_j, a_i; a_i^+, a_j^+ \right| \psi \right\rangle g(i',j') \geq 0 \qquad (10.13)$$

The Q-condition is given by:

$$\begin{aligned} Q(i,j;i',j') &= \left\langle \psi \left| a_j, a_i; a_i^+, a_j^+ \right| \psi \right\rangle \\ &= \Gamma(i,j';j,i') - \delta(i,i')\gamma(j,j') - \delta(j,j')\gamma(i,i') \\ &\quad + \delta(i,j')\gamma(j,i') + \delta(j,i')\gamma(i,j') - \delta(i,j')\gamma(j,i') \end{aligned} \qquad (10.14)$$

10.3.1 G-Condition (Garrod) and Percus

If the operator $A = \sum_{i,j} g(i,j) a_i^+ a_j$ (g is any function of the two indices) due to the positive semidefinite property of the operator A^+A, $\langle \psi | A^+A | \psi \rangle \geq 0$. Then, the G-condition states that:

$$G(i,j;i',j') = \langle \psi | A^+A | \psi \rangle \tag{10.15}$$

It depends linearly on 1-RDM and 2-RDM and can be written as:

$$G(i,j;i',j') = \langle \psi | A^+A | \psi \rangle = \Gamma(i,j';j,i') + \delta(i,i')\gamma(j'j) \tag{10.16}$$

10.3.2 T-Conditions (Erdahl)

For an arbitrary, totally antisymmetric function $g(i,j,k)$, the operators A^+A and AA^+ are both positive semidefinite, where $A = \sum_{i,j,k} g(i,j,k) a_i a_j a_k$. We can express this in terms of the RDM expressions similar to the derivations of Q or G conditions. Separately taking $\langle \psi | A^+A | \psi \rangle$ and $\langle \psi | AA^+ | \psi \rangle$, we can see that each term contains 3-RDM, which is defined as $\langle \psi | a_i^+, a_j^+, a_k^+, a_k, a_j, a_i | \psi \rangle$ keeping the opposite sign. Hence, in the sum of these functions, $\langle \psi | A^+A + AA^+ | \psi \rangle$ only the 1-RDM and 2-RDM will be present. Of course, this sum is nonnegative as well. The result is that $T1$ is a positive semidefinite matrix.

The hermitian matrix $T1$ is given by Eq. 10.17:

$$T1(i,j,k;i',j',k') = \langle \psi | a_i^+, a_j^+, a_k^+, a_k a_j, a_i + a_i, a_j, a_k, a_k^+, a_j^+, a_i^+ | \psi \rangle \tag{10.17}$$

It is related to 1-RDM and 2-RDM by the equation:

$$T1(i,j,k;i',j',k') = A[i,j,k|A|i',j',k']$$
$$\left[\frac{1}{6}\delta(i,i')\,\delta(j,j')\,\delta(k,k') - \frac{1}{2}\delta(i,i')\,\delta(j,j')\,\gamma(k,k') \right.$$
$$\left. + \frac{1}{4}\delta(i,i')\,\Gamma(j,k;j'k') \right] \tag{10.18}$$

10.3.3 T2 Condition

The $T2$ condition follows in a similar way from the positive semidefinite property of the operator $A^+A + AA^+$ where $A = \sum_{i,j,k} g(i,j,k) a_i^+ a_j a_k$. If $g(i,j,k)$ is antisymmetric with respect to (j,k) only the result will make $T2$ into a positive semidefinite property. The hermitian matrix $T2$ is defined by:

$$T2(i,j,k;i',j',k') = \left\langle \psi \left| a_k^+, a_j^+, a_i, a_i^+ a_j, a_k + a_i^+, a_j, a_k, a_k^+, a_j^+, a_i \right| \psi \right\rangle \quad (10.19)$$

$$T2(i,j,k;i',j',k') = A[j,k\,|A|\,j',k']$$

$$\left[\frac{1}{2}\delta\,(j,j')\,\delta\,(k,k')\,\gamma\,(i,i') + \frac{1}{4}\delta\,(i,i')\,\Gamma\,(j',k';j,k) \right.$$

$$\left. - \delta\,(j,j')\,\Gamma\,(i,k';i'k) \right] \quad (10.20)$$

10.4 Computations Using the RDM Method

Following the clear statement of the RDM approach and of the most important N-representability conditions, the first significant computational results came in the 1970s. Kijewski applied the RDM method to doubly ionized carbon ($N = 4$), C++, using a basis of 10 spin orbitals ($r = 10$). Garrod and his co-authors were the first ones to actually solve the semidefinite programming, imposing the P, Q and G conditions, by which they obtained very accurate results for atomic beryllium ($N = 4$ and $r = 10$).

10.5 The SDP Formulation of the RDM Method

Let C, A_p ($p = 1,2,\ldots,m$) be given block diagonal symmetric matrices with prescribed block sizes, and $c, a_p \in R^s$ ($p = 1,2,\ldots,m$) be given s-dimensional vectors. A diagonal matrix with elements a can be represented by $Diag(a)$.

The objective function to be maximized is:

$$\langle C, X \rangle + \langle Diag(c), Diag(x) \rangle$$

Subject to:

$$\langle A_p, X \rangle + \langle Diag(a_p), Diag(x) \rangle = b_p(p = 1,2..,\ldots,m)$$
$$X \succeq 0, x \in R^s \quad (10.21)$$

Its dual is:
Minimize $b^T y$

Subject to:

$$S = \sum_{p=1}^{m} A_p y_p - C \succeq 0$$

$$\sum_{p=1}^{m} Diag(ap)yp = Diag(c)^y \in R^m \tag{10.22}$$

where $(X;x)$ are the primal variables and $(S;y)$ are the dual variables. Primal-dual interior-point methods and their variants are the most established and efficient algorithms to solve general semidefinite programming. Reduced density matrix with $(P,Q,G,T1,T2)$ N-representability conditions can be treated as an SDP. 1-RDM variational variable Γ_1 and its corresponding Hamiltonian H_1 are two index matrices; the 2-RDM variational variable Γ_2, and the corresponding Hamiltonian H_2, Q and G are four index matrices. T_1 and T_2 are six index matrices. Map each pair (i,j) or triple (i,j,k) to a composite index for these matrices, resulting in symmetric matrices of order $r(r-1)/2 \times r(r-1)/2$ for Γ_2, H_2 and Q, a symmetric matrix of order $r(r-1)(r-2)/6 \times r(r-1)(r-2)/6$ for T_1, and a symmetric matrix of order $r^2(r-1)/2 \times r^2(r-1)/2$ for $T2$. For example, the four-index element $\Gamma_2 (i,j;i',j')$ with $1 \le i < j \le r$ and $1 \le i' < j' \le r$ can be associated with the two-index element $\Gamma_2(j-i+(2r-i)(i-1)/2 j' - i' + (2r-i')(i'-1)/2)$. We assume, henceforth, that all matrices have their indices mapped to two indices, and we keep the same notation for simplicity also, due to the antisymmetry property of the 2-RDM, Γ_2, and of the N-representability conditions Q, $T1$ and $T2$, and also due to the spin symmetry. Let us define a linear transformation svec:

$$S^n \rightarrow R^{n(n+1)/2} \tag{10.23}$$
$$U \in S^n$$

$$svec(U) = \left(U_{11}, \sqrt{2}U_{12}, U_{22}, \sqrt{2}U_{13}, \sqrt{2}U_{23}, U_{23}, \dots, \sqrt{2}U_{1n}, \dots U_{nn}\right)^T \tag{10.24}$$

To formulate the RDM method with (P,Q,G,T_1,T_2) conditions as the dual SDP, we define:

$$y = \left[svec(\Gamma_1)^T, svec(\Gamma_2)^T\right]^T \in R^m \tag{10.25}$$

$$b = \left[svec(H_1)^T, svec(H_2)^T\right]^T \in R^m \tag{10.26}$$

Now, express the N-representability conditions through the dual slack matrix variable S by defining it as having the following diagonal blocks: $(\Gamma_1,(I-\Gamma_1),\Gamma_2,Q,G,T_1,T_2)$. Then, the ground state energy can be computed with the dual linear function:

$$E = \min_{y} b^t y \tag{10.27}$$

10.6 Comparison of Results

Zhengji Zhao et al. computed the ground state energies of 47 molecules by the RDM method, imposing the (P,Q), (P,Q,G), $(P,Q,G,T1)$, $(P,Q,G,T2)$, and $(P,Q,G,T1,T2)$ conditions. These results are compared with results obtained by other, more familiar methods, such as singly and doubly substituted configuration interaction (SDCI), Brueckner doubles (with triples) (BD(T)) and coupled cluster singles and doubles with perturbational treatment of triples (CCSD(T)-CCSD(T), which is arguably the most accurate single method available in Gaussian 98). The RDM method provides a lower bound for the full CI result in the same model space, and it gives exact solutions for the cases $N = 2$ and $N = r - 2$ using only the P and Q conditions. Previous numerical results of Nakata et al. suggest that adding the G condition to the P and Q conditions is essential to obtain a solution that is competitive at least with the Hartree-Fock approximation. This generalization is again confirmed by this research. In certain cases (LiH, BeH, BH^+, CH^-, NH, NH^-, OH^+, OH, OH^-, HF^+, HF, SiH^-, HS^+) the difference between the result of the RDM method using P, Q, and G conditions, RDM (P,Q,G) and the full CI result is around 0.1 milli Hartree (mH). In those cases, the accuracy also compares favorably with the CCSD(T), BD(T), and SDCI approximations. The RDM (P,Q,G) errors are found to be much more; still, it is well below the Hartree-Fock error in magnitude. The results of the RDM method are improved by the inclusion of the $T1$ condition, and improved spectacularly by adding both the $T1$ and $T2$ conditions (or even $T2$ alone). They found that the RDM method with P, Q, G, $T1$, and $T2$ conditions gives almost the exact full CI values for the ground state energies, with an error around 0.1 mH or less.

When the $T1$ and $T2$ conditions are added, the dipole moment error falls to around 0.0001 a.u. or less for most of the molecules. Once the energy is obtained with a high accuracy, the dipole moment calculation also reaches a high accuracy. This is another advantage of the RDM method over the other, traditional variational methods, in which a first order error in the trial wavefunction results in a second order error in the energy, so a poor trial function may produce amazingly good results on the ground state energy, but not on the other ground state properties.

10.7 Research in RDM

Appreciating the level of accuracy that the RDM method can attain, the present trend is to make computations using this method. A number of research papers in this regard are available. Some of them are mentioned below.

Gidofalvi and Mazziotti used variational RDM theory to evaluate the strength of Hamiltonian-dependent conditions. A theory for the absorption line shape of molecular aggregates in condensed phase is formulated based on a reduced density-matrix approach by Yang and Mino. They illustrated the applicability of the present theory by calculating the line shape of a dimer (a pair consisting of a donor and an acceptor of an energy transfer).

Entropy maximization has proven effective in treating certain aspects of the phase problem of X-ray diffraction. Entropy on an N-representable one-particle density matrix is well defined by D. M. Collins.

Reduced density matrix descriptions were developed by Jacobs, Verne et al. for linear and non-linear electromagnetic interactions of moving atomic systems, considering the applied magnetic fields. Atomic collision processes are treated as environmental interactions. Applications of interest include electro-magnetically induced transparency and related pump-probe optical phenomena in atomic vapors.

10.8 Exercises

1. A harmonic oscillator is brought to thermal equilibrium at a temperature T and then is disconnected from the reservoir and coupled to a two state system in such a way that the two-state system is in a $\sigma_3 = +1$ state if the level of the oscillator is even, and $\sigma_3 = -1$ if it is odd. Write the reduced density matrix if one is interested only in the two-state system. Use the density operator to compute $\langle \sigma_3 \rangle$.

2. For a 2-state system, write down the most general form of the density matrix. (finding all the constraints on the coefficients).

3. Consider two systems: 1 and 2, each in the states: $|\psi_1\rangle = \frac{1}{\sqrt{2}}(|a\rangle_1 + |b\rangle_1)$ and $|\psi_2\rangle = \frac{1}{\sqrt{2}}(|a\rangle_2 + |b\rangle_2)$. Write down the density matrix for each system. Write down the combined state for the two systems. Find the density matrix for the combined systems. Find the reduced density matrix for system 2.

References

1. Gidofalvi G, Mazziotti D (2004) Variational reduced-density-matrix theory: strength of Hamiltonian-dependent positivity conditions. Chem Phys Lett 398:4-6
2. Yang M (2005) A reduced density-matrix theory of absorption line shape of molecular aggregate. J Chem Phys 123:12 pp 124705–124706
3. Collins DM (1993) Entropy on charge density: making the quantum mechanical connection. Acta Cryst D49 pp 86-89
4. Fukuda M et al. (2007) Large-scale semidefinite programs in electronic structure calculation. Math Prog 109:2–3
5. McRae WB, Davidson ER (1972) Linear inequalities for density matrices II. J Math Phys 13:1527

6. Garrod C, Fusco MA (1976) Role of Model System in Few-Body Reduction of N-Fermion Problem. Int J Quant Chem 10:495
7. Vandenberghe L, Boyd S (1996) Semidefinite programming. SIAM 38: 49
8. Wolkowicz H, Saigal R, Vandenberghe L (2000) Handbook of Semidefinite Programming: Theory, Algorithms, and Applications. Kluwer Academic, Norwell, MA
9. Nakata M, Nakatsuji H, Ehara M, Fukuda M, Nakata K, Fujisawa K (2001) Variational calculations of fermion second-order reduced density matrices by a semidefinite programming algorithm. J Chem Phys 114:8282
10. Jacobs V et al. (2007) Advanced Optical and Quantum Memories and Computing IV. Proc SPIE 6482 pp 64820X

Chapter 11
Molecular Mechanics

11.1 Introduction

Molecular mechanics (MM) computes the structure and energy of molecules based on nuclear motions. In this method, electrons are not considered explicitly, but rather it is assumed that they will find their optimum distribution once the positions of the nuclei are known. This assumption is based on the Born-Oppenheimer approximation that nuclei are much heavier than electrons and their movement is negligibly small compared to the movement of electrons. Nuclear motions such as vibrations and rotations can be studied separately from electrons. The electrons are supposed to move fast enough to adjust to any movement of the nuclei. In a very general sense, MM treats a molecule as a collection of weights connected with springs, where the weights represent the nuclei and the springs represent the bonds. Based on this treatment, molecular properties can be well studied. The method is based on the following assumptions:

1. Nuclei and electrons are lumped together and treated as unified atom-like particles.
2. Atom-like particles are treated as spherical balls.
3. Bonds between particles are viewed as springs.
4. Interactions between these particles are treated using potential functions derived from classical mechanics.
5. Individual potential functions are used to describe different types of interactions.
6. Potential energy functions rely on empirically derived parameters that describe the interactions between sets of atoms.
7. The potential functions and the parameters used for evaluating interactions are termed a *force field*.
8. The sum of interactions determines the conformation of atom-like particles.

A comparative study of the three major computational chemistry techniques can be made as given in Table 11.1.

K. I. Ramachandran et al., *Computational Chemistry and Molecular Modeling*
DOI: 10.1007/978-3-540-77304-7, ©Springer 2008

Table 11.1 Comparative study of ab initio, semiempirical and molecular mechanics techniques

Ab initio	Semi-empirical	Molecular mechanics
Counting all electrons	Ignoring some electrons (simplification)	Ignoring all electrons. Only nuclei are taken into consideration
Limited to tens of atoms and best performance using a supercomputer	Limited to hundreds of atoms	Molecules containing thousands of atoms
Can be applied to inorganics, organics, organo-metallics and molecular fragments (the catalytic components of an enzyme)	Can be applied to inorganics organics, organo-metallics and small oligomers (peptide, nucleotide, saccharide)	Can be applied to inorganics, organics, oligonucleotides, peptides, saccharides, metallo-organics and inorganics
Extended to a vacuum or implicit solvent environment	Extended to a vacuum or implicit solvent environment	Extended to a vacuum, implicit, or explicit environment
Applicable to ground, transition, and excited states	Applicable to ground, transition, and excited states	Applicable to the ground state only. Thermodynamics and kinetics via molecular dynamics properties

11.2 Triad Tools

Molecular mechanics depends upon three tools-force fields, parameter sets, and minimizing algorithms, together sometimes called *triad tools* (Fig. 11.1).

A force field is a set of functions and constants used to find the potential energy of the molecule. In general, the potential energy of the system can be represented as sum of the force field functions (Eq. 11.1):

$$E = \sum_{ij} k_{ij} x_i x_j + \sum_{ijk} k_{ijk} x_i x_j x_k \qquad (11.1)$$

Here, k_{ij} is a constant depending up on the bond length (the distance between x_i and x_j) and k_{ijk} is a constant depending upon the bond angle (the bond angle between x_i, x_j and x_k). However, the molecular mechanics energies will not be confused with absolute quantities. The only difference in energy between two or more conformations, states, or levels will have meaning. In most cases, in MM or its tool, the empirical force field (EFF, or simply, force field, FF), the data determined experimentally for small molecules can be extrapolated to larger molecules (transferable). It is aimed at quickly providing energetically favorable conformations for large systems.

Parameters included in the parameter set define the reference points and force constants allowing for the calculation of different levels of potential energy calculations, which are caused due to the inclusion of attractive or repulsive interactions between atoms.

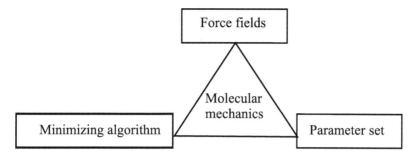

Fig. 11.1 MM triad tools

Algorithms to calculate new geometrical positions from an initial guess to provide geometry optimization use the so-called optimizers or minimizers. Different methods such as the steepest descent, the conjugate gradient, Powel, Newton-Raphson, BFGS, line searches, etc. are available in this step. Different techniques to overcome local-global minima problem are provided. Geometry optimization requires the global minimum to be achieved.

The force fields generally take the form of $E_{\text{total}} = E_r + E_\theta + E_\phi + E_{\text{nb}} +$ [special terms], where the total energy (E_{total}) is expressed as the sum of energies associated with bond stretching (E_r), bond angle bending (E_θ), bond torsion (E_ϕ), nonbond interactions (E_{nb}), and specific terms such as hydrogen bonding (E_{hb}) in biochemical systems. Most MM equations are similar in the types of terms they contain. However, there are some differences in the forms of the equations that can affect the choice of force field and parameters for the systems of interest. We need quantum mechanics to describe bonding accurately but can approximate bonding with simple physical models.

11.3 The Morse Potential Model

The Morse potential (Philip M. Morse), is a fitting model for the potential energy of diatomic molecules such as dihydrogen. It is suitable for the vibrational structure of the molecule, as it explicitly includes the effects of bond breaking, such as the existence of unbound states. It also accounts for the anharmonicity of real bonds and the non-zero transition probability for overtone and combination bands.

The potential is represented by the function:

$$V(r) = D_e \left(1 - e^{-a(r-r_e)}\right)^2 \tag{11.2}$$

Here, is the distance between the atoms, is the equilibrium bond distance, is the well depth (defined relative to the dissociated atoms), and a controls the "width" of the potential. The dissociation energy of the bond can be calculated by subtracting the zero point energy from the depth of the well. The force constant of the bond can be found by taking the second derivative of the potential energy function, from which it can be shown that the parameter, a, is:

$$a = \sqrt{k_e/2D_e} \tag{11.3}$$

11.4 The Harmonic Oscillator Model for Molecules

The harmonic oscillator is a simple mechanical model of a moving mass fixed to a wall with the help of a spring. A similar model can be considered for a small atom such as hydrogen connected to a large atom or molecule. The large molecule can be considered as stationary relative to the fast motions of the small hydrogen.

Hooke's Law gives the relationship between the force applied to an unstretched spring and the amount the spring is stretched when the force is applied. In physics, Hooke's law of elasticity is an approximation that states that the amount by which a material body is deformed (the strain) is linearly related to the force causing the deformation (the stress). Materials for which Hooke's law is a useful approximation are known as linear-elastic or "Hookean" materials. For such materials, the extension produced is directly proportional to the load:

$$F = -kx \tag{11.4}$$

where x is the distance by which the material is elongated, F is the restoring force exerted by the material, and k is the force constant (or spring constant). The negative sign indicates that the force exerted by the spring is in the direction opposite to the direction of displacement. It is called a "restoring force," as it tends to restore the system to equilibrium (Fig. 11.2). But by Newtonian mechanics, the force, $F = ma$, where m is the mass of the body and a the acceleration. From Eq. 11.2, we can write the expression as:

$$F = ma = m\frac{d^2x}{dt^2} = -kx. \tag{11.5}$$

The force to compress a spring varies from $F_{ext} = F_0 = 0$ at $x_i = 0$ to $F_{ext} = F_x = kx$ (at $x_f = x$). Since force increases linearly with x, the average force that must be applied is:

$$F_{average} = F_{ext} = \frac{1}{2}(F_0 + F_x) = \frac{1}{2}kx \tag{11.6}$$

The work done by F_{ext} is:

$$W = F_{ext}x = \frac{1}{2}kx^2 \tag{11.7}$$

The potential energy stored in the compressed (or stretched) spring will be the calculated work required to compress (or stretch) the spring. Hence, the potential

Fig. 11.2 Harmonic oscillator in one dimension

energy is:

$$E_{pe} = \frac{1}{2}kx^2 \tag{11.8}$$

and is stored in the spring as potential energy.

Solving the differential equation (Eq. 11.5), we obtain:

$$x(t) = A\cos\left(\sqrt{\frac{k}{m}}t\right) \pm B\sin\left(\sqrt{\frac{k}{m}}t\right) \tag{11.9}$$

where $\sqrt{\frac{k}{m}} = 2\pi v = \omega$, the oscillation frequency. The angular frequency in radians is related to the frequency in cycles per second (Hertz) by Eq. 11.8:

$$v = \frac{1}{2\pi}\sqrt{\frac{k}{m}} \tag{11.10}$$

Substituting the value of ω in Eq. 11.9, we get:

$$x(t) = A\cos(\omega t) \pm B\sin(\omega t) \tag{11.11}$$

If we assume an initial condition $x(t=0) = A$ and $\frac{dx}{dt}(t=0) = 0$, then the solution is reduced to:

$$x(t) = A\cos(\omega t) \tag{11.12}$$

The potential energy can be derived from this equation as shown in Eq. 11.13:

$$E_{pe} = V = -\int_0^x F\,dx = -\int_0^x (-kx)\,dx = \frac{kx^2}{2} \tag{11.13}$$

11.5 The Comparison of the Morse Potential with the Harmonic Potential

The Morse potential is more accurate than the harmonic potential; still, it is not widely used as it is computationally expensive. The Morse potential allows a bond to stretch to an unrealistic length. By this model, for a structure with long bonds

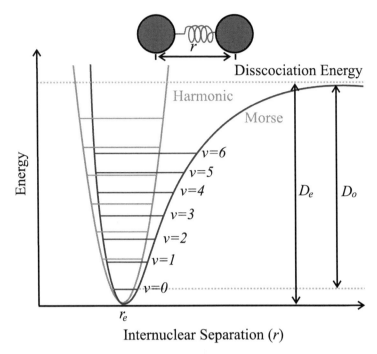

Fig. 11.3 The Morse potential and the harmonic oscillator potential

there would be almost no force pulling the atoms together. Hence, convergence in this method might be problematic or nonphysical results might be obtained. The major defect of the harmonic potential is that the force is estimated as very high even at a very high distance. This may destroy some important structural features. A graphical comparison of these two potentials is illustrated in Fig. 11.3. Unlike the energy levels of the harmonic oscillator potential, which are evenly spaced, the Morse potential level spacing decreases as the energy approaches the dissociation energy. The dissociation energy D_e is larger than the true energy required for dissociation D_o due to the zero point energy of the lowest ($v = 0$) vibrational level.

11.6 Two Atoms Connected by a Bond

We can transform the "two body" problem, with the masses connected to a spring as a "single body" problem with masses of two bodies replaced by a single reduced mass μ vibrating with respect to a stationary center of mass x_c as shown in Fig. 11.4. In diatomic covalently bonded molecules like dihydrogen a similar formulation can be made. The reduced mass is calculated from Eq. 11.12.

$$\mu = \frac{m_1 m_2}{m_1 + m_2} \tag{11.14}$$

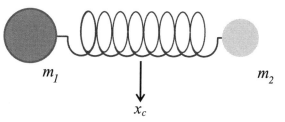

Fig. 11.4 Two masses connected together by a spring (bond)

The vibrating frequency v expression will become automatically:

$$v = \frac{1}{2\pi}\sqrt{\frac{k}{\mu}} \tag{11.15}$$

11.7 Polyatomic Molecules

In polyatomic molecules, each atom is kept in its position by one or more chemical bonds. Each chemical bond may be modeled as a harmonic oscillator in a space defined by its potential energy as a function of the degree of stretching or compression of the bond along its axis (Fig. 11.5).

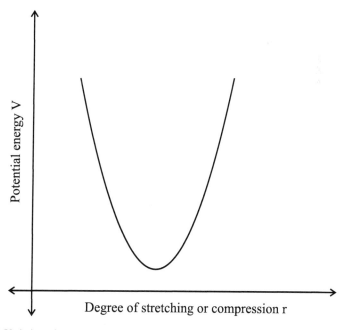

Fig. 11.5 Variation of potential energy with degree of stretching or compression

11.8 Energy Due to Stretching

Bond stretching or compression from the natural bond position is associated with an increase in potential energy. The corresponding energy change is described by an equation similar to Hooke's law for a spring, with a cubic term instead of square term in the expression. This cubic term helps to keep the energy from rising too sharply as the bond is stretched.

$$V_{\text{stretching}} = 143.88 \frac{k_s}{2} (l - l_o)^2 (1 - 2(l - l_o)) \tag{11.16}$$

where k_s is the stretching force constant in $mdyn.A^{-1}$, l_o is the natural bond length in A, l is the actual bond length in A, and 143.88 is to convert the unit to kcal.mol^{-1}.

11.9 Energy Due to Bending

The bending of bonds will also be associated with an increase in energy. The potential energy expression associated with bending is given by:

$$E_\theta = 0.21914k_\theta (\theta - \theta_o)^2 \left(1 + 7 \times 10^8 (\theta - \theta_o)^4\right) \tag{11.17}$$

where k_θ is the force constant associated with bending in $mdyn.A^{-1}rad^{-2}$, θ is the actual bond angle in degrees, θ_o is the natural bond angle in degrees, and 0.21914 is the conversion factor. This potential function works very well for bends of up to about 10 degrees. To handle special cases, such as cyclobutane, special atom types and parameters are used in the force field.

11.10 Energy Due to Stretch-Bend Interactions

When a bond angle is dropped, the two bonds forming the angle will stretch to alleviate the strain. To include such phenomena, cross term (multiple) potential functions are introduced. Cross term potential functions take into account at least two terms such as bond stretching and bond bending. The potential energy expression for this change is given as:

$$E_{s\theta} = 2.51124k_{s\theta} (\theta - \theta_o) \left[(l - l_o)_a + (l - l_o)_b\right] \tag{11.18}$$

$k_{s\theta}$ is the corresponding force constant in $mdyn.A^{-1}rad^{-1}$, a and b represents bonds to a common atom, and 2.51124 is the conversion factor.

11.11 Energy Due to Torsional Strain

Intramolecular rotations (rotations about torsion or dihedral angles) require energy. For example, the conversion of a chair conformer to a boat conformer is endothermic. The torsion potential is a Fourier series that accounts for all 1–4 through-bond relationships:

$$E_{tor} = \frac{V_1}{2}\left(1 + \cos\omega\right) + \frac{V_2}{2}\left(1 + \cos 2\omega\right) + \frac{V_3}{2}\left(1 + \cos 3\omega\right) \tag{11.19}$$

where V_1, V_2 and V_3 are force constants in the Fourier series in $kcal.mol^{-1}$, and ω is the torsion angle from $0°$ to $180°$.

11.12 Energy Due to van der Waals Interactions

The van der Waals radius of an atom is its effective size. As two non-bonded atoms are brought together, the van der Waals attraction between them increases (the van der Waals force and the corresponding energy are inversely proportional to distance). When the distance between them equals the sum of the van der Waals radii the attraction is at a maximum. If the atoms are brought still closer together there is a strong van der Waals repulsion (a sharp increase in energy). The energy expression takes the form of:

$$E_{vdW} = \varepsilon\left[2.90 \times 10^5 \exp\left(-12.50\frac{r_0}{r_v}\right) - 2.25\left(\frac{r_v}{r_0}\right)^6\right] \tag{11.20}$$

where ε is the energy parameter, which determines the depth of the potential energy well (for C–C it is 0.044 while for C–H it is 0.046), r_v is the sum of the van der Waals radii of the interacting atoms, and r_0 is the distance between the interacting centers.

11.13 Energy Due to Dipole-Dipole Interactions

In some force fields electrostatic interactions are accounted for by atomic point charges. In other force fields, such as MM2 and MMX, bond dipole moments are used to represent electrostatic contributions. One can readily see that the equation below stems from Coulomb's law. The energy is calculated by considering all dipole-dipole interactions in a molecule. If the molecule has a net charge, (e.g., NH_4^+) charge-charge and charge-dipole calculations must also be carried out.

$$E_{dipole} = \frac{\mu_i \mu_j}{D\left(r_{ij}\right)^3}\left(\cos\chi - 3\cos\alpha_i \times \cos\alpha_j\right) \tag{11.21}$$

D is the dielectric constant of the system, χ the angle between the dipoles, and μ_i and μ_j the corresponding charges, α_i and α_j the angles between the dipoles and a vector connecting the dipoles. r_{ij} is the distance between the dipoles (Fig. 11.6).

11.14 The Lennard-Jones Type Potential

Real fluids have a continuous intermolecular potential, which can be approximated by the following equation:

$$V(r) = \varepsilon \left[\left(\frac{m}{n-m} \right) x^{-n} - \left(\frac{n}{n-m} \right) x^{-m} \right] \tag{11.22}$$

Here, n and m are constants, $x = r/r_m$, and r_m is the separation corresponding to the minimum energy. The most common form of the Lennard-Jones potential (the LJ-12-6 potential) is obtained when $n = 12$ and $m = 6$. This expression clearly supports the decay of dispersion forces.

11.15 The Truncated Lennard-Jones Potential

It is customary to model the repulsive interactions between hard spheres by a truncated Lennard-Jones potential defined by:

$$V(r) = 4\varepsilon \left[\left(\frac{\sigma}{r} \right)^{12} - \left(\frac{\sigma}{r} \right)^{6} \right] + \varepsilon \text{ if } r \le 2^{1/6}\sigma \tag{11.23}$$

$$V(r) = 0 \text{ if } r > 2^{1/6}\sigma \tag{11.24}$$

The advantage of this potential is that it provides a more realistic representation of repulsive interaction than assuming an infinitely steep potential.

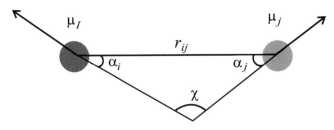

Fig. 11.6 Dipole-dipole interaction

11.16 The Kihara Potential

The Kihara spherical core potential (Maitland et al., 1981) is a slightly more complicated alternative to the LJ potential. The formulation is as follows:

$$V(r) = \infty \text{ if } r \le d \tag{11.25}$$

$$V(r) = 4\varepsilon \left[\left(\frac{\sigma - d}{r - d} \right)^{12} - \left(\frac{\sigma - d}{r - d} \right)^{6} \right] \text{ if } r > d \tag{11.26}$$

where d is the diameter of an impenetrable hard core at which $V(r) = \infty$. The Kihara potential can also be applied to non-spherical molecules by using a convex core of any shape.

11.17 The Exponential -6 Potential

The exponential decay of the intermolecular repulsion can be effectively explained through this potential. The potential is:

$$V(r) = \infty \text{ if } r \le \lambda r_m \tag{11.27}$$

$$V(r) = \frac{\varepsilon}{\left(1 - \frac{6}{\alpha} \right)} \left\{ \frac{6}{\alpha} \exp \left[\alpha \left(1 - \frac{r}{r_m} \right) \right] - \left(\frac{r_m}{r} \right)^{6} \right\} \text{ if } r > \lambda r_m \tag{11.28}$$

where α is the repulsive-wall steepness parameter, ε is the maximum energy of attraction occurring at a separation of r_m, and λr_m is the distance at which the potential goes through a false maximum. The value of λ can be obtained (Hirschfelder et al., 1954) by finding the smallest root of the following equation:

$$\lambda^{7} \exp \left[\alpha \left(1 - \lambda \right) \right] - 1 = 0 \tag{11.29}$$

The false maximum is an unsatisfactory feature of the exp-6 potential. At $r = 0$, the exponential term has a finite value allowing the dispersion term to dominate at very small intermolecular separation. Consequently, the potential passes through a maximum and then tends to $-\infty$ at $r \to 0$. Therefore, the condition that $V(r) = \infty$ when $r \le \lambda r_m$ must be imposed to use the use the potential especially in a simulation. Alternatively, damping functions for the dispersion term have been proposed, which overcome this problem.

11.18 The BFW Two-Body Potential

This is an atom-specific potential, which is applicable to a specific atom or class of atom. For example, the Barker-Fisher-Watts potential for argon is:

$$V(r) = \varepsilon \left[\sum_{i=0}^{5} A_i (x-1)^i \exp\left[\alpha(1-x) \right] - \sum_{j=0}^{2} \frac{C_{2j+6}}{\delta + x^{2j+6}} \right] \tag{11.30}$$

where $x = r/r_m$ and the other parameters are obtained by fitting the potential to experimental data for molecular beam scattering, and long range interaction coefficients. The contribution from s repulsion has an exponential dependence on intermolecular separation and the contribution to dispersion of the C_6, C_8, and C_{10} coefficients are included.

11.19 The Ab Initio Potential

A two body potential can be obtained by fitting a carefully chosen function to data obtained from ab initio calculations. For example, Eggenberger et al. used ab initio calculations to obtain the following potential for the interaction between neon atoms:

$$V(r) = a_1 \exp\left[-a_2 (r/a_0)^2 \right] + a_3 \exp\left[-a_4 (r/a_0)^2 \right] + a_5 \exp\left[-a_6 (r/a_0)^2 \right]$$
$$+ a_7 (r/a_0)^{-10} + a_8 (r/a_0)^{-8} + a_7 (r/a_0)^{-6} \tag{11.31}$$

where a_0 is the Bohr radius and the remaining parameters do not have any physical meaning.

It is interesting to compare the functional similarity of the potential with accurate empirical two-body potential such as the BFW potential. We can observe that all of these potentials have an exponential term and contributions from r^{-6}, r^{-8}, and r^{-10} intermolecular separations.

11.20 The Ionic and Polar Potential

Molecules are associated with permanent multipole moments or charges which result in electrostatic interactions. The application of Coulomb's law of electrostatic interaction between charges q, dipole moments μ, and quadrupole moment Q between molecules a and b yields:

$$V^{(q,q)}(r) = \frac{q_a q_b}{r} \tag{11.32}$$

$$V^{(q,\mu)}(r) = \frac{q_a \mu_b \cos \theta_b}{r^2} \tag{11.33}$$

$$V^{(q,Q)}(r) = \frac{q_a Q_b \left(3 \cos^2 \theta_b - 1\right)}{4r^3} \tag{11.34}$$

$$V^{(\mu,\mu)}(r) = \frac{\mu_a \mu_b \left(2 \cos \theta_a \cos \theta_b - \sin \theta_a \sin \theta_b \cos (\phi_a - \phi_b)\right)}{r^3} \tag{11.35}$$

$$V^{(\mu,Q)}(r) = \frac{3\mu_a Q_b}{4r^4} \left[\cos \theta_a \left(3 \cos^2 \theta_b - 1\right) - 2 \sin \theta_a \sin \theta_b \cos \theta_b \cos (\phi_a - \phi_b)\right] \tag{11.36}$$

$$V^{(Q,Q)}(r) = \frac{3Q_a Q_b}{16r^5} \left[1 - 5 \cos^2 \theta_a - 5 \cos^2 \theta_b - 15 \cos^2 \theta_a \cos^2 \theta_b\right.$$
$$\left. + 2\left[\sin \theta_a \sin \theta_b \cos (\phi_a - \phi_b) - 4 \cos \theta_a \cos \theta_a\right]^2\right] \tag{11.37}$$

where θ_a, θ_b, ϕ_a, and ϕ_b define the various orientation angles between the molecules.

11.21 Commonly Available Force Fields

Some of the commonly available force fields are mentioned below.

11.21.1 MM2, MM3, and MM4

The MM family of force fields (MM2, MM3, and MM4) was introduced by Allinger et al. [2, 3]. and are widely used for the computations of small molecules. The force field can identify sp, sp^2 and sp^3 hybridized carbon atoms, organic intermediates such as free radical and carbocation, the carbonyl functional group, and cyclohydrocarbons such as cyclopropane and cyclopropene (Leach, 2001). The MM family was parameterized to fit values obtained through electron diffraction, which provide mean distances between atoms averaged over vibrational motion at room temperature. The bond stretching potential is represented by the classic Hooke's law expansion:

$$V_{(l)} = \frac{k}{2} (l - l_0)^2 \left[1 - k'(l - l_0) - k''(l - l_0) - k'''(l - l_0) - \ldots\right] \tag{11.38}$$

In MM2, expansion is made up to cubic terms, which may cause the cubic function to pass through a maximum that is far from the reference value. This has lead to disastrous expansion of bonds in some experiments. This defect is overcome in MM3 by limiting the use of the cubic contribution only when the structure is sufficiently close to its equilibrium geometry and is inside the actual potential well. Leach includes a quartic term in MM3, which eliminates the inversion problem and leads to an even better description. MM2 has a similar defect with bond bending and is corrected in MM3. Most of the force fields agree to a point-charge electrostatic model, where the point of origination of a charge is assigned to a particular atom. Hence, the MM family assigns dipoles to the bonds in the molecule. The electrostatic energy is then given by a sum of dipole-dipole interaction energies. This approach can be irresistible for molecules (ions) that have a formal charge and which require charge-charge and charge-dipole terms to be included in the energy expression. The MM family of force fields is often regarded as the "gold standard" as these force fields have been painstakingly derived and parameterized based on the most comprehensive and highest quality experimental data. In MM4, computational problems are negligibly small compared to MM2 and MM3.

11.21.2 AMBER

AMBER (Cornell et al., 1995 [5]) was originally parameterized for a limited number of organic systems and it has been widely used for proteins and nucleic acids. Like other force fields developed for use in modeling proteins and nucleic acids, it uses more specific atom types – specifically, according to Leach, the carbon atom at the junction between a six-and a five-membered ring is assigned an atom type that is different from the carbon atom in an isolated five-membered ring such as histidine, which in turn is different from the atom type of a carbon atom in a benzene ring (Leach, 2001). AMBER can be used for polymers and small molecules with some additional parameters. It generally gives reasonable results for gas-phase model geometries, solvation free energies, vibrational frequencies, and conformational energies. It should be noted that AMBER employs a united atom representation – there does exist an all atom representation of AMBER as well – which differs from an all atom representation in that non-polar hydrogen atoms are not represented explicitly, but are coalesced into the description of the heavy atoms to which they are bonded. This results in significant additional speed in calculations based on AMBER compared to other force fields. AMBER also includes a hydrogen-bond term which augments the value of the hydrogen-bond energy derived from the dipole-dipole interaction of the donor and acceptor groups. However, the contribution of the hydrogen-bond term is only approximately $0.5\,\text{kcal.mol}^{-1}$. It uses general torsion parameters. According to Leach, the energy profile for rotation about a bond that is described by a general torsion potential depends solely upon the atom types of the two atoms that comprise the central bond and not upon the atom types of the terminal atoms. AMBER takes a position midway between those force fields that consistently use more terms for all torsions and those force fields that only use a single term in the torsion expansion. United atom force fields such as AMBER usually use improper torsion terms to maintain stereochemistry at chiral centers. The MM family is an example of a force field that consistently uses more than one term to define the torsion expansion – specifically, it uses three terms. The potential field expression is as follows:

$$V = \sum_{\text{bonds}} \frac{k_l}{2}(l-l_0)^2 + \sum_{\text{angles}} \frac{k_\theta}{2}(\theta-\theta_0)^2 + \sum_{\text{dihedral}} \sum_{n} \frac{V_n}{2}[1+\cos(n\tau-\gamma)]$$

$$+ \sum_{i<j} 4\varepsilon_{ij}\left[\left(\frac{\sigma_{ij}}{r_{ij}}\right)^{12}-\left(\frac{\sigma_{ij}}{r_{ij}}\right)^{6}\right] + \frac{1}{vdW_{\text{scale}}} \sum_{i<j}^{1,4\text{terms}}\left[\left(\frac{\sigma_{ij}}{r_{ij}}\right)^{12}-\left(\frac{\sigma_{ij}}{r_{ij}}\right)^{6}\right]$$

$$+ \sum_{\text{Hbonds}} \left[\frac{C_{ij}}{r_{ij}^{12}}-\frac{D_{ij}}{r_{ij}^{10}}\right] + \sum_{i<j} \frac{q_i q_j}{\varepsilon r_{ij}} + \frac{1}{EE_{\text{scale}}} \sum_{i<j}^{1,4\text{terms}} \frac{q_i q_j}{Dr_{ij}} \qquad (11.39)$$

11.21.3 CHARMM

CHARMM (Chemistry at Harvard Macromolecular Mechanics, developed by Mackerell and Karplus, et al., 1995) was parameterized by experimental data. It has been used widely for simulations ranging from small molecules to solvated complexes of large biological macromolecules. CHARMM performs well over a broad range of calculations and simulations, including the calculation of interaction and conformation energies, geometries, local minima, time-dependent dynamic behavior, and barriers to rotation, vibrational frequencies, and free energy. CHARMM uses a flexible and comprehensive energy function:

$$E_{(\text{pot})} = E_{\text{bond}} + E_{\text{torsion}} + E_{\text{oop}} + E_{\text{elect.}} + E_{\text{vdW}} + E_{\text{constraint}} + E_{\text{user}} \qquad (11.40)$$

where the out-of-plane (OOP) angle is an improper torsion. The van der Waals term is derived from rare-gas potentials, and the electrostatic term can be scaled to mimic solvent effects. Hydrogen-bond energy is not included as a separate term as in AMBER. Instead, hydrogen-bond energy is implicit in the combination of van der Waals and electrostatic terms.

11.21.4 Merck Molecular Force Field

The Merck molecular force field (MMFF) (Halgren, 1996) is similar to MM3 but differs in focus on its application to condensed-phase processes in molecular dynamics. It achieves MM3-like accuracy for small molecules and is applicable to proteins and other systems of biological significance. It is designed to be a transferable force field for pharmaceutical compounds that accurately treats conformational energetics and nonbonded interactions. This force field is adequate for both gas phase and condensed phase calculations. It has a large number of cross terms, which is the major reason for its transferability. The internal bonded terms used in this force field are bonds, angles, stretch-bend, out-of-plane bending, and dihedrals. Nonbonded terms include van der Waals and electrostatic. Energy expressions based on these terms are given below.

11.21.4.1 Bond

$$E_{\text{bond}} = k_{\text{bond}} \left(r_{ij} - r_{ij}^0 \right)^2 \cdot \left(1 + cs \left(r_{ij} - r_{ij}^0 \right) + \frac{7}{12} cs^2 \left(r_{ij} - r_{ij}^0 \right)^2 \right) \qquad (11.41)$$

where k_{bond} is the force constant, r_{ij} is the bond length between atoms i and j, and cs is the cubic stretch constant.

11.21.4.2 Angle Bending

$$E_{\text{angle}} = k_\theta \left(\theta_{ijk} - \theta_{ijk}^0 \right)^2 \cdot \left(1 + cb \left(\theta_{ijk} - \theta_{ijk}^0 \right) \right) \qquad (11.42)$$

where k_θ is the force constant, θ_{ijk} is the bond angle between I, j, and k, and cb is the cubic bent constant $\left(-0.007^{0-1} \right)$.

11.21.4.3 The Near Linear/Linear Angle

$$E_{\text{angle,linear}} = k_{ijk\text{linear}} \left(1 + \cos \theta_{ijk} \right) \qquad (11.43)$$

11.21.4.4 Stretch-Bend

$$E_{\text{stretch-bending}} = \left(k_{ijk} \left(r_{ij} - r_{ij}^0 \right) + k_{kji} \left(r_{kj} - r_{kj}^0 \right) \right) \left(\theta_{ijk} - \theta_{ijk}^0 \right) \qquad (11.44)$$

Here, k_{ijk} and k_{kji} are the force constants coupling the ij and kj stretches to the ijk angle.

11.21.4.5 OOP Bending

$$E_{\text{oop}} = k_{\text{oop}} \left(\chi_{ijk;l} \right)^2 \qquad (11.45)$$

$\left(\chi_{ijk;l} \right)$ is known as the Wilson wag, which is the angle between the bond jl and the plane ijk, where j is the central atom. A typical example of OOP bending is at the tricoordinate centers (e.g., the benzene ring).

11.21.4.6 Dihedral/Torsional

$$E_{\text{torsion}} = 0.5 \left(V_1 \left(1 + \cos \phi \right) + V_2 \left(1 + \cos 2\phi \right) + V_3 \left(1 + \cos 3\phi \right) \right) \qquad (11.46)$$

Here, the V terms are the force constants for the terms in the Fourier series and ϕ is the dihedral angle.

11.21.4.7 van der Waals (Buffered 14-7)

$$E_{vdW} = \varepsilon_{ij} \left(\frac{1.07R_{ij}^*}{R_{ij} + 0.07R_{ij}^*} \right)^7 \left(\frac{1.12R_{ij}^{*7}}{R_{ij}^7 + 0.07R_{ij}^{*7}} - 2 \right) \tag{11.47}$$

Here, R_{ij} is the distance between atoms i and j, R_{ij}^* is the minimum interaction energy distance between the atoms (based on parameterized atomic polarizability), ε_{ij} is the well depth between the atoms (based on the Slater-Kirkwood expression, including the polarizability and the number of electrons).

11.21.4.8 Electrostatic

$$E_{electrostatic} = \frac{q_i q_j}{D(R_{ij} + \delta)^n} \tag{11.48}$$

Here, D is the Dielectric constant, δ is the electrostatic buffering constant $(= 0.05)$, and q_i and q_j are the charges on atoms i and j. The charge on any atom is given by:

$$q_i = q_i^0 + \sum \omega_{ki} \tag{11.49}$$

q_i^0 is the formal atomic charge (usually 0) and ω_{ki} terms are bond charge increments summed over all the covalent bonds to the atom i.

11.21.4.9 Internal Parameters Used in MMFF

1. MP2/6-31G* optimized conformations encompassing 360 compounds and later tested on a set of ca. 700 conformations.
2. Geometries of molecules.
3. Vibrational spectra.
4. Conformational energetics (relative energies if minima).
5. Nonbond parameters.
6. VdW terms optimized based on high level ab initio dimer calculations.
7. (MP4(SDTQ) with Sadlej's "medium polarized" basis set ($10s$, $6p$, $4d/5s$, $4p$) contracted to ($5s$, $3p$, $2d/3s$, $2p$).
8. Electrostatic terms based on 70 dimer interaction energies and geometries at the HF/6-31G* level.

11.21.5 The Consistent Force Field

The consistent force field (CFF) family (Maple and Hagler, 1994) was developed by Halgren and the Biosym Consortium. These force fields have anharmonic and cross term enhancements. Furthermore, these force fields are derived at their core from ab initio methods rather than from purely experimental data. They were developed to mimic peptide and protein properties. The CFF force fields use quartic polynomials for bond stretching and angle bending. For torsions they use a three-term Fourier expansion. The van der Waals interactions are represented by using an inverse 9th-power term for repulsive behavior instead of the more customary 12th-power term. Hagler, precursory to the development of the CFF force field, showed that no explicit hydrogen bond term is required to accurately model hydrogen-bonding interactions, as the combination of electrostatic and van der Waals calculations sufficiently captured the hydrogen-bonding contributions. This enabled significant simplification in deriving many recently developed force fields. The development of CFF was the first major force field developed based upon ab initio quantum mechanical calculations on small molecules, although not as broadly applied as for the more recent MMFF94. The quantum mechanics calculations were performed on structures distorted from equilibrium in addition to the expected calculations on structures at equilibrium. This yielded a wealth of data for fitting and parameterization [1].

11.22 Some Other Useful Potential Fields

1. GROMACS – This force field is optimized in the package of the same name.
2. GROMOS – A force field that comes as part of the GROMOS (GROningen MOlecular Simulation package), a general-purpose molecular dynamics computer simulation package for the study of biomolecular systems. The GROMOS force field (A-version) has been developed for application to aqueous or apolar solutions of proteins, nucleotides, and sugars. However, a gas phase version (B-version) for simulation of isolated molecules is also available.
3. OPLS-aa, OPLS-ua, OPLS-2001, OPLS-2005 – Members of the OPLS family of force fields developed by William L. Jorgensen at the Yale University Department of Chemistry.
4. ENZYMIX – A general polarizable force field for modeling chemical reactions in biological molecules. This force field is implemented with the empirical valence bond (EVB) method and is also combined with the semimacroscopic PDLD approach in the program in the MOLARIS package.
5. ECEPP/2 – The first force field for polypeptide molecules, developed by F. A. Momany, H. A. Scheraga and colleagues.
6. QCFF/PI – A general force field for conjugated molecules.
7. CFF/ind and ENZYMIX – The first polarizable force field which has subsequently been used in many applications to biological systems.

8. PFF (Polarizable Force Field) – Developed by Richard A. Friesner and coworkers.
9. DRF90 – Developed by P.Th. van Duijnen and coworkers.
10. SP-basis Chemical Potential Equalization (CPE) approach – Developed by R. Chelli and P. Procacci.
11. CHARMM polarizable force field – Developed by B. Brooks and coworkers.
12. The SIBFA (Sum of Interactions Between Fragments Ab initio computed) force field for small molecules and flexible proteins – Developed by Nohad Gresh (Paris V, René Descartes University) and Jean-Philip Piquemal (Paris VI, Pierre & Marie Curie University). SIBFA is a molecular mechanics procedure formulated and calibrated on the basis of ab initio supermolecule computations.
13. AMOEBA force field – Developed by Pengyu Ren (University of Texas at Austin) and Jay W. Ponder (Washington University).
14. ORIENT procedure – Developed by Anthony J. Stone (Cambridge University) and coworkers.
15. Non-Empirical Molecular Orbital (NEMO) procedure – Developed by Gunnar Karlström and coworkers at Lund University.
16. Gaussian Electrostatic Model (GEM) – A polarizable force field based on density fitting developed by Thomas A. Darden and G. Andrés Cisneros at NIEHS, and Jean-Philip Piquemal (Paris VI University).
17. Polarizable procedure – Based on the Kim-Gordon approach developed by Jürg Hutter and coworkers (University of Zurich)
18. ReaxFF – A reactive force field developed by William Goddard and coworkers. It is fast, transferable, and is the computational method of choice for atomistic-scale dynamical simulations of chemical reactions.
19. EVB (empirical valence bond) – This reactive force field, introduced by Warshel and coworkers, is a reliable way of using force fields in modeling chemical reactions in different environments. The EVB facilitates calculations of actual activation free energies in condensed phases and in enzymes.
20. VALBOND – A function for angle bending that is based on the valence bond theory and works for large angular distortions, hypervalent molecules, and transition metal complexes.

11.23 The Merits and Demerits of the Force Field Approach

The power of the force field approach can be listed as follows:

1. Force field-based simulations can handle large systems, and are several orders of magnitude faster (and cheaper) than quantum-based calculations.
2. The analysis of the energy contributions can be done at the level of individual or classes of interactions.
3. The modification of the energy expression can be done to bias the calculation.

Table 11.2 Information available from computational methods

Data item	Molecular mechanics	Semi-empirical	Ab initio
Heat of formation	YES	YES	YES
Entropy of formation	YES	YES	YES
Free energy of formation	YES	YES	YES
Heat of activation	NO	YES	YES
Entropy of activation	NO	YES	YES
Free energy of activation	NO	YES	YES
Heat of reaction	YES	YES	YES
Entropy of reaction	YES	YES	YES
Free energy of reaction	YES	YES	YES
Strain energy	YES	NO	NO
Vibrational spectra	NO	YES	YES
Dipole moment	NO	YES	YES
Geometry optimization	YES	YES	YES
Electronic bond order	NO	YES	YES
Electronic distribution	NO	YES	YES
Mulliken population analysis	NO	YES	YES
Transition state location	NO	YES	YES

Applications beyond the capability of classical force field methods include:

1. Electronic transitions (photon absorption).
2. Electron transport phenomena.
3. Proton transfer (acid/base reactions).

A comparison of the computing facility of MM methods with ab initio and semi-empirical methods can be seen in Table 11.2.

11.24 Parameterization

In addition to the functional form of the potentials, a force field defines a set of parameters for each type of atom. For example, a force field would include distinct parameters for an oxygen atom in a carbonyl functional group and in a hydroxyl group. The typical parameter set includes the following.

1. Atomic mass.
2. van der Waal's radii.
3. Partial charge for individual atoms.
4. Bond length.
5. Bond angle.
6. Dihedral angles for pairs, triplets, and quadruplets of bonded atoms.
7. Effective spring constant for each force constant.

Most current force fields use a "fixed-charge" model by which each atom is assigned a single value for the atomic charge that is not affected by the local electrostatic environment; proposed developments in next-generation force fields incorporate models for polarizability, in which a particle's charge is influenced by electrostatic interactions with its neighbors. For example, polarizability can be approximated by the introduction of induced dipoles; it can also be represented by Drude particles, or massless, charge-carrying virtual sites attached by a spring like harmonic potential to each polarizable atom. The introduction of polarizability into force fields in common use has been inhibited by the high computational expense associated with calculating the local electrostatic field.

Parameter sets and functional forms are defined by force field developers to be self-consistent. Because the functional forms of the potential terms vary extensively between even closely related force fields (or successive versions of the same force field), the parameters from one force field should never be used in conjunction with the potential from another.

11.25 Some MM Software Packages

A number of software packages are available for MM studies; the most important among them are listed in Table 11.3.

11.26 Exercises

1. If the O–H bond distance calculated from the MM3 parameter set for a water molecule is 94.7 pm and the H–O–H bond has an angle of 105°, compute the distance between the nuclei of hydrogen atoms of water in the gas phase. Calculate the moment of inertia of water molecule about the principal axis.

2. What is the MM3 standard enthalpy of formation at 298.15 K of styrene? Is the minimum-energy structure planar, or does the ethylene group move out of the plane of the benzene ring?

3. Cyclopentadiene (Fig. 11.7) dimerises to produce specifically the endo dimer (2) rather than the exo dimer (1). The hydrogenation of this dimer proceeds to give initially one of the dihydro derivatives (3) or (4). Only after prolonged hydrogenation is the tetrahydro derivative formed. Compute the geometries and energies of all four species (1–4). Compare their thermodynamic functions. (The relative stabilities of the pairs of compounds 1/2 and 3/4 should indicate which of each pair is the less strained and/or hindered in a thermodynamic sense). The observed reactivity towards cyclodimerisation and hydrogenation can of course be due to either thermodynamic (i.e., product stability) or kinetic (i.e., transition state stability) factors. In pericyclic reactions in particular, stereoselectivity is

Table 11.3 Important software for MM studies

Package name	Creator
AMBER	Peter Kollman, University of California, San Francisco
AMMP	Rob Harrison, Thomas Jefferson University, Philadelphia
ARGOS	Andy McCammon, University of California, San Diego
BOSS	William Jorgensen, Yale University
BRUGEL	Shoshona Wodak, Free University of Brussels
CFF	Shneior Lifson, Weizmann Institute
CHARMM	Martin Karplus, Harvard University
CHARMM/GEMM	Bernard Brooks, National Institutes of Health, Bethesda
DELPHI	Bastian van de Graaf, Delft University of Technology
DISCOVER	Molecular Simulations Inc., San Diego
DL_POLY	W. Smith & T. Forester, CCP5, Daresbury Laboratory
ECEPP	Harold Scheraga, Cornell University
ENCAD	Michael Levitt, Stanford University
FANTOM	Werner Braun, University of Texas, Galveston
FEDER/2	Nobuhiro Go, Kyoto University
GROMACS	Herman Berendsen, University of Groningen
GROMOS	Wilfred van Gunsteren, BIOMOS and ETH, Zurich
IMPACT	Ronald Levy, Rutgers University
MACROMODEL	Schodinger, Inc., Jersey City, New Jersey
MM2/MM3/MM4	N. Lou Allinger, University of Georgia
MMC	Cliff Dykstra, Indiana Univ. and Purdue Univ. at Indianapolis
MMFF	Tom Halgren, Merck Research Laboratories, Rahway
MMTK	Konrad Hinsen, Inst. of Structural Biology, Grenoble
MOIL	Ron Elber, Cornell University
MOLARIS	Arieh Warshal, University of Southern California
MOLDY	Keith Refson, Oxford University
MOSCITO	Dietmar Paschek & Alfons Geiger, University of Dortmund
NAMD	Klaus Schulten, University of Illinois, Urbana
OOMPAA	Andy McCammon, University of California, San Diego
ORAL	Karel Zimmerman, INRA, Jouy-en-Josas, France
ORIENT	Anthony Stone, Cambridge University
PCMODEL	Kevin Gilbert, Serena Software, Bloomington, Indiana
PEFF	Jan Dillen, University of Pretoria
Q	Johan Åqvist, Uppsala University
SIBFA	Nohad Gresh, INSERM, CNRS, Paris
SIGMA	Jan Hermans, University of North Carolina
Tinker	Jay William ponder, Washington University School of Medicine

controlled by the electronic properties of the molecules (stereoelectronic control), and hence can only be understood in terms of molecular wavefunction. On the basis of the results obtained from the molecular mechanics technique, predict whether the cyclodimerisation of cyclopentadiene and the hydrogenation of the dimer is kinetically or thermodynamically controlled [4].

4. The PCBs are a family of chlorinated biphenyls that are claimed to have all sorts of evil properties, none of which have been proven for humans. Of particular interest is 2,3,4,3',4'-pentachlorobiphenyl, which is referred to by biologists

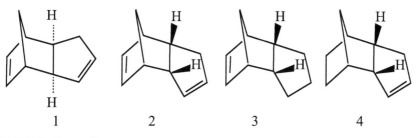

Fig. 11.7 Cyclopentadiene

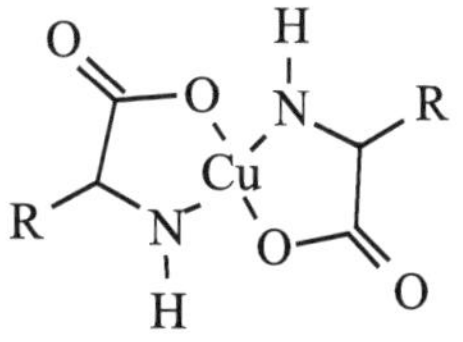

Fig. 11.8 Copper (II) complex of amino acid

as a "coplanar biphenyl", and argued, as a consequence of its coplanarity, to have toxicity comparable to dioxins. Is it coplanar? If not, what would be the energetic cost of making it coplanar? What happens to the coplanarity if you remove some of the chlorines? Follow MM modeling techniques to make the computation.

5. Copper (II) complexes of amino acids have the general structure as shown in Fig. 11.8. Make a computational chemistry study to predict whether the ligands around copper are placed in a square planar or tetrahedral manner. Does this depend upon the nature of amino acid coordinates? With two stereogenic centers, this kind of complex can exist in diastereomeric forms. Can both be formed from a single enantiomer of the amino acid? What is the energy difference between them (this is important because such complexes are sometimes used to resolve racemic amino acids)?

References

1. ChemShell, a Computational Chemistry Shell. See http://www.chemshell.org
2. Allinger NL, Yuh YH, Lii J-H (1989) J Am Chem Soc 111: 8551
3. Allinger NL, Kuohsiang C, Lii J-H (1996) J Comp Chem 17 pp 642–668
4. Beachy MD, Chasman D, Murphy RB, Halgren TA, Friesner RA (1997) J Am Chem Soc 119 pp 5908–5920
5. Cornell WD, Cieplak P, Bayly CI, Gould IR, Merz KM Jr, Ferguson DM, Spellmeyer DC, Fox T, Caldwell JW, Kollman PA (1995) J Am Chem Soc 117 pp 5179–5197

Chapter 12
The Modeling of Molecules Through Computational Methods

12.1 Introduction

Performing a geometry optimization is the primary step in studying a molecule using computational techniques. Geometry optimizations classically attempt to locate a minimum on the potential energy surface in order to foretell the equilibrium structures of molecular systems. They may also be used to locate transition structures or intermediate structures. Moreover, the geometry of a molecule determines many of its physical and chemical properties. We know that the energy of a molecule changes with its structure. Hence, understanding the methods of geometry optimization is the major requirement for energy minimization. It is essential to understand the geometry of a molecule before running computations.

12.2 Optimization

Optimization modeling can be carried out by identifying the objectives, the design variables, and the constraints, and by using an algorithm to find the solution to the problem. Optimality conditions will help us to determine whether we have indeed reached our goal of an optimum solution.

12.2.1 Multivariable Optimization Algorithms

Optimization problems, which we come across in molecular modeling, are multivariable problems, where the objective functions have more than a single variable on which the given function depends on. If we consider a "two variable problem", say, $f(x) = x_1^2 + x_2^2$, and say $x_1 = 3$ and $x_2 = 4$, then every x_1 and x_2 has a function value (i.e., height). This function can be represented by a surface (Fig. 12.1). We

K. I. Ramachandran et al., *Computational Chemistry and Molecular Modeling*   229
DOI: 10.1007/978-3-540-77304-7, ©Springer 2008

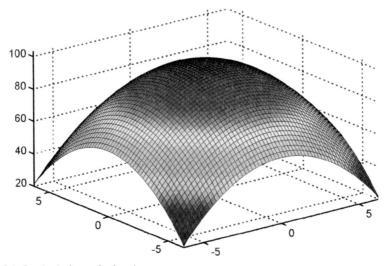

Fig. 12.1 Quadratic form of a function

have to find the minimum value of the function and at what values of x_1 and x_2 is the minimum value attained. For example, the minimum occurs at $f(x) = 0$ and occurs when $x_1 = x_2 = 0$. We can also put constraints such as $x_1 + x_2 = 5$, in which case the solution must lie on the line of constraint. However, we will be discussing only unconstrained problems now.

12.2.2 Level Sets, Level Curves, and Gradients

The function values under study are represented as contour maps with circles representing each function value (Fig. 12.2). Any function $f(x) = C$ is a level set, which is a set of points having the same height. These contours are called *level sets* or *level curves*. At any point on the circle or curve, the function value will be the same (Fig. 12.3). The outermost contour will have the highest function value and the inner circles will progressively have smaller and smaller values.

At the bottommost point, the function will have zero value and is said to be the minimum at that point. At each point on the curve, there are gradients, given by $\nabla f(x)$, pointing to the steepest direction. The direction of steepest descent is given by $-\nabla f(x)$, which we get by searching in the opposite direction. The contour map is a vector field, with gradients at every point. The gradients are always tangent to the level surface. There is a plane tangential to the point, and the gradient will always be orthogonal to the plane. Maximum and minimum points in the contour map are shown by Fig. 12.4.

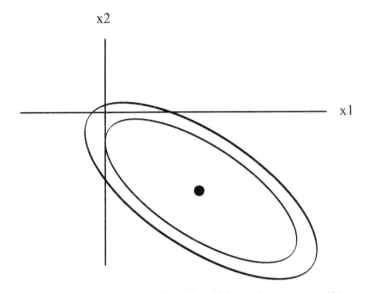

Fig. 12.2 Contours of the quadratic form. Each ellipsoidal curve has a constant $f(x)$

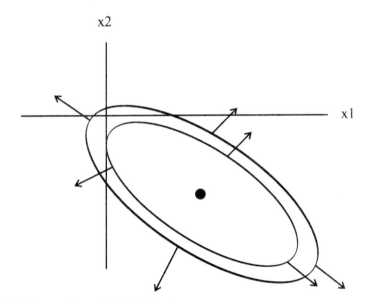

Fig. 12.3 Vectors are tangent to the level surface

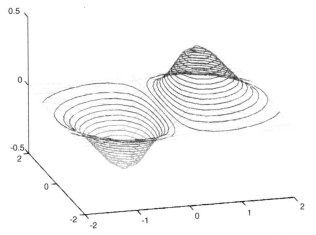

Fig. 12.4 Maximum and minimum points on the contour map (generated from MATLAB)

12.2.3 Optimality Criteria

The optimality criteria for multivariable functions are different, as compared to uni-variable functions (although the definition of local, global, and inflection points still hold). The gradient function is a vector quantity and not a scalar quantity as in uni-variable functions. We derive the optimal criteria by using the definition of the local optimal point and using the Taylor Series expansion of the multivariable function.

The objective function is a function on N variables, represented by $x_1, x_2, \ldots, x_n$ The gradient vector at any point x is represented by $\nabla f(x)$, which is an N-dimensional vector given as follows:

$$\nabla f(x) = \text{Partial derivatives of } f(x) \text{ with respect to } x_1, x_2, \ldots, x_n$$

For a two-dimensional case, the gradient (first derivative) of $f(x) = x_1^2 + x_2^2$ will be:

$$\begin{bmatrix} \dfrac{\partial f}{\partial x_1} \\[2mm] \dfrac{\partial f}{\partial x_2} \end{bmatrix} = \begin{bmatrix} 2x_1 \\ 2x_2 \end{bmatrix}. \tag{12.1}$$

The first order partial derivatives are calculated using the central difference method. The second order derivatives in multivariable functions form a matrix $\nabla f(x)$, better known as the Hessian matrix. A point $x*$ is a stationary point if $\nabla f(x) = 0$ and the point is a minimum, maximum or an inflection point if $\nabla^2 f(x)$ is positive-definite, negative-definite or otherwise.

A matrix $\nabla^2 f(x)$ is defined to be positive-definite if, for any point y in the search space, the quantity $y^T \nabla^2 f(x) y > 0$ (or $y^T A y > 0$), where A is a symmetric, positive definite matrix. The matrix is said to be positive definite if all the eigenvalues of the matrix or all the principal derivatives are positive. In our case, we are interested in the matrix A being positive definite and our principles and calculations are based on this assumption.

A matrix $\nabla^2 f(x)$ is defined to be negative definite if, for any point y in the search space, the quantity $y^T \nabla^2 f(x)y \leq 0$ (or $y^T Ay < 0$) where A is a symmetric, positive definite matrix. The negative definiteness can also be verified by testing the positive definiteness of $-A$.

If the matrix A is positive or negative definite at only some points, but not uniformly across, then it is neither positive definite nor negative definite.

12.2.4 The Unidirectional Search

We use the successive unidirectional search along each component of a vector to find the minimum along a search direction. It is a one-dimensional search, performed by comparing the function values along a specific direction. The search is performed, for a point x^t, along a search direction s^t. Only points lying on a N-Dimensional line, passing through x^t and oriented along the search direction s^t, are considered. The derivative of this function is called a directional derivative. Any point on this line can be expressed as:

$$x(\alpha) = x^t + \alpha S^t \tag{12.2}$$

α is a scalar quantity which specifies the distance of $x(\alpha)$ from x^t, $x(\alpha)$ is a vector specifying all the design variables $x_i(\alpha)$. α can be positive or negative; when $\alpha = 0$, $x(\alpha) = x^t$.

12.2.5 Finding the Minimum Point Along S^t

To find the minimum point along s^t, the following steps are used.

1. Rewrite the multivariable function in terms of a single variable.
2. Substitute each x_i by $x_i(\alpha)$, as given in the above equation.
3. Use single variable search methods to find the minimum along this line. (Generally the binding phase method is used for bracketing and the golden search method is used to find the specific minimum).
4. Once we find the optimum α^*, we can find the point $x(\alpha)$, using Eq. 12.2.

Multivariable optimizations can be done with the help of algorithms which makes use of two types of methods: direct search methods and the gradient-based methods. In the optimization problems of computational chemistry, the latter is found to be more reliable due to the following reason: direct search methods need many function evaluations to converge to a solution. Hence, gradient based methods are faster than direct search methods. Thus, our discussion in this regard will be limited to gradient-based methods only.

12.2.6 Gradient-Based Methods

These methods use the derivative values of the objective functions in the algorithms. Many objective functions are not differentiable, so, the derivatives cannot be applied directly. We cannot apply the algorithms to discrete or discontinuous functions. Efficient algorithms can be used if the derivative is available. The gradients can also be calculated numerically. These concepts are very complex to be applied directly, especially for multivariate functions, where there are many interactions between the variables.

The algorithms require first derivative, second derivative, or sometimes both values. The derivative values are calculated at neighboring points only using the central difference theorem.

By definition, the derivative $\nabla f(x)$ at x^t is the direction of maximum increase (steepest ascent) of the function $f(x)$. So, to find the minimum, we need to travel in a direction opposite to that of the maximum descent, which is the steepest descent direction given by $-\nabla f(x)$. The function value will decrease rapidly, as we move in that direction.

A search direction d^t, is a descent direction at a point x^t, if the condition $\nabla^2 f(x^t).d^t < 0$ is satisfied in the vicinity of point x^t. There are several ways by which we can approach the problem, using gradient methods. Some of them are listed below.

1. Cauchy's steepest descent method (the algorithm).
2. Newton's method.
3. Marquardt's method.
4. The conjugate gradient method.
5. The steepest descent method.
6. The conjugate directions method.

Conjugate gradient methods are iterative methods used in the solution of equations of the type:

$$Ax = b \qquad\qquad (12.3)$$

where A is a known symmetric, positive definite or indefinite matrix, and b is a known vector. The same problem can be expressed as a convex scalar quadratic equation, of the form:

$$f(x) = \frac{1}{2}x^T Ax - b^T x + c \qquad\qquad (12.4)$$

12.2.6.1 Major Definitions Used in Derivations

Inner products: $x^T y = \sum x_i y_i$

$$x^T y = y^T x$$

$x^T y = 0$ if x, y are orthogonal to each other.

$$(AB)^T = B^T A^T$$
$$(AB)^{-1} = B^{-1} A^{-1}$$

Expressions that reduce to a 1-by-1 matrix such as $x^T A x$ are scalar quantities. Matrix A is positive definite if, for every non zero vector x, we have:

$$x^T A x > 0 \qquad (12.5)$$

We use contours of the quadratic form. Ellipsoidal curves have a constant $f(x)$, as shown in Fig. 12.3.

$\nabla f(x)$ is the first derivative of the quadratic form. For every point x, the gradient points in the direction of the steepest increase lead to an increase in $f(x)$. Gradients are perpendicular to contour lines. The gradient, given by the first derivative, $\nabla f(x)$, is a vector field, where each vector points towards the direction of the steepest increase of $f(x)$. The gradient at the bottom of this field is zero. So, to minimize $f(x)$, set the gradient $\nabla f(x) = 0$. Integrating Eq. 12.3, we get $\nabla f(x)$ (steepest increase direction) and $-\nabla f(x)$ (steepest descent direction):

$$\nabla f(x) = \frac{1}{2} A^T x + \frac{1}{2} A x - b = A x - b \qquad (12.6)$$
$$(\text{using } A^T = A, \text{ since } A \text{ is symmetric})$$

So, at a minimum, set the first derivative to be zero. Hence:

$$\nabla f(x) = A x - b = 0 \qquad (12.7)$$

We need to solve the equation $A x = b$. If A is a symmetric, positive definite matrix, the solution is a minimum of $f(x)$. Even if A is not symmetric, still we will have $\frac{1}{2}(A^T + A)$ in the formula, which makes it into a symmetric matrix. The solution of $A x = b$ is a critical point of $f(x)$. So, the minimum point of the function is the solution to the set of problems of type $A x = b$. The solution of the function lies in the intersection point of n-hyper planes, each of dimension $(n-1)$. For the two-dimensional case, the solution is the intersection of two lines. In summary, to solve $A x = b$, find an x that minimizes $f(x)$.

12.2.7 The Method of Steepest Descent

In this method, we start at some arbitrary point x_0 and proceed to move towards the minimum point, in the direction of steepest descent, $-\nabla f(x)$. We start our trial at x_0, then slide to x, the minimum point. Take steps $x_1, x_2, \ldots, x_n$ until we are close to x. Take steps in the direction of steepest descent which is $-\nabla f(x) = b - A x$. The definitions used are as follows:

1. *Error* – It tells us how far the current point is away from the real optimum point. It can be computed from Eq. 12.8.

$$e_i = x_i - x \qquad (12.8)$$

2. *Residual* – It tells us how far we are from the correct value b. It is computed from Eq. 12.9.

$$r_i = b - Ax_i$$
$$= -Ae_i$$
$$= -\nabla f(x) \tag{12.9}$$

This is the direction of the steepest descent:

$$\nabla f(x) = -r_i \tag{12.10}$$

The residual r_i is the error transformed by A into the same space as b. In the first trial, we make the movement from x_0 to x_1 where:

$$x_1 = x_0 + \alpha r_0 \tag{12.11}$$

We need to find α by using the line search method and choose α to minimize f along the line of steepest descent. The path is given by a line created by the intersection of a plane and a paraboloid. So, α minimizes f and is computed by finding the first derivative of $f(x)$ and is set to zero (Fig. 12.5).

According to the chain rule the first derivative of the function with respect to α is:

$$df(x)/d\alpha = d/dx[f(x)^T]d/dx[f(x_1)] \tag{12.12}$$

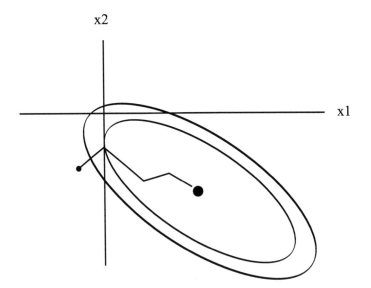

Fig. 12.5 Method of steepest descent

In the differentiating equation (Eq. 12.11) with respect to α we get r_0 for the last term:

$$df(x)/d\alpha = f'(x_1)Tr_0 \qquad (12.13)$$

So, we need to choose α so that $f(x)^T r_0$ are perpendicular. Therefore, the derivative at the new point x_1 is perpendicular to r_0, the residual at x_0. Now, we need to find the value of α. We need to express α in terms of r_0 values, since r_0 is known. We have seen that $f'(x_1)r_0 = 0$, $\nabla f(x) = f'(x_1)^T = -r_i$.

$$\text{But: } r_1^T r_0 = 0 \qquad (12.14)$$

(multiplying both sides by -1):

$$(b - Ax_1)^T r_0 = 0 \qquad (12.15)$$

(Expanding r_i from residual value)

$$[b - A(x_0 - \alpha r_0)]^T r_0 = 0 \qquad (12.16)$$

(Expanding x_i from Eq. 12.12)

$$\left[(b - Ax_0)^T r_0 - \alpha (Ar_0)^T r_0\right] = 0 \qquad (12.17)$$

$\left(\text{Applying the transpose rule for } (AB)^T\right)$:

$$(b - Ax_0)^T r_0 = \alpha (Ar_0)^T r_0 \qquad (12.18)$$

$$\alpha = r_0^T r_0 / r_0^T Ar_0 \qquad (12.19)$$

So, putting it all together, the method of steepest descent method can be generalized as computing:

$$r_i = b - Ax_i \qquad (12.20)$$

$$\alpha_i = r_i^T r_i / r_i^T Ar_i \text{ and} \qquad (12.21)$$

$$x_{i+1} = x_i + \alpha_i r_i . \qquad (12.22)$$

Here, we have to calculate values of r_i and α_i for each x_{i+1}. So, for each new value of x, we need to compute r_i, which has one matrix-vector multiplication (Ax_i) and to compute α_i, which has another matrix vector (Ar_0) multiplication.

In order to reduce the number of matrix-vector multiplications, we multiply by A and add b to Eq. 12.21. Although Eq. 12.20 still needs to compute r_0, the Ar_i in Eq. 12.20 and in Eq. 12.23 needs to be computed only once for each iteration.

$$-A(x_{i+1}) = -Ax_i + \alpha_i Ar_i \qquad (12.23)$$

Adding b, we have:

$$b - A(x_{i+1}) = b - Ax_i + \alpha_i Ar_i \qquad (12.24)$$

$$r_{1+1} = r_1 - \alpha_i Ar_i \qquad (12.25)$$

Points to note:

1. The convergence pattern is zigzag. Each gradient is perpendicular to the previous gradient.
2. The cost of computation is two matrix vector multiplications per iteration.
3. The algorithm is dominated by matrix-vector products.
4. We can eliminate one A by pre-multiplying by $-A$ and, adding b to both sides, we get:

$$r_{i+1} = r_i - \alpha_i A r_i \tag{12.26}$$

Then, $A r_i$ will be calculated only once per iteration and used in Eqs. 12.21 and 12.24. However, the major disadvantages of this reduction in computation are:

1. The absence of feedback on x_i.
2. The accumulation of a floating point round off error. This causes x_i to converge near x.

We can minimize these disadvantages by using Eq. 12.20 periodically to recompute the correct residual. The steepest descent converges to the exact solution on the first iteration either if the error term is an eigenvector or error values are all equal.

12.2.8 The Method of Conjugate Directions

The steepest descent method takes search steps in the same direction more than once. Instead of that, if we had orthogonal search directions, then that would have the following advantages:

1. This takes only one step per direction.
2. Proceed in that direction only, which reduces the computation time. For example, in a two-dimensional problem, only two steps will be required.

For each step choose a point:

$$x_{i+1} = x_i + \alpha_i d_i \tag{12.27}$$

While computing α_i, make sure that error e_i is perpendicular to d_i:

$$\text{So } d_i^T e_{i+1} = 0 \tag{12.28}$$

$$d_i^T (e_i + \alpha_i d_i) = 0 \tag{12.29}$$

$$d_i^T e_i + d_{i+1}^T \alpha_i d_i = 0 \tag{12.30}$$

$$\alpha_i = \frac{-d_i^T e_i}{d_i^T d_i} \tag{12.31}$$

This expression requires e_i to be known to us. This complexity can be avoided by choosing A-orthogonal (conjugate) directions (Fig. 12.6).

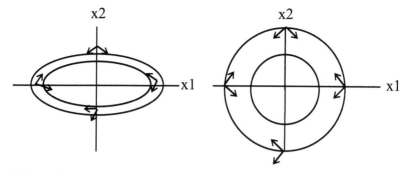

Fig. 12.6 A-orthogonal vectors

A set of non-zero vectors $(d_0, d_1, \ldots)$ is said to be conjugate, with a symmetric positive definite matrix A, if $d_i^T A d_j = 0$ for all $i \neq j$. So, given $x_0 \varepsilon R^n$ and a set of conjugate directions $d_0, d_1, \ldots$, we can generate a sequence $\{x_i\}$ by setting:

$$x_{i+1} = x_i + \alpha d_i \tag{12.32}$$

By making use of conjugacy we can minimize $f(x)$ in n steps, by successively minimizing it along the individual directions in the conjugate set.

If α_i is a 1-D minimizer of the quadratic function $f(x)$, along $x_i + \alpha d_i$ given by:

$$\alpha_i = \frac{-d_i^T e_i}{d_i^T d_i} = \frac{d_i^T r_i}{d_i^T A d_i} \tag{12.33}$$

the sequence $\{x_i\}$ generated by this algorithm converges to solution $x*$ in "n" steps. Successive minimizations along the co-ordinate directions will minimize with a diagonal Hessian in "n" iterations.

If A is a diagonal, the contours of the quadratic functions are aligned with the coordinate directions. So, we can find the minimum by performing the one-dimensional minimizations along the co-ordinate directions $e_1, e_2, \ldots, e_n$ in n steps. So, the new requirement is that e_{i+1} must be A-orthogonal to d_i. This is equivalent to finding a minimum point along the search line, as we have seen in steepest descent. So, as before, this is achieved by differentiating Eq. 12.26 with respect to α:

$$\frac{df(x_{i+1})}{d\alpha} = f'(x_{i+1})^T \frac{d(x_{i+1})}{d\alpha} = 0 \tag{12.34}$$

$$-r_{i+1}^T d_i = 0 \tag{12.35}$$

Since $r_{i+1} = -A e_{i+1}$, the equation becomes:

$$d_i^T A e_{i+1} = 0 \tag{12.36}$$

With A-orthogonal vectors, we make the α_i equation as:

$$\alpha_i = \frac{-d_i^T A e_i}{d_i^T A d_i} \tag{12.37}$$

Since $Ae_i = -r_i$, the equation becomes:

$$\alpha_i = \frac{-d_i^T r_i}{d_i^T A d_i} \qquad (12.38)$$

With this expression, we can compute Eq. 12.26 without knowing the error e_i. If $d_i = r_i$, (the search vector is the same as residual), then α_i formula for A-orthogonal search directions will be the same α_i formula used for steepest descent. Hence, the method of conjugate directions converges in N steps. The procedure is summarized as follows:

1. Select some d_0.
2. Choose a minimum point x_i such that the corresponding e_i is A-orthogonal to d_0.
3. Compute the initial error e_0 as the sum of A-orthogonal components.
4. Each step of the conjugate directions eliminates one of the components.
5. Choose a minimum point x_i, such that e_i is A-orthogonal to d_0.

12.2.9 The Gram-Schmidt Conjugation Method

We have seen that use of A-orthogonal directions $\{d_i\}$ eliminates e_i. The *Gram-Schmidt method* takes n linearly independent vectors $(u_0, u_1, \ldots, u_{n-1})$ and constructs d_i from the u_i (Fig. 12.7).

In order to construct the d_i, we take the u_i and subtract out any component that are not A-orthogonal to the previous d vectors. Set $d_0 = u_0$.

$$\text{For } k < i > 0, \text{ set } d_i = u_i + \sum_{k=0}^{i-1} \beta_{ik} d_k \qquad (12.39)$$

Here, i stands for values which are already known and k stands for values to be computed. Post-multiplying by Ad_j:

$$d_i^T A d_i = u_i^T A d_i + \sum_{k=0}^{i-1} \beta_{ik} d_k^T A d_j \qquad (12.40)$$

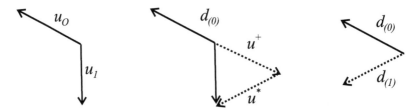

Fig. 12.7 Gram-Schmidt conjugation of two vectors

This is for the terms except when $k = j$ (A orthogonal).

So, we have,

$$u_i^T A d_j + \beta_{ij} d_j^T A d_j = 0 \text{ for } i > j \tag{12.41}$$

$$\beta_{ij} = \frac{-u_i^T A d_j}{d_j^T A d_j} \text{ for } i > j \tag{12.42}$$

All other terms for which $k \neq j$ it becomes zero. We know that $d_i^T A d_j = 0$ and $d_i^T r_j = 0$ and $u_i^T r_j = 0$ when $i < j$ (where $i =$ previous directions and j are current and future directions):

$$d_i = u_i + \sum_{k=0}^{i-1} \beta_{ik} d_i \tag{12.43}$$

$$d_i^T A e_j = -u_i^T A e_j + \sum \beta_{ik} d_k^T A e_j \tag{12.44}$$

where the sigma terms becomes zero for $j > 1$:

$$A e_j = -r_j, \text{ i.e., } d_i^T r_j = 0 \text{ and } u_i^T r_j = 0$$

For $j = i$:

$$d_i^T A e_i = -u_i^T A e_i$$

or:

$$d_i^T r_i = -u_i^T r_i \tag{12.45}$$

The disadvantages are as follows:

1. Using Gram-Schmidt conjugation in conjugate directions requires that all search vectors must be kept in memory to construct new ones.
2. n^3 operations generate the full set.

One method of conjugate directions, namely the *conjugate gradient method*, solves this problem for us.

12.2.10 The Conjugate Gradient Method

This is a method of "conjugate directions" where the search directions are constructed by conjugation of the residual, i.e., setting $u_i = r_i$, the crucial step that was mentioned earlier (Fig. 12.8).

We have:

$$r_{j+1} = -A e_{i+1} = -A(e_j + \alpha_i d_j) = r_j - \alpha_i A d_j \tag{12.46}$$

$$r_i^T r_{j+1} = r_i^T r_j - r_i^T \alpha_i A d_j \tag{12.47}$$

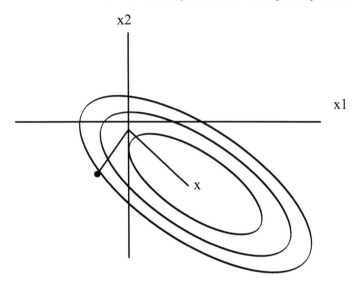

Fig. 12.8 Conjugate gradient method – converges in N steps

Instead of u's we have chosen r_0, r_1 etc . . .

$$r_i^T r_i = 0 \text{ for } i \neq j \qquad (12.48)$$

Referring Eq. 12.42, where we have the value for β_{ij}, and if we use r_i in Eq. 12.41, instead of u_i, we get:

$$\beta_{ij} = \frac{-r_i^T A d_j}{d_j^T A d_j} \qquad (12.49)$$

$$\text{So, } \beta_{ij-1} = (1/\alpha_{i-1}) \left(\frac{-r_i^T r_i}{d_{i-1}^T A d_{i-1}} \right) = 2^{\text{nd}} \text{ term for } i = j+1 \qquad (12.50)$$

$$\beta_{ij} = 0 \text{ for } i > j+1 \qquad (12.51)$$

So, substituting for u_i and β_{jk}, we get:

$$d_i = r_i - \left[\frac{r_i^T r_i}{\alpha_{i-1} d_{i-1}^T A d_{i-1}} \right] d_{i-1} \qquad (12.52)$$

Now we come to a set of A orthogonal directions d_i with which we can work to reach the optimum in N steps.

12.3 Potential Energy Surfaces

A potential energy surface is often represented by illustrations, as given in Fig. 12.9. These surfaces specify the way in which the energy of a molecular system varies with small changes in its structure. In this way, a potential energy surface is a mathematical relationship linking the molecular structure and the resultant energy.

For example, for a diatomic molecule, the potential energy surface can be represented by a two-dimensional plot with the internuclear separation on the x-axis and the energy at that bond distance on the y-axis; in this case, the potential energy surface is a curve. For larger systems, the surface has as many dimensions as there are degrees of freedom within the molecule. The potential energy surface illustration considers only two of the degrees of freedom within the molecule, and plots the energy above the plane defined by them, creating a surface. Each point represents a particular molecular structure, with the height of the surface at that point corresponding to the energy of that structure. Our illustrated example surface contains three minima: a minimum is a point at the bottom of a valley, from which motion in any direction leads to a higher energy. Two of them are local minima, corresponding to the lowest point in some limited region of the potential surface, and one of them is the global minimum, the lowest energy point anywhere on the potential surface. Different minima correspond to different conformations or structural isomers of the molecule under investigation. The illustration also shows two maxima and a saddle point (the latter corresponds to a transition state structure) [7].

At both minima and saddle points, the first derivative of the energy, known as the gradient, is zero. Since the gradient is the negative of the forces, the forces are also

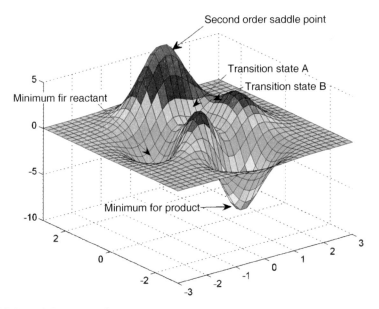

Fig. 12.9 Potential energy surface

zero at such points. *A point on the potential energy surface where the forces are zero is called a stationary point.* All successful optimizations locate a stationary point, although not always the one that was intended. Geometry optimizations usually locate the stationary point closest to the geometry from which they started. When you perform a minimization, intending to find the minimum energy structure, there are several possibilities as to the nature of the results: you may have found the global minimum, you may have found a local minimum but not the global minimum, or you may have located a saddle point.

12.3.1 Convergence Criteria

Convergence criteria set up for the potential energy surface in different software may be slightly different. In most cases, computational cutoff values will be set up initially for tools such as forces, root-mean-square of forces, displacement, and root-mean-square of displacement. Values below these predefined cutoff values will be considered as zero during computation. Major criteria for convergence can be summarized as follows:

1. Forces must be zero.
2. The root-mean-square of the forces should be zero.
3. The computed displacement for the next step of optimization should be zero or less than a predefined cutoff value.
4. The root-mean-square of the displacement for the next step should be zero or less than a cutoff value.

However, for large molecules, if the forces are $(1/100)^{th}$ of the cutoff value, even though other criteria are not satisfied, the molecule can be considered as having attained geometry minimization.

The output files of the geometry optimization of ethene with GAUSSIAN 03 W and SPARTAN '02 using the 6-31G(d) basis set are included in the URL. The relevant values from the output are included in Table 12.1.

Table 12.1 Convergence criteria satisfied in the geometrical optimization of ethene

Item	Value	Threshold	Converged?
Maximum force	0.000177	0.000450	YES
RMS force	0.000118	0.000300	YES
Maximum displacement	0.001119	0.001800	YES
RMS displacement	0.000602	0.001200	YES

Table 12.2 Geometry optimization and frequency

Computational search	Frequency	Inference
Geometry minimization	No imaginary frequency	Attained geometry minimization
Geometry minimization	One or more imaginary frequencies	The structure is a saddle point

12.3.2 Characterizing Stationary Points

A geometry optimization alone cannot determine the nature of the stationary point that it attains. In order to characterize a stationary point, it is necessary to perform a frequency calculation on the optimized geometry. Electronic structure programs such as GAUSSIAN are able to carry out such calculations (you can even perform an optimization followed by a frequency calculation at the optimized geometry in a single job).

In order to distinguish a local minimum from the global minimum, it is necessary to perform a conformational search. We might begin the computation by altering the initial geometry slightly and then performing another minimization. Note that modifying the dihedral angles is often a good place to start. There are also a variety of conformational search tools that can help with this task. We will focus here on distinguishing between minima and saddle points via frequency calculations. The completed frequency calculation will include a variety of results such as frequencies, intensities, the associated normal modes, the zero point energy of the structure and various thermochemical properties. In identifying whether there are any frequency values less than zero, these frequencies are known as imaginary frequencies. The number of imaginary frequencies indicates the sort of stationary point to which the given molecular structure corresponds (Table 12.2). By definition, a structure which has n imaginary frequencies is an nth order saddle point. Thus, the minimum will have zero imaginary frequencies, and an ordinary transition structure will have one imaginary frequency since it is a first order saddle point.

12.4 The Search for Transition States

Transition states correspond to saddle points on the potential energy surface. Strictly speaking, a transition state (Fig. 12.10) of a chemical reaction is a first order saddle point. Like minima, the first order saddle points are stationary points with all forces zero. Unlike minima, one of the second derivatives in the first order saddle is negative. The eigenvector with the negative eigenvalue corresponds to the reaction coordinate. *A transition state search thus attempts to locate stationary points with one negative second derivative.*

The energy state of the activated complex should be located at a first-order saddle point on the potential energy surface, i.e., a point which is a maximum in one

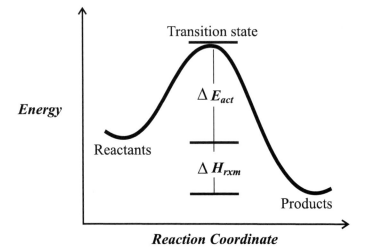

Fig. 12.10 Transition state of a reaction

direction and a minimum in all other directions. The structure associated with the first-order saddle point will exhibit one imaginary frequency and the normal mode of vibration associated with this frequency should emulate the motion of the atoms along the reaction coordinate. We will consider some typical computational problems solved with softwares like GAUSSIAN, Spartan, etc.

12.4.1 Computing the Activated Complex Formation

Let us compute the activated complex formation during hydroboration of ethylene. The reaction is given by Fig. 12.11.

We here illustrate the computation using Spartan. For further details refer to the Spartan manual. The build input for ethylene is shown in Fig. 12.11 and BH_3 is shown with the sp^2 hybridization icon. One procedure used to build an activated complex is the Reaction icon. This procedure utilizes the linear synchronous transit method and is activated by clicking a button in the tool bar.

To optimize the geometry of the activated complex, click on Setup in the tool bar and select Calculations from the pop-up menu. Pick Transition State Geometry, Semi-Empirical, and MNDO. Check the boxes next to Frequencies and Vibration

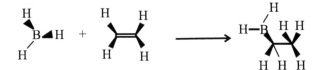

Fig. 12.11 Computing activated complex

Modes. Click the OK button to close the Setup Calculations window and select Submit from the Setup menu. When the Save As window appears, create the Transition Spartan file in the folder (refer to the URL for details).

The energy state of the activated complex should be located at a first-order saddle point on the potential energy surface, i.e., a point which is a maximum in one direction and a minimum in all other directions. The structure associated with the first-order saddle point will exhibit one imaginary frequency (here it is -225.29) and the normal mode of vibration associated with this frequency should imitate the motion of the atoms along the reaction coordinate.

To confirm that the energy state of our structure is located at a first-order saddle point, click Display in the tool bar and select Vibrations from the pop-up menu. The Vibrations List window which appears contains the frequencies of the normal modes of vibration for the structure. The imaginary frequencies have an "i" in front of the number and appear at the beginning of the list. To determine if the motion of the atoms in the normal mode of vibration associated with the imaginary frequency is consistent with the formation of products in the forward direction and reactants in the reverse direction, click the box next to the imaginary frequency in the Vibrations List window and observe the animation. Does the structure appear to move toward the product in one direction and reactants in the other direction?

Now, let us predict the intermediate structure formed during the transformation of cis-C_3H_5Cl to trans-C_3H_5Cl. Here, we use GAUSSIAN. First of all, let us assume that the intermediate is formed by the dihedral rotation of $H-C-C-H$ (the 2nd and 3rd carbon atoms). To draw the structures and get the required input files for calculation, we use Gaussview GUI. The dihedral angle is rotated by 180° to get the structure of the suggested intermediate (Fig. 12.12).

All these models have been subjected to geometry optimization with the route terms "#T RHF/6-31G(d) Opt Freq Test" (refer to the URL for details). The frequency computation of the second structure produces an imaginary frequency suggesting that this conformation could be an intermediate structure. However, the difference in energy between trans(0) and trans(180) conformers is only 0.000517144 Hartrees or 0.324512828735 kcal/mol. This energy is much less than the energy required for the rotation of the C=C double bond. Hence, it cannot be considered as a transition structure of cis and trans forms. Moreover, the symmetry A of the imag-

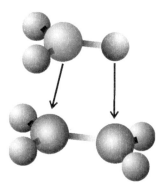

Fig. 12.12 Spartan input diagram for identifying the transition state

inary frequency suggests that it is a symmetry breaking mode. This small imaginary frequency (-220.8853) could be due to some modest geometry distortion. In the eigenvector of the Hessian, giving the displacements for the normal mode corresponding to the imaginary frequency, significant values are from D1 to D6 (Table 12.3). On comparing these values with typical methyl rotation values (included in the output corresponding to methyl rotation of ethane), the suggested structure can be considered as obtained by the motion corresponds to methyl rotation.

Now, we assume the transition state to be obtained by rotating the Cl–C–C–H dihedral angle [5]. By changing the dihedral angles, we obtain a structure with the Z-matrix as given (Fig. 12.13 and Fig. 12.14). With this input, the model is again subjected to geometry optimization with the same route terms (refer to the C_3H_3Cl transition state file of the URL).

The results show that this transition state has got a high value of imaginary frequency. In the Hessian, angles A8 and A9, the dihedral angles D6 to D10 are significantly corresponding to the transition. The energy level diagram (Fig. 12.15) reads an energy difference of 110.5665 kcal/mol between the cis and transition forms and

Table 12.3 GAUSSIAN output eigenvector of the Hessian

Variable	Displacement
D1	0.40739
D2	0.39336
D3	0.41850
D4	0.40447
D5	0.41850
D6	0.40447

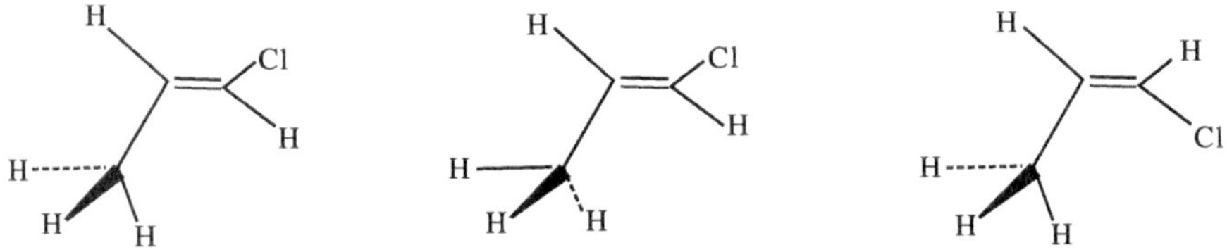

Trans (0)   Trans (180)   Cis (0)

Fig. 12.13 Transformation of cis-C_3H_5Cl to trans-C_3H_5Cl

Z-matriz of C_3H_5Cl

C								
H	1	1.0880						
C	1	1.5656	2	111.2790				
H	1	1.0880	3	1075901	3	-117.0269	0	
H	1	1.0937	3	116.0383	4	-119.9580	0	
H	3	1.1073	1	103.9960	4	66.9831	0	
C	3	1.4493	1	104.3350	4	169.9561	0	
H	7	1.1296	3	130.8670	1	-54.1900	0	
Cl	7	1.7600	3	121.9200	1	125.8000	0	

Fig. 12.14 Z-matrix of C_3H_5Cl

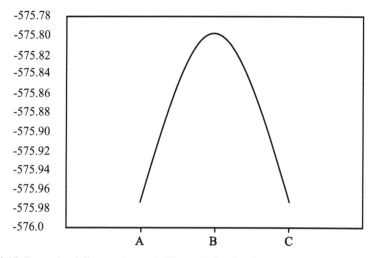

Fig. 12.15 Energy level diagram (energy in Hartrees) showing the cis (*A*), trans (*C*), and the intermediate (*B*) for C_3H_3Cl

108.4774 kcal/mol between the trans and transition forms. This suggested structure can be identified as a transition state. Similarly, we can change the dihedral angle and find other transition states. For an accurate modeling of the rotation with respect to a double bond, higher level of theory like CASSCF is used rather than Hartree-Fock (HF).

12.5 The Single Point Energy Calculation

The single point energy (SPE) calculation is a basic molecular modeling calculation where the energy of the molecule at a specific molecular geometry is computed. This type of computation helps to obtain basic information about the molecule, to obtain a consistency check on the geometry of the molecule, to predict properties related to the energy changes, and so on. The calculation can be set at any level of theory as is required for the computation. We shall see some typical computations carried out using SPE calculation.

Let us make the SPE calculation of water with different basis sets and levels of computation, starting from a lower level to a higher level. In each higher level of computation, the structure from the lower level is taken so that each computation modifies the SPE to attain the theoretical one (Table 12.4).

SPEs, sometimes known simply as molecular energies, are typically in units of Hartrees, which can be converted to more common energy terms such as kilojoules mol^{-1} (kJ mol^{-1}), kilocalories mol^{-1} (kcal mol^{-1}), or electron volts (eV). An HTML set up for the interconversion of energy units is included in the URL. Any change in a molecular geometry will require that a new SPE calculation be performed.

Table 12.4 SPE of water

Route	Sp energy (Hartree)
opt hf/6-31+g(d)	− 76.0171125670
hf/6-31+g(d) sp	− 76.0177423002
b3lyp/6-31+g(d) sp	− 76.4217149983

12.6 The Computation of Solvation

12.6.1 The Theory of Solvation

Solvation is associated with the interaction between solvent and solute molecules, which will lead to changes in energy, stability, and molecular orientation (distribution). Hence, those properties, which will depend upon energy such as vibrational frequency, spectrum, etc. will also change along with solvation. Moreover, changes in stability may change the optimization criteria [3].

The space occupied by the solute molecules dispersing the solvent molecules is said to be the *solvent cavity*. The energy required to push aside solvent molecules is known as the *cavitation energy*. This is thermodynamically balanced by the solvent-solute interaction. The interaction between the solvent and the solute is given by Eq. 12.53:

$$E = \frac{q_i q_j}{\kappa r_{ij}} \qquad (12.53)$$

where κ is the dielectric constant, and $q_i q_j$ the charges separated by r_{ij}.

The solvent molecules reorient to provide maximum interaction leading into structural distortions. Solvent energy modeling by considering the cleavage of solvent-solvent bonds and setting up of solvent-solute bonds is called the linear solvent energy relationship (LSER). Solvation modeling is broadly classified into the following types.

12.6.1.1 The Group Additivity Method

The contribution of each group or atom to solvation is set up (QSPR). Then, using a fitting technique, the total solvation effect of the molecule is determined. This technique is known as the group additivity method.

12.6.1.2 The Continuum Method

In this method, the solvent is considered as a continuum with a given dielectric constant.

12.6.2 The Solvent Accessible Surface Area

The surface area of the solvent accessible to solute molecules is known as the solvent accessible surface area (SASA). The maximum interaction will be in the region close to the solute molecules. The solvation free energy, ΔG_s^0, is given by:

$$\Delta G_s^0 = \sum_i \sigma_i A_i , \qquad (12.54)$$

where σ_i is the surface tension associated with a region i and A_i the surface area. In this equation we are not considering the difference in energy contributions by different solvent sites.

12.6.3 The Onsager Model

In this model, the solvation system is considered as a molecule with a multipole moment inside a spherical cavity surrounded by a continuum dielectric. This method is helpful in predicting the solvation effect, even if the molecule is with zero dipole moment.

12.6.4 The Poisson Equation

Electrostatic interaction between an arbitrary charge density $\rho(r)$ and a continuum dielectric with the dielectric permitivity ε is given by Poisson's potential equation:

$$\nabla^2 \phi = -\frac{4\pi\rho(r)}{\varepsilon} . \qquad (12.55)$$

12.6.5 The Self-Consistent Reaction Field Calculation

The self-consistent field calculation (SCRF) is an adaptation of the Poisson method for ab initio calculations. This method models the systems in a non-aqueous medium. Different types of calculations can be set up on the basis of the difference in the shape of the solvent cavity and the difference in the description of the solute such as dipole, multipole, etc. Some of these types are mentioned below.

12.6.5.1 The Onsager Model (SCRF=Dipole)

In this model, the solute is considered as occupying a fixed spherical cavity of a radius a_0 within the solvent field. A dipole in the solute molecule will induce a dipole (induced dipole) on the medium. The solvent and the solute are stabilized by the interaction between the solute dipoles and solvent induced dipoles. The systems with a zero dipole moment will not exhibit solvation by this model.

12.6.5.2 The Tomasi Polarized Continuum Model (SCRF=PCM)

The Tomasi polarized continuum model (PCM) differs in the cavity. The cavity is considered as a union of a series of interlocking atomic spheres. The effect of polarization of the solvent continuum is calculated numerically by integration rather than by approximation.

12.6.5.3 The Isodensity Surface Model (SCRF=IPCM)

In this method, the cavity is considered as an isodensity surface. It is calculated by an iterative procedure. The isodensity surface has a very natural intuitive shape, corresponding to the reactive shape of solute molecules to provide maximum interaction. It is not a predefined shape such as a sphere.

12.6.6 The Self-Consistent Isodensity Polarized Continuum Model

In the self consistent isodensity polarized continuum model (SCI-PCM), the isosurface and the electron density are effectively coupled. The procedure solves for the electron density, which minimizes the energy, including the solvation energy. This, in turn, depends upon the cavity and electron density. This accounts for the full coupling between the cavity and electron density.

Route words used to make SCRF calculations with GAUSSIAN are included in Table 12.5.

Table 12.5 Running SCRF calculations using GAUSSIAN

Sl.no	Model	Required input
1	SCRF=Dipole	a_0 and ε
2	SCRF=PCM	ε
3	SCRF=IPCM	ε
4	SCRF=SCIPCM	ε

12.7 The Population Analysis Method

The population analysis in computational chemistry stands for estimating the partial atomic charges or orbital electronic density from calculations carried out, particularly those based on the linear combination of the atomic orbitals molecular orbital method. The Mulliken population analysis is the most common type of this computation.

12.7.1 The Mulliken Population Analysis Method

Due to its simplicity, the Mulliken population analysis has become the most familiar method to count electrons associated with an atom in a molecule. The total number of electrons in a closed shell system is given by the integral over the electron density as:

$$N = \int dr \rho(r) = 2 \sum_{1=1}^{N/2} \int dr \psi_i^*(r) \psi_i(r) \tag{12.56}$$

If the coefficients of the basis functions b_μ^* and b_v in the molecular orbital are $C_{\mu i}^*$ and C_{vi} in the ith molecular orbital:

$$N = 2 \sum_{i=1}^{N/2} \sum_{\mu=1}^{K} \sum_{v=1}^{K} C_{\mu i}^* C_{vi} \int dr b_\mu^*(r) b_v(r)$$

$$= 2 \sum_{i=1}^{N/2} \sum_{\mu=1}^{K} \sum_{v=1}^{K} C_{\mu i}^* C_{vi} S_{\mu v} \tag{12.57}$$

where $S_{\mu v}$ is the overlap integral. Introducing the density matrix:

$$P_{\mu v} = 2 \sum_{i=1}^{N/2} C_{\mu i}^* C_{vi} \tag{12.58}$$

N assumes the following simplified form:

$$N = \sum_{\mu=1}^{K} \sum_{v=1}^{K} P_{v\mu} S_{\mu v} = \sum_{\mu=1}^{K} (PS)_{\mu\mu} = Tr(PS) \tag{12.59}$$

$(PS)_{\mu\mu}$ can be interpreted as the charge to be associated with basis function b_μ. The partial trace:

$$\rho M(A) = \sum_{\mu \in A} (PS)_{\mu\mu} \tag{12.60}$$

with the sum running over all basis functions that are centered at the atom with position R_A is called the Mulliken charge of that atom. It is seen here that the definition of the Mulliken charge is only meaningful if the basis set consists of basis functions that can be associated with an atomic site.

The Mulliken spin density ρ_s is defined as the difference of the Mulliken charges of spin-up and spin-down electrons. The sum over the Mulliken charges of all atoms equals the total number of electrons in the system. Likewise, the sum over the Mulliken spin densities equals the total spin of the system. It is noted here that the Mulliken spin density is in fact not a spin density but an integrated spin density, i.e., a spin. It nevertheless persists with the common notation.

Molecular orbitals and their energies can be computed with the keyword Pop = Reg in the route section of GAUSSIAN input. The required data will be obtained in the output. The atomic contributions for each atom in the molecule are given for each molecular orbital numbered in order of increasing energy. It includes the following:

1. The molecular orbital and orbital energies.
2. The symmetry of the orbitals.
3. The nature of the orbital – occupied or virtual
4. The relative magnitude of each orbital
5. The gross orbital population
6. The atomic contributions
7. The Mulliken population analysis
8. The density matrix

12.7.2 The Merz-Singh-Kollman Scheme

In the Merz-Singh-Kollman (MK) scheme, atomic charges are fitted to reproduce the molecular electrostatic potential (MEP) at a number of points around the molecule. At first, the MEP is calculated at a number of grid points located on several layers around the molecule. The layers are constructed as an overlay of van der Waals spheres around each atom. (Fig. 12.16). The points located inside the van der Waals volume are neglected [1].

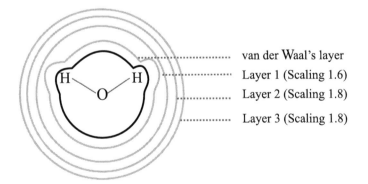

Fig. 12.16 MK scheme

The b results are achieved by sampling points not too close to the van der Waals surface and the van der Waals radii are therefore modified through scaling factors. The smallest layer is obtained by scaling all radii with a factor of 1.4. The default MK scheme then adds three more layers constructed with scaling factors of 1.6, 1.8, and 2.0.

After computing the MEP at the valid grid points located on all four layers, atomic charges are derived that reproduce the MEP as closely as possible. The additional constraint in the fitting procedure is that the sum of all atomic charges is considered as equal to that of the overall charge of the system. An input file for calculating the MK charges for water at the Becke3LYP/6-31G(d) level of theory is: #P Becke3LYP/6-31G(d) pop=MK scf=(direct,tight) (using Gaussian 03).

12.7.3 Charges from Electrostatic Potentials Using a Grid-Based Method (CHELPG)

This method is similar to the MK method. In this method, atomic charges are fitted to reproduce the MEP at a number of points around the molecule. As a first step of the fitting procedure, the MEP is calculated at a number of grid points spaced 3.0 pm apart and distributed regularly in a cube. The dimensions of the cube are chosen such that the molecule is located at the center of the cube, adding 28.0 pm headspace between the molecule and the end of the box in all three dimensions. All points falling inside the van der Waals radius of the molecule are discarded from the fitting procedure.

After evaluating the MEP at all valid grid points, atomic charges are derived that reproduce the MEP in the most optimum way. The additional constraint in the fitting procedure is that the sum of all atomic charges equals that of the overall charge of the system. Gaussian input file for calculating the CHELPG charges for water is: #P HF/STO-3G pop=chelpg scf=(direct,tight).

12.7.4 The Natural Population Analysis Method

The analysis of the electron density distribution in a molecular system based on the orthonormal natural atomic orbitals is known as natural population analysis (NPA). Natural populations $n_i(A)$ are the occupancies of the natural atomic orbitals. These rigorously satisfy the Pauli's exclusion principle $0 < n_i(A) < 2$. The population of an atom $n(A)$ is the sum of natural populations:

$$n(A) = \sum n_i(A) \tag{12.61}$$

A distinguishing feature of the NPA method is that it largely resolves the basis set dependence problem encountered in the Mulliken population analysis method.

12.8 Shielding

A nucleus with a resultant nuclear magnetic moment (μ) zero provides an excellent search of the magnetic fields inside a sample. When it is exposed to a static homogeneous magnetic field, the nuclear magnetic moment will process around the direction of the magnetic field with a frequency directly proportional to the magnitude of the magnetic field. The frequency, and thus the magnetic field at the nuclear site, can be detected by nuclear magnetic resonance (NMR) experiments [2].

When a static homogeneous magnetic field H is applied, the electronic system reacts to it by producing currents. These currents in turn give rise to an additional magnetic field ΔH at the nuclear site. The chemical shielding tensor of that nucleus can be defined as follows:

$$\sigma_{\alpha\beta} = -\frac{\Delta H_\alpha}{H_\beta} \tag{12.62}$$

$$\text{Where} \qquad \alpha, \beta \in \{x, y, z\} .$$

The chemical shielding tensor, in fact, depends upon the chemical surrounding of a nucleus. Hence, $\sigma_{\alpha\beta}$ is unique for a chemical environment. It differs for a nucleus in an atom being covalently or ionically bonded to its neighbors. NMR spectroscopy has become a standard tool to characterize chemically different sites of an ion in a molecule or in a crystal. The total magnetic field at the nucleus is the sum of the external magnetic field and the nuclear magnetic field. This leads into an energy splitting of:

$$\Delta E = -\mu.H^{\text{total}} = -\mu (1 - \sigma) H \tag{12.63}$$

Therefore, σ can be identified as a mixed second derivative of the ground state energy in the presence of both a nuclear magnetic moment and an external magnetic field with respect to these two parameters. By a Taylor expansion:

$$E (H, \mu) = E_0 + \ldots + \sum_{i,j} H_i \frac{\partial^2 E(H, \mu)}{\partial H_i \partial \mu_j} \mu_j + \ldots \tag{12.64}$$

To calculate the chemical shielding the electronic Hamiltonian operator is expanded to include the external magnetic field and the magnetic field of the nuclear magnetic moments. This is done applying the minimal substitution:

$$p \rightarrow p + (e/c) A^{(\text{tot})}(r) \tag{12.65}$$

of the momentum operator where $A^{(\text{tot})} = A + A^{\text{nucleus}}$ is the vector potential of the above contributions to the total magnetic field. The ground state energy is then evaluated using the usual Rayleigh-Schrödinger many body perturbation theory and the above mixed derivative yields the chemical shielding tensor.

The two vector potentials (for a nucleus at R) are given by:

$$A(r) = \frac{1}{2} H r \tag{12.66}$$

$$A^{\text{nucleus}}(r) = \frac{\mu(r-R)}{|r-R|^3} \tag{12.67}$$

The extended Hamiltonian operator includes the following terms:

$$\hat{H}_{(H,r)} = \hat{H}_{\text{electron}} + \sum_a H_a \hat{H}_a^{(1,0)} + \sum_a \mu_a \hat{H}_a^{(0,1)} + \sum_{a\beta} H_a \hat{H}_{a\beta}^{(1,1)} \mu_\beta$$

$$+ \frac{1}{2} \sum_{a\beta} H_a \hat{H}_{a\beta}^{(2,0)} H_\beta \tag{12.68}$$

The various contributions are:

$$\hat{H}_\alpha^{(1,0)} = -\frac{i}{2c} \sum_{j=1}^N (r_j \times \nabla_j)_\alpha \tag{12.69}$$

$$\hat{H}_\alpha^{(0,1)} = -\frac{i}{c} \sum_{j=1}^N \frac{\left[(r_j - R)\nabla_j\right]_\alpha}{|r-R|^3} \tag{12.70}$$

$$\hat{H}_\alpha^{(1,1)} = -\frac{1}{2c^2} \sum_{j=1}^N \frac{\left[r_j(r_j - R)\delta_{\alpha\beta} - r_{j\alpha}\left(r_{j\beta} - R_\beta\right)\right]}{|r-R|^3} \tag{12.71}$$

$$\hat{H}_\alpha^{(2,0)} = -\frac{1}{4c^2} \sum_{j=1}^N \left(r_j^2 \delta_{\alpha\beta} - r_{j\alpha} r_{j\beta}\right) \tag{12.72}$$

Evaluating the expression for the shielding constant using the Hellmann-Feynman theorem, one arrives at:

$$\sigma^{\alpha\beta} = -\left\langle \psi^{(0)} \left| \hat{H}_\alpha^{(1,1)} \right| \psi^{(0)} \right\rangle - \left[\frac{\partial}{\partial H_\beta} \left\langle \psi^{(H_\beta)} \left| \hat{H}_\alpha^{(0,1)} \right| \psi^{(H_\beta)} \right\rangle \right]_{H_\beta} = 0 \tag{12.73}$$

Here, $\psi^{(0)}$ is the unperturbed wavefunction and $\psi^{(H_\beta)}$ is the wavefunction in the presence of the external magnetic field. The two terms represent the diamagnetic and paramagnetic contribution to the shielding tensor. It should be noted that the diamagnetic contribution depends only on the unperturbed wavefunction, whereas the paramagnetic contribution is determined solely by the perturbed wavefunction.

To calculate the perturbed wavefunction in the presence of a magnetic field, it is sufficient to use the Hamiltonian $\hat{H}' = \hat{H}_0 + \sum H_\alpha \hat{H}_\alpha^{(1,0)}$ and to solve the associated Roothaan equations, $F'C = SC\varepsilon$, where the one electron part of the Fock operator receives an additional field dependent term. Note that in this case the expansion coefficients are allowed to become complex to accommodate for the perturbation.

12.9 Electric Multipoles and Multipole Moments

Multipole moments are the coefficients of a series expansion of a potential due to either continuous or discrete sources. A multipole moment usually involves powers (or inverse powers) of the distance to the origin, as well as some angular dependence. In principle, a multipole [4] expansion provides an exact description of the potential and generally converges under two conditions:

1. if the sources (e.g., charges) are localized close to the origin and the point at which the potential is observed is far from the origin.
2. the reverse, i.e., if the sources (e.g., charges) are located far from the origin and the potential is observed close to the origin. In the first (more common) case, the coefficients of the series expansion are called either exterior multipole moments, or simply multipole moments, whereas in the second case they are called interior multipole moments. The zeroth-order term in the expansion is called the monopole moment, the first-order term is denoted as the dipole moment, and the third, and fourth terms are denoted as quadrupole and octupole moments, etc.

12.9.1 The Quantum Mechanical Dipole Operator

Consider a set of N electric point charges $Q_1, Q_2, \ldots, Q_n$ at position vectors r_1, $r_2, \ldots, r_n$. For instance, this collection may be a molecule consisting of electrons and nuclei. The physical quantity (observable) *dipole* has the *quantum mechanical operator*:

$$P_e = \sum_{i=1}^{N} Q_i r_i \tag{12.74}$$

It is a vector quantity with components along the x, y, and z axes:

$$(P_e)_x = \sum_{i=1}^{N} Q_i x_i \tag{12.75}$$

$$(P_e)_y = \sum_{i=1}^{N} Q_i y_i \tag{12.76}$$

$$(P_e)_z = \sum_{i=1}^{N} Q_i z_i \tag{12.77}$$

If two equal and opposite charges, $+Q$ and $-Q$ are separated by d, then the electric dipole moment has magnitude Qd and pointing towards the direction of a vector from the negative charge to the positive charge.

Electric second moments will be given by six independent terms,

$$\sum_{i=1}^{N} Q_i x_i x_i, \sum_{i=1}^{N} Q_i x_i y_i, \sum_{i=1}^{N} Q_i x_i z_i, \sum_{i=1}^{N} Q_i y_i z_i, \sum_{i=1}^{N} Q_i y_i y_i, \sum_{i=1}^{N} Q_i z_i z_i \; .$$

This is normally represented by a 3×3 symmetric matrix:

$$\begin{bmatrix} \sum_{i=1}^{N} Q_i x_i x_i & \sum_{i=1}^{N} Q_i x_i y_i & \sum_{i=1}^{N} Q_i x_i z_i \\ \sum_{i=1}^{N} Q_i y_i x_i & \sum_{i=1}^{N} Q_i y_i y_i & \sum_{i=1}^{N} Q_i y_i z_i \\ \sum_{i=1}^{N} Q_i z_i x_i & \sum_{i=1}^{N} Q_i z_i y_i & \sum_{i=1}^{N} Q_i z_i z_i \end{bmatrix} \tag{12.78}$$

The *quadrupole moment* of a system is defined as:

$$\Theta_{ij} = \sum q \left(3 x_i x_j - r^2 \delta_{ij} \right) \tag{12.79}$$

The corresponding potential is:

$$V(R) = \frac{1}{4\pi\varepsilon_0} \sum \frac{\Theta_{ij}}{2R^3} n_i n_j \tag{12.80}$$

where R is a vector with origin in the system of charges and n is the unit vector in the direction of R. The matrix representation of the quadrupole moment is:

$$\begin{bmatrix} \sum_{i=1}^{N} Q_i \left(3 x_i x_i - r_i^2 \right) & 3 \sum_{i=1}^{N} Q_i x_i y_i & 3 \sum_{i=1}^{N} Q_i x_i z_i \\ 3 \sum_{i=1}^{N} Q_i y_i x_i & \sum_{i=1}^{N} Q_i \left(3 y_i y_i - r_i^2 \right) & 3 \sum_{i=1}^{N} Q_i y_i z_i \\ 3 \sum_{i=1}^{N} Q_i z_i x_i & 3 \sum_{i=1}^{N} Q_i z_i y_i & \sum_{i=1}^{N} Q_i \left(3 z_i z_i - r_i^2 \right) \end{bmatrix} \tag{12.81}$$

This matrix has zero trace (the sum of the diagonal elements).

The quadrupole moment gives a measure of deviation from spherical symmetry. The properties of the electric quadrupole matrix are normally investigated with the matrix in its principal axis system.

12.9.2 The Dielectric Polarization

Dielectric polarization stands for the charge separation in a small unit volume $d\tau$. The charge separation is again equivalent to a dipole moment.

The electric dipole induced dp_e is directly proportional to the volume $d\tau$.

$dp_e = P d\tau$, where the proportionality constant P is known as the dielectric polarization.

The applied field on a system induces the dipole moment on all the molecules. The dependence of the dipole moment p_e on the external electrostatic field E is given by the expression:

$$P_e(E) = P_e(E = 0) + \alpha E + \dots \tag{12.82}$$

where $P_e(E = 0)$ is the permanent electric dipole moment, αE is the product of dipole polarizability α and the applied field. The higher terms stand for hyperpolarizabilities [6].

P_e and E are vectors and α is a tensor quantity which can be represented as:

$$\alpha = \begin{bmatrix} \alpha_{xx} & \alpha_{xy} & \alpha_{xz} \\ \alpha_{yx} & \alpha_{yy} & \alpha_{yz} \\ \alpha_{zx} & \alpha_{zy} & \alpha_{zz} \end{bmatrix} \tag{12.83}$$

which by proper transformation results in a diagonal matrix of the following type:

$$\alpha = \begin{bmatrix} \alpha_{aa} & 0 & 0 \\ 0 & \alpha_{bb} & 0 \\ 0 & 0 & \alpha_{cc} \end{bmatrix} \tag{12.84}$$

α_{aa}, α_{bb} and α_{cc} are the principal values of polarizability. For symmetrical molecules, the principal axes of polarizability correspond to symmetry axes. The one-third sum of diagonal elements is known as the mean polarizability $\langle \alpha \rangle$.

12.10 Vibrational Frequencies

For a system with a reduced mass μ and a spring constant k, the allowed vibrational energies are given by:

$$\varepsilon_{\text{vib}} = \frac{h}{2\pi} \sqrt{\frac{k}{\mu}} \left(v + \frac{1}{2} \right) \tag{12.85}$$

Quantum mechanically, the normalized vibrational wavefunctions are given by:

$$\psi_v(\xi) = \left(\frac{\sqrt{\beta/\pi}}{2^v v!} \right)^{1/2} H_v(\xi) \exp(-\xi^2/2) \tag{12.86}$$

where $\beta = 2\pi\sqrt{\mu k}/h$ and $\xi = \sqrt{\beta} x$. The polynomials H_v are known as the Hermite polynomial (given in Table 12.6).

The smallest allowed value of vibrational energy is known as the zero point energy. It is given by:

$$E_{\text{zpe}} = (h/2\pi) \sqrt{k/\mu} \left(0 + \frac{1}{2} \right) \tag{12.87}$$

Table 12.6 Hermite polynomials

v (Energy quantization number)	$H_v(\xi)$
0	1
1	2ξ
2	$4\xi^2 - 2$
3	$8\xi^3 - 12\xi$
4	$16\xi^4 - 48\xi^2 + 12$
5	$32\xi^5 - 160\xi^3 + 120\xi$

It is the correction to the electronic energy of the molecule to compensate the effect of vibration, even at zero K.

The vibration of molecules is best described by the quantum mechanical approach. But, in practice, molecules need not behave like a harmonic oscillator description, which is used in this method. Bond stretching is better described by a Morse potential and conformational changes have a sine wave type behavior. However, the harmonic oscillator description is used as an approximate treatment for low vibrational quantum numbers. Frequencies computed with ab initio methods and a quantum harmonic oscillator approximation tend to be about 10%, due to the difference between a harmonic potential and the true potential. For the very low frequencies, the computed frequency may be far from the experimental values. Many studies are done carried out using ab initio methods and multiplying the resulting frequencies by about 0.9 to get a good estimate of the experimental results.

Vibrational frequencies from semiempirical calculations tend to be qualitative. The density functional theory (DFT) methods give frequencies with this same level of accuracy, but with a somewhat smaller deviation from the experimental results. It is possible to compute vibrational frequencies using ab initio methods without using the harmonic oscillator approximation. For a diatomic molecule, the quantum harmonic oscillator energies can be obtained by knowing the second derivative of energy with respect to the bond length at the equilibrium geometry. For a non-harmonic oscillator energy, the entire bond dissociation curve must be computed, which requires far more computer time. Likewise, computing anharmonic frequencies for any molecule requires computing at least a sampling of all possible nuclear motions. Due to the enormous amount of time necessary to compute all of these energies, this sort of calculation is very seldom done.

Another method for computationally describing molecules is the molecular mechanics (MM) method. The forces acting on the atoms are modeled as simple algebraic equations such as harmonic oscillators, Morse potentials, etc. All of the constants for these equations are usually obtained from experimental results. A suitable force field can be designed to describe the geometry of the molecule only or specifically created to describe the motions of the atoms. The calculation of the vibrational frequencies by determining the geometry using a harmonic oscillator approximation can yield usable results, if the force field was designed to reproduce the vibrational frequencies. MM does not perform so well if the structure is significantly different from the compounds in the parameterization set.

Another technique built around MM is a dynamics simulation. In a dynamics simulation, the atoms move around for a period of time following Newton's equations of motion. This motion is a superposition of all of the normal modes of vibration and the frequencies cannot be determined directly from this simulation. However, the spectrum can be determined by doing a Fourier transform on these motions. The motion corresponding to a peak in this spectrum is determined by taking just that peak and doing the inverse Fourier transform to see the motion. This technique can be used to calculate anharmonic modes, very low frequencies, and frequencies corresponding to conformational transitions. However, a fairly large amount of computer time may be necessary to get enough data from the dynamics simulation to get a good spectrum.

Another related issue is the computation of the intensities of the peaks in the spectra. Peak intensities depend upon the probability that a particular wavelength photon will be absorbed or Raman scattered. These probabilities can be computed from the wavefunction by first computing the transition dipole moment. Some types of transitions turn out to have a zero probability due to the molecule's symmetry or the spin of the electrons. This is where the spectroscopic selection rules come from.

12.11 Thermodynamic Properties

Consider an ideal gas composed of diatomic molecules AB; in the limit of absolute zero temperature, all the molecules are in the ground state of electronic and vibrational motion. The ground state dissociation energy of a molecule is the energy needed to dissociate the molecule into its ground vibrational state to atoms in their ground states.

$$AB \rightarrow A_{(g)} + B_{(g)} \tag{12.88}$$

$$D_0 = D_e - E_{zpe} \tag{12.89}$$

If zero point vibrational energy is considered as the zero point energy, then:

$$D_0 = D_e - \frac{1}{2}h \sum_{k=1}^{3N-6} v_k \tag{12.90}$$

For the processes of the gas-phase molecule to gas phase atoms, the change in the internal energy is given by $D_0 N_A$, where N_A is the Avogadro number. Hence, for the process change in internal energy:

$$\Delta U_0^o = D_0 N_A . \tag{12.91}$$

In the limit of absolute zero, the change in internal energy is equal to change in enthalpy. Thus:

$$\Delta U_0^o = \Delta H_0^o = D_0 N_A \tag{12.92}$$

Table 12.7 Computed enthalpy using CCSD(T)

Molecule	Computed enthalpy (kcal/mol)	Experimental enthalpy (kcal/mol)	Zero point energy (kcal/mol)
CH	141.7 ± 0.3	141.2 ± 4.2	4.04
CH_2	92.8 ± 0.4	92.2 ± 1.09	10.55
CH_3	35.8 ± 0.6	35.6 ± 0.2	18.6
CH_4	-15.9 ± 0.7	-16.0 ± 0.1	27.71
CH_2O	-25.0 ± 0.3	-25.0 ± 0.1	16.53
HCO	9.8 ± 0.3	10.3 ± 1.9	7.69
CO	-27.4 ± 0.2	-27.2 ± 0.04	3.10

We can calculate ΔH_0^o for a reaction by knowing the theoretical or computational atomization energy of the product and the experimental atomization energy of the reactant. ΔH_0^o =(experimental atomization energy of the reactant – Computational atomization energy of the product).

For example, the geometry optimization of water with HF/6-31G* (UHF) computes the internal energy $U_e = -76.010746$ Hartree. The ground state atomic energies of H and O are, respectively, -0.498233 Hartree and -74.783931 Hartree. The predicted D_e for the change is:

$$H_2O \rightarrow 2H_{(g)} + O_{(g)} \tag{12.93}$$

$$2(-0.498233) + 1(-74.783931) - (-76.010746) = 0.23035 \text{ Hartree} = 6.27 \text{ eV}.$$

With HF/6-31G*, computed fundamental frequencies are 3643, 1634, and 3748 cm^{-1} and $E_{zpe} = 0.56$ eV and $D_0 = 5.71$ eV. The experimental value of D_0 obtained from chemical data is 9.51 eV.

The gas phase experimental atomization energy of water is 219.4 kcal/mol while the predicted atomization energy is only 132 kcal/mol ($D_0 N_A$). The inclusion of electron correlation terms minimizes the error.

Table 12.7 clearly illustrates the efficiency of computation using correlation functions (CCSD(T)).

12.12 Molecular Orbital Methods

Molecular orbital (MO) methods are trying to combine MO approximations describing the active part of the modeling system (ab initio, density functional to semi-empirical) with either some lower level MO methods or MM describing the inactive parts [8]. They are as follows:

1. *IMOMM:* (Integrated MO + MM – Maseras and Morokuma, 1995). In this method, the active part of the system is treated by some MO methods while the nonactive part is treated only by an MM method.

2. *IMOMO:* (Integrated MO + MO – Humbel et al., 1996). In this method we treat the active part of our system with a sophisticated MO method, whereas the nonactive part is treated with some lower level MO method.

3. *ONIOM:* (Our own N-layered integrated molecular orbital + molecular mechanics – Svensson et al., 1996). The ONIOM method divides the system into n layers like an onion. For example, the ONIOM3 method divides the system into 3 parts. With ONIOM3, we can use high level MO methods to describe the active part, some lower level MO method to describe the semiactive part and MM to describe the inactive part of the system. An example could be CCSD (T) on the active part, HF or MP2 on the semiactive part and MM on the inactive part of the system. The ONIOM facility in commercial software such as Gaussian 03, Spartan, etc. provides substantial performance gains for geometry optimizations. ONIOM calculations enable both the steric and electrostatic properties of the entire molecule to be taken into account, when modeling the processes in the high accuracy layer. These techniques yield molecular structures and properties that are in very good agreement with the experiments. Refer to Gaussian and Spartan manuals for details.

12.13 Input Formats for Computations

There are a number of input formats which are taken up by different computational chemistry environments. The Z-matrix input is the general representation of molecules.

12.13.1 The Z-Matrix Input as the Common Standard Format

The Z-matrix format is a matrix representation of the molecule giving the entire data required for computations and is of the following form;

$$[group[,]]atom, p_1, r, p_2, \alpha, p_3, \beta, J$$

or, alternatively, $[group[,]]atom, p_1, x, y, z$. The elements of this form are described as follows:

group   This stands for the atomic group number and is optional. It can be used if different basis sets are used for different atoms of the same kind. The basis set is then referred to by this group number and not by the atomic symbol.

atom   This includes the chemical symbol of the new atom placed at position p_0. This may optionally be appended (without a blank) by an integer, which can act as a sequence number, e.g., C1, H2, etc. Dummy centers with no charge and basis functions are denoted either as Q or X, optionally appended by a number, e.g., Q1; note that the first atom in the z-matrix must not be called X, since this may be confused with a symmetry specification (use Q instead).

p_1 This stands for the atom to which the present atom is connected. This may be either a number n, where n refers to the nth line of the Z-matrix, or an alphanumeric string as specified in the *atom* field of a previous card, e.g., C1, H2, etc. The latter form works only if the atoms are numbered in a unique way.

r This is the distance of new atom from p_1. This value is given in Bohr, unless "ANG" has been specified directly before or after the symmetry specification.

p_2 A second atom needed to define the angle $\alpha(p_0, p_1, p_2)$. The same rules hold for the specification as for p_1.

α Internuclear angle $\alpha(p_0, p_1, p_2)$. This angle is given in degrees and must be in the range $0 < \alpha < 180°$.

p_3 A third atom needed to define the dihedral angle $\beta(p_0, p_1, p_2, p_3)$. Only applies if $J = 0$ (see below).

β The dihedral angle $\beta(p_0, p_1, p_2, p_3)$ in degree. This angle is defined as the angle between the planes defined by (p_0, p_1, p_2) and (p_1, p_2, p_3) $(-180° \leq \beta \leq 180°)$. Only applies if $J = 0$ (see below).

J If this is specified and nonzero, the new position is specified by two bond angles, rather than a bond angle and a dihedral angle. If $J = \pm 1$, β is the angle $\beta(p_0, p_1, p_3)$. If $J = 1$, the triple vector product $(p_1 - p_0)$. $[(p_1 - p_2) \times (p_1 - p_3)]$ is positive, while this quantity is negative if $J = -1$.

x, y, z Cartesian coordinates of the new atom. This form is assumed if $p_1 \leq 0$; if $p_1 < 0$, the coordinates are frozen in geometry optimizations.

All atoms, including those related by symmetry transformations, should be specified in the Z-matrix. Note that for the first atom, no coordinates need to be given, for the second atom only p_1, r are needed, while for the third atom p_3, β, J may be omitted.

12.13.2 Multipurpose Internet Mail Extensions

Multipurpose Internet Mail Extensions (MIME) is an Internet standard that extends the format of e-mail to support the following:

- Text in character sets other than US-ASCII
- Non-text attachments
- Multi-part message bodies
- Header information in non-ASCII character sets

MIME is also a fundamental component of communication protocols such as HTTP, which requires that data be transmitted in the context of email-like messages, even though the data might not fit this context. For UNIX/LINUX there is a tar.gz file available which registers chemical MIME types on your system. Programs can

Table 12.8 File extensions used in computational chemistry

File extension	MIME type	Proper name
alc	chemical/x-alchemy	Alchemy format
csf	chemical/x-cache-csf	CAChe MolStruct CSF
cbin, cascii, ctab	chemical/x-cactvs-binary	CACTVS format
cdx	chemical/x-cdx	ChemDraw eXchange file
cer	chemical/x-cerius	MSI Cerius II format
c3d	chemical/x-chem3d	Chem3D format
chm	chemical/x-chemdraw	ChemDraw file
cif	chemical/x-cif	Crystallographic information file, Crystallographic information framework
cmdf	chemical/x-cmdf	CrystalMaker data format
cml	chemical/x-cml	Chemical markup language
cpa	chemical/x-compass	Compass program of the Takahashi
bsd	chemical/x-crossfire	Crossfire file
csm, csml	chemical/x-csml	Chemical style markup language
ctx	chemical/x-ctx	Gasteiger group CTX file format
cxf, cef	chemical/x-cxf	Chemical eXchange format
emb, embl	chemical/x-embl-dl-nucleotide	EMBL nucleotide format
spc	chemical/x-galactic-spc	SPC format for spectral and chromatographic data
inp, gam, gamin	chemical/x-gamess-input	GAMESS input format
fch, fchk	chemical/x-gaussian-checkpoint	Gaussian checkpoint format
cub	chemical/x-gaussian-cube	Gaussian cube (wavefunction) format
gau, gjc, gjf	chemical/x-gaussian-input	Gaussian input format
gcg	chemical/x-gcg8-sequence	Protein sequence format
gen	chemical/x-genbank	ToGenBank format
istr,ist	chemical/x-isostar	IsoStar library of intermolecular interactions
jdx, dx	chemical/x-jcamp-dx	JCAMP spectroscopic data exchange format
kin	chemical/x-kinemage	Kinetic (protein structure) images
mcm	chemical/x-macmolecule	MacMolecule File Format
mmd, mmod	chemical/x-macromodel-input	MacroModel molecular mechanics
mol	chemical/x-mdl-molfile	MDL molfile
smiles, smi	chemical/x-daylight-smiles	Simplified molecular input line entry specification
sdf	chemical/x-mdl-sdfile	Structure-data file

then register as a viewer, editor, or processor for these formats, so that full support for chemical MIME types is available. All other common input file extensions used in computational chemistry are listed in Table 12.8.

12.13.3 Converting Between Formats

OpenBabel and JOELib are open source tools specifically designed for converting between file formats. We have used OpenBabel here to illustrate a format conversion among common computational environments by taking water as an example.

12.13.3.1 GAUSSIAN *Z*-Matrix Format

```
0 1
O
H   1          B1
H   1          B2 2         A1
B1 0.96000000
B2 0.96000000
A1 109.50000006
```

12.13.3.2 Alchemy Format

```
3 ATOMS, 2 BONDS, 0 CHARGES

1 O3 0.0000   0.0000   0.1140 0.0000
2 H  0.0000   0.7808  -0.4562 0.0000
3 H  0.0000  -0.7808  -0.4562 0.0000
1 2 1 SINGLE
2 3 1 SINGLE
```

12.13.3.3 GAUSSIAN 03 Format

```
0 1
O 0.00000   0.00000   0.11404
H 0.00000   0.78084  -0.45615
H 0.00000  -0.78084  -0.45615
```

12.13.3.4 GAMESS Input Format (INP)

```
$CONTRL COORD=CART UNITS=ANGS $END
$DATA

Put symmetry info here
O 8.0 0.00000   0.00000   0.11404
H 1.0 0.00000   0.78084  -0.45615
H 1.0 0.00000  -0.78084  -0.45615
$END
```

12.13.3.5 MOPAC Cartesian Format (MOPCRT)

```
PUT KEYWORDS HERE

O 0.00000 1   0.00000 1   0.11404 1
H 0.00000 1   0.78084 1  -0.45615 1
H 0.00000 1  -0.78084 1  -0.45615 1
```

12.13.3.6 SMILES FIX Format (FIX)

```
O
0.000 0.000 0.114
```

12.13.3.7 XYZ Cartesian Coordinate Format (XYZ)

```
3
Energy: -47430.8699204
O 0.00000   0.00000   0.11404
H 0.00000   0.78084 -0.45615
H 0.00000 -0.78084 -0.45615
```

12.13.3.8 Protein Data Bank Format (PDB)

```
COMPND UNNAMED
AUTHOR GENERATED BY OPEN BABEL 2.0.2
HETATM 1 O HOH 1 0.000   0.000   0.114 1.00 0.00 O
HETATM 2 H HOH 1 0.000   0.781 -0.456 1.00 0.00 H
HETATM 3 H HOH 1 0.000 -0.781 -0.456 1.00 0.00 H
CONECT 1 2 3
CONECT 2 1
CONECT 3 1
MASTER 0 0 0 0 0 0 0 0 3 0 3 0
END
```

12.14 A Comparison of Methods

We shall make a comparison of different methods to identify the most suitable method for a required computation.

12.14.1 Molecular Geometry

Molecular geometry can be computed at any level. Ab initio HF/STO-3G calculations give acceptable predictions of bond distances and quite good predictions of bond angles. However, there are some exceptions to this statement: an error of 0.72 A.U. in the Na_2 bond length and of 0.23 .U. for NaH. HF/STO-3G bond lengths for molecules with only first-row elements are more accurate than for second-row molecules. An improved result can be achieved by using a bigger basis set. The order of trials STO-3G, 3-21G, 3-21G(*), and 6-31G* normally gives improved results. The experimental results conducted by Hehre et al. showing the variation of average absolute errors with an increase in size of basis set is included in Table 12.9. The computation of the dihedral angle is better with ab initio HF methods.

Table 12.9 Average absolute errors in the bond length (A.U.) and bond angle

Method	AH_n-length	AB single bonds in H_mABH_n	AB multiple bonds in H_mABH_n	AB length in hypervalent species	Angle in H_mABH_n
HF/STO-3G	0.054	0.082	0.027		2.0°
HF/3-21 G	0.016	0.067	0.017	0.125	1.7°
HF/3-21 G*	0.017	0.040	0.018	0.015	1.8°
HF/6-31 G*	0.014	0.030	0.023	0.014	1.5°

The following results were reported earlier regarding computation of geometries.

1. The predicted dihedral angle for hydrogen peroxide is 180° against the actual 112° with 3-21 basis set. Computation with HF/6-31 G* improves the results.

2. Conformational angles of cyclobutane and cyclopentane are better estimated by HF/6-31G*.

3. The average absolute errors in a sample of 73 bond lengths in H_mABH_w type molecules reduced from 0.021 A with HF/6-31G* and 0.013 A with MP2/6-31G* (Hehre et al., pp. 156-161).

4. Feller and Peterson [9] conducted a study of 184 small molecules examining the effect of various frozen-core correlation methods using the basis sets aug-cc-pVTZ, aug-cc-pVDZ, and aug-cc-pVQZ. The results with aug-cc-pVTZ are included in Table 12.10. The HF errors increased with increase in the basis set size for these three sets. MP4 results for AB lengths were less accurate than MP2 results.

5. The DFT method gives promising results with 6-31G* or larger basis sets. The DFT method should not be done with basis sets smaller than 6-31G* as the method does not include correlation. Average absolute errors in bond lengths and bond angles for a sample of 108 molecules containing two to eight atoms were reported by Scheiner, Baker, and Andzelm [10]. This result is included in Table 12.11. The B3PW91 hybrid functional gave the best results of the four functionals studied. The same team of scientists conducted DFT calculations with five different basis sets and found that as the basis set size increased, the errors in DFT geometries decreased significantly.

6. Dihedral angle computation provided an average absolute error [11]: of 3.8° with HF/6-31G*, 3.6° with MP2/6-31G*, and 3.4° with BP86 for a basis set that is TZP on nonhydrogens and DZP on hydrogens.

Table 12.10 Comparison of results with the aug-cc-pVTZ basis set

Average error	HF	MP2	MP4	CCSD	CCSD(T)
A-H bond length (A.U.)	0.014	0.011	0.007	0.009	0.009
A-B bond length (A.U.)	0.028	0.022	0.030	0.011	0.016
Bond angle (Degrees)	1.6	0.3	0.3	0.3	0.4

Table 12.11 Absolute average error with DFT methods

HF/ 6-31G**	MP2/ 6-31G**	SVWN/ 6-31G**	BLYP/ 6-31G**	BPW91/ 6-31G**	B3PW91/ 6-31G**
0.021A.U. 1.3°	0.015 A.U. 1.1°	0.016 A.U. 1.1°	0.021 A.U. 1.2°	0.017 A.U. 1.2°	0.011 A.U. 1.0°

Table 12.12 Average error semiempirical methods

Property	MNDO	AM1	PM3
Bond length in A.U.	0.055	0.051	0.037
Bond angle in degrees	4.3	3.8	4.3

Table 12.13 RMS error and MM methods

Property	MMFF94	MM3	UFF	CHARMm
Length (A.U.)	0.014	0.010	0.021	0.016
Angle (degrees)	1.2	1.2	2.5	3.1

7. Semiempirical methods usually give satisfactory bond lengths and angles. The results will be normally less accurate than that obtained by ab initio or DFT methods. For compounds containing H, C, N, O, F, Al, Si, P, S, Cl, Br, and I, average absolute errors in 460 bond lengths and 196 bond angles were reported by Stewart [12]. The result is included in Table 12.12.

   MNDO, AMI, and PM3 do not include d orbitals and are not particularly accurate for geometries of molecules with elements from the second and later rows. For such molecules, MNDO/d can be effectively used.

8. The performance of semiempirical methods for dihedral angles is not satisfactory.

9. MM force fields usually give good results for geometries for the kinds of molecules for which the field has been properly parameterized. In the experiment conducted by Halgren [13] for 30 organic compounds with MMFF94, MM3, UFF, and CHARMm, RMS errors for bond length and bond angle are given in Table 12.13.

12.14.2 Energy Changes

The result of experimentation conducted by Scheiner, Baker and Andzelm [10] is included in Table 12.14. They took 108 atomization energies (atom), 66 bond dissociation energies (BD), 73 hydrogenation enthalpies (HE) and 29 combustion energies (CE). The average absolute errors in kcal/mol were included.

For DFT, hybrid functionals are found to be very effective.

Table 12.14 Energy computation comparison

Method	Atom	BD	HE	CE
HF/6-31G**	119.2	58.8	8.5	44.5
MP2/6-31G**	22.0	8.8	7.0	11.2
SVWN/6-31G**	52.2	22.1	11.3	21.8
BPW91/6-31G**	7.4	5.9	10.1	27.6
BPW91/TZ2P	7.3	5.5	5.5	15.9
B3PW91/6-31G**	6.8	5.6	6.8	26.2
B3PW91/TZ2P	6.5	5.1	3.9	14.4

Holder et al. [14] conducted experiments on the standard enthalpy of formation using AM1, PM3, and SAM1 for molecules containing H, C, N, O, F, Cl, Br, and I. Average errors were found to be respectively 6.4, 5.3 and 4.0 kcal/mol. Stewart [12] conducted a semiempirical computation with 886 compounds of H, C, N, O, F, Al, Si, P, S, Cl, Br, and I. The average absolute error with in MNDO, AM1 and PM3 were 23.7, 14.2 and 9.6 kcal/mol.

Thiel and Voityuk [15] conducted computations for 99 S-containing compounds with semiempirical methods. As MNDO/d and SAM include d orbitals, these were found to give improved results. They did the computations with MNDO, AM1, PM3, SAM1, SAM1d, and MNDO/d. The average absolute gas phase errors in kcal/mol is found to be, respectively, 48.4, 10.3, 7,5, 8.3, 7.9, and 5.6.

MM2 and MM3 usually give gas-phase heats of formation with 1 kcal/mol accuracy for compounds similar to those used in the parameterization. For example, the average absolute MM3 error in the standard enthalpy of formation for a sample of 45 alcohols and ethers is 0.6 kcal/mol [16]. Many MM programs do not include provision for calculation of heats of formation.

12.14.3 Dipole Moments

Hehre reported the following average absolute errors for a sample of 21 small molecules HF/STO-3G–0.65D, HF/3-21G*–0.34 D and HF/6-31G*–0.30 D. The STO-3G basis set is not very reliable here. For a sample of 108 compounds, the average absolute errors with the 6-31G** basis set were [10]: HF–0.23 D, MP2–0.20 D, SVWN–0.23 D, BLYP–0.20 D, BPW91–0.19 D, B3PW91–0.16 D. Extremely accurate dipole moments were obtained with a gradient corrected functional and a very large basis set (an uncontracted version of the aug-cc-pVTZ set); for BLYP, the average absolute error was only 0.06 D. Semiempirical methods give reliable dipole moments. For 125 compounds of H, C, N, O, F, Al, Si, P, S, CI, Br, and I, average absolute errors are: MNDO–0.45 D, AMI–0.35 D, PM3–0.38 D [17]. For 196 compounds of C, H, N, O, F, CI, Br, and I, average absolute errors are: AMI–0.35 D, PM3–0.40 D, SAM1–0.32 D [18].

12.14.4 Generalizations

1. The overall reliability of the EH, CNDO, and INDO methods for calculating molecular properties is poor.
2. The ab initio SCF MO method is usually reliable for ground-state, closed-shell molecules, provided one uses a basis set of suitable size.
3. The STO-3G basis set is not generally reliable, and this basis set is little used nowadays.
4. MP2 perturbation theory usually substantially improves calculated properties, as compared with HF results.
5. DFT with gradient-corrected functionals (and especially hybrid functionals) usually performs substantially better than the HF method.
6. The AMI and PM3 methods are significantly less reliable than HF calculations with basis sets of suitable sizes.
7. MM is usually reliable for those kinds of molecules for which the method has been properly parameterized, but some existing MM force fields are not very reliable. For small and medium organic compounds, MM2, MM3, MM4, and MMFF94 are generally reliable.
8. MMFF94, OPLS, and AMBER force fields are found to be giving reliable structure predictions.
9. MMFF94 and OPLS fields give the best energy predictions. The comparisons of this section consider only compounds of H-Ar.
10. For compounds involving transition metals, ab initio SCF MO calculations often do not give good results.

The density-functional method may well be useful for transition-metal compounds.

12.15 Exercises

1. Find the energy difference between the trans and gauche conformations of dichloro ethane in the environments cyclohexane and gas phase using HF, MP2 (Onsager) and B3LYP. (#T B3LYP/6-31+G(d) SCRF(IPCM) SCF=Tight Test 6D).
2. Find the vibrational frequencies of formaldehyde in acetonitrile using the Onsager SCRF model and the SCIPCM model.
3. Predict the energy difference between the gauche and the trans conformers of dichloroethane in its liquid state ($e = 10.1$) and in acetonitrile ($e = 35.9$).
4. Compute the frequency associated with carbonyl stretch in a solution with acetonitrile for formaldehyde, acetaldehyde acetone, acrolein, formamide, acetyl chloride, and methyl acetate.
5. Use GaussView to draw carbon monoxide and set up an input file to perform a HF geometry optimization and frequency calculation with the 6-31+g(d) basis set. Use GaussView to visualize the results.

6. Find the single point energy calculation of water with #T RHF/6-31G(d) Pop = Full Test. At the end of this tutorial you should have the following:

    a. A printout of the HOMO.
    b. A printout of the LUMO.
    c. A printout containing the thermo-chemistry (enthalpy, entropy, free energy, thermal corrections, zero-point energy) and the archive.

7. Find the NMR shielding constants of methane. (The key word in the route section will be #T RHF/6-31G(d) NMR Test).

8. Run a single point energy calculation on propene and determine the following information from the output:

    a. What is the standard orientation of the molecule? In what plane do most of the atoms lie?
    b. What is the predicted HF energy?
    c. What is the magnitude and direction of the dipole moment of propene?
    d. Describe the general nature of the predicted charge distribution. (The key word is #T RHF/6-31G(d) Test).

9. Make a table of energies and dipole moments of three stereo isomers of 1,2-dichloro-1,2-difluoro ethane. You will be required to run the HF/6-31G(d) single point energy calculation for each. (Ref. Exer.2_02a(RR),2_02b(SS) and 2_02c(meso)).

10. Acetone and acetaldehyde are functional group isomers. Calculate the difference in the HF energy and dipole moments of these two.

11. Ethylene and formaldehyde are iso-electronic. Compare the dipole moment of these two. Compare the HOMO and LUMO in both.

12. Compare the NMR properties of butane, trans-2-butene and 2-butyne. (Ref. Exer.2_05a 2_05b and 2_05c) Run with HF/6-31G(d) and B3LYP/6-31G(d)).

13. Calculate the magnetic shielding of nitrogen in pyridene and compare it to its saturated cyclohexane analogue.

14. Fullerene compounds have received a lot of attention in recent years. Predict the energy of C-60 and look at its HOMO predicted at the HF level with the 3-21G basis set. Include SCF=Tight in the route section.

15. Run the geometry optimization of ethylene.(#T RHF/6-31G(d) Opt Test).

16. Find the energy difference between the trans and gauche conformations of dichloro ethane in the environments cyclohexane and gas phase using HF, MP2 (Onsager) and B3LYP. (#T B3LYP/6-31+G(d) SCRF(IPCM) SCF=Tight Test 6D).

17. Find the vibrational frequencies of formaldehyde in acetonitrile using the Onsager SCRF model and the SCIPCM model.

18. Perform frequency calculations of ethylene, chloro ethylene, vinyl alcohol, propene, and vinyl amine and study the vibrational and energy effects of these substitution on ethylene.

19. An amino acid can be present in two forms: the unionized form, $H_2NCHRCO_2H$, and the Zwitterion, $^+H_3NCHRCO_2^-$. You will explore the properties and the rel-

ative stability of the two forms in the case of the simplest amino acid glycine. An ab initio calculation with a good basis set will be used to obtain good energies.

20. Minimize the energy of the structure using MM with the Merck molecular force field. Comment on the structure of the conformer produced by the minimization.

21. Calculations on glycine in the unionized form, $H_2NCH_2CO_2H$. The unionized species is conformationally more flexible. Perform a search of low energy conformers.

22. A barrier to internal rotation of the amide bond: One can argue for a considerable double bond character in the amide bond using either valence bond or MO arguments. This double bond character was first noted by Pauling and makes an important contribution to the structure of proteins. The goal of this section of the exercise is twofold: 1) verify that the anti or s-trans conformer is more stable than the s-cis, and 2) determine the barrier to internal rotation.

References

1. Brown RD, Boggs JE, Hilderbrandt R, Lim K, Mills IM, Nikitin E, Palmer MH (1996) Pure Appl Chem 68:387
2. Stanton JF, Gauss J, Watts JD, Lauderdale WJ, Bartlett RJ, (1992) Int J Quant Chem S26: 879
3. Bader RFW (1990) Atoms in Molecules: A Quantum Theory. Clarendon, Oxford
4. Wiberg KB, Rablen PRJ (1993) Comp Chem 14:1504
5. Becke ADJ (1993) Chem Phys 98:5648
6. Stephens PJ, Devlin FJ, Chabalowski CF, Frisch M (1994) J Phys Chem 98: 11623
7. Pople JA, Head-Gordon M, Fox DJ, Raghavachari K, Curtiss LA (1989) J Chem Phys 90:5622
8. Curtiss LA, Raghavachari K, Trucks GW, Pople JA (1989) J Chem Phys 94:7221
9. Feller D, Peterson KA (1998) J Chem Phys 108:154
10. Scheiner AC, Baker J, Andzelm JW (1997) J Comp Chem 18:775
11. St-Amant A et al. (1995) J Comp. Chem 16:1483
12. Stewart JJP (1991) J Comp Chem 12:320
13. Halgren TA (1996) J Comp Chem 17:553
14. Holder AJ et al. (1994) Tetra 50:627
15. Thiel WR, Voityuk AA (1996) J Phys Chem 100:616
16. Allinger NL et al. (1990) J Am Chem Soc 112:8293
17. Stewart JJP (1989) J Comp Chem 10:221
18. Dewar MJS et al. (1993) Tetra 49:5003

Chapter 13
High Performance Computing

13.1 Introduction – Supercomputers vs. Clusters

Supercomputer is a term that people use to represent enormous processing capacity. The machines of CRAY class have made people think about these in a totally different way – a huge computer having multiples of processors on the single board. Traditionally, supercomputers have only been built by a selected number of vendors. A company or organization that required the performance of such a machine had to have a huge budget required for its supercomputer. People started thinking of some other better alternative which they could afford. The concept of cluster computing was introduced when people first tried to spread different jobs over more computers and then gather back the data from these systems. With the development of the personal computing (PC) platform, the performance gap between a supercomputer and a cluster of multiple personal computers became smaller. So, today when we say supercomputer, it is just a term to mean a huge processing capacity and any machine giving specific gigaflops speed may be considered to have supercomputing facilities.

13.2 Clustering

In general, clustering refers to technologies that allow multiple computers to work together to solve common computing problems. To anyone who has worked as a network or system administrator, some of the benefits of clustering will be immediately apparent. The increased processing speed offered by performance clusters, increased transaction or response speed offered by load-balancing clusters, or the increased reliability offered by high availability clusters can be vital in a variety of applications and environments [1].

Take, for example, the modeling of a macromolecule or a polymer or a biopolymer like protein. This requires massive amounts of data and very complex calculations. By combining the power of many workstation-class or server-class machines,

K. I. Ramachandran et al., *Computational Chemistry and Molecular Modeling*   275
DOI: 10.1007/978-3-540-77304-7, ©Springer 2008

performance levels can be made to reach supercomputer levels, and that even for a much lower price than the traditional supercomputer. Most people consider *clustering* or *server clustering* as a high performance group of computers used for scientific research. However, this is just one of the types of clustering available. The basic idea behind the "performance clustering" approach is to make a large number of individual machines act like a single, very powerful machine. This type of cluster is best applied to large and complex problems that require huge computing horsepower. Applications such as molecular dynamic simulations, the modeling of polymers, computational drug designing, and quantum mechanical modeling are prime areas of computational chemistry for high-performance clusters.

A second type of clustering technology allows a network of servers to share the load of traffic from clients. By load balancing the traffic across an array of servers, access time improves and the reliability of computation increases. Moreover, since many servers are handling the work, the failure of one system will not cause a catastrophic breakdown.

Another type of clustering involves the servers to act as live backups of each other. This is called *high availability clustering* (HA clustering) or *redundancy clustering*. By constantly tracking the performance and stability of the other servers, a high availability cluster allows for greatly improved system uptimes. This can be crucial in high traffic simulation sites. Load balancing and high availability clusters share many common components, and some clustering techniques make use of both types of clustering.

13.3 How Clusters Work

At its core, clustering technology has two basic parts. The first component of clustering consists of a customized operating system (such as the kernel modifications made to Linux) with special compiler programs to take full advantage of clustering. The second component is the hardware interconnection (interconnects) between machines (nodes) in the server cluster. These interconnects are often highly dedicated interfaces. In some cases, the hardware will be designed specifically for the clustered systems. However, in most common Linux cluster implementations, this interconnect is handled by a dedicated fast *Ethernet* or *gigabit Ethernet network*. The assignment of tasks, status updates, and program data can be shared between machines across this interface, while a separate network is used to connect the cluster to the outside world. The same network infrastructure can often be used for both of these functions. However, this simplification may affect the performance of computing, especially when the network traffic is high [2]. By splitting the problem into tasks that can be executed in a parallel manner, computation is carried out fast.

Performance clustering works in a similar manner to traditional symmetric multiprocessor (SMP) servers. The most widely known high-performance clustering solution for Linux is *Beowulf* (Fig. 13.1). It grew out of research at NASA and can provide supercomputer-class processing power for the cost of run-of-the-mill PC hardware. By connecting those PCs through a fast Ethernet network, the computing power is increased to the level of a supercomputer. Probably the best-known type of Linux-based cluster may be a Beowulf cluster. A Beowulf cluster consists of multiple machines connected to one another on a high speed LAN. In order to extract the computing resources of clusters, special cluster-enabled applications must be written using clustering libraries. The most popular clustering libraries are PVM and the message passing interface (MPI). By using the clustering libraries, programmers can design applications that can span across an entire cluster computing resources rather than being confined to the resources of a single machine. For many applications, PVM and MPI allow computing problems [3, 4] to be solved at a rate that scales almost linearly relative to the number of machines in the cluster.

The servers of a high availability cluster do not normally share the processing load, unlike performance cluster servers. Nor do they share the traffic load as load-balancing clusters do. Instead, they keep themselves ready to take over the computational charge for a failed or defective server instantaneously. Although we will not get the performance increased from a high availability cluster, due to their increased flexibility and reliability, they have been made necessary in today's information-intensive computational environment. High availability clustering also allows easier server maintenance. One machine from a cluster of servers can be taken out, shut down, upgraded, reloaded after sometime, or allowed to work without collecting information from it.

13.4 Computational Clusters

Computational/high performance Linux clusters started back in 1994 when Donald Becker and Thomas Sterling built a cluster for NASA. This cluster was made up of 16 DX4 processors connected by 10 Mbit Ethernet and was named Beowulf. Since then, the Beowulf project has been joined by other software projects trying to provide useful solutions to turning commercial off the shelf (COTS) hardware into clusters capable of supercomputing speed [5, 6, 18].

13.5 Clustering Tools and Libraries

MPI is a library specification for message-passing, proposed as a standard by the industry consortium of vendors, implementers, and users. It has many free and commercial implementations, but because MPI is an open standard, while any person or company can twist MPI to optimize it for his or their own use, the calling structure and API must remain unchanged. All manufacturers of commercial supercomputers provide a version of MPI with their systems.

Fig. 13.1 The full Perseus Beowulf cluster

LAM/MPI is a high-quality open-source implementation of MPI, including all of MPI-1.2 and much of MPI-2. LAM/MPI has a rich set of features for system administrators, parallel programmers, application users, and parallel computing researchers.

From its beginnings, LAM/MPI was designed to operate on heterogeneous clusters. With support for Globus and Interoperable MPI, LAM/MPI can span clusters of clusters. Several transport layers, including Myrinet, are supported by LAM/MPI. With TCP/IP, LAM imposes virtually no communication overhead, even at gigabit Ethernet speeds. New collective algorithms exploit hierarchical parallelism in SMP clusters. Some of the useful MPI formulations are listed below.

LAM (local area multicomputer) is an MPI programming environment and development system introduced at the Ohio Supercomputer Center and Notre Dame University, now being developed and maintained by a group at Indiana University. It is freely available for download.

MP-Lite is a lightweight message passing library designed to deliver the maximum performance to applications in a portable and user-friendly manner.

MPICH is a portable implementation of MPI, developed at Argonne National Laboratory. It is freely available, and an extremely vanilla implementation of MPI, which makes it easy for porting to various Unix modifications. There is also a Windows NT version available.

13.6 The Cluster Architecture

It has become widely accepted that cluster setup and management is extremely tedious and error-prone, due to the inherent autonomy of the nodes in a cluster. Hence, using a cluster is much more difficult than using a traditional supercomputer. These

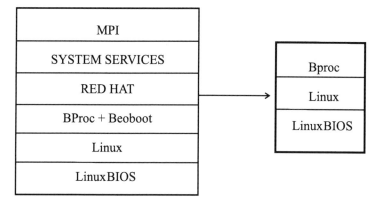

Fig. 13.2 For this modified cluster architecture, only the front end is made as a fully loaded system. The cluster nodes themselves have installed only LinuxBIOS. They receive the kernel (BProc + Linux) from the front end

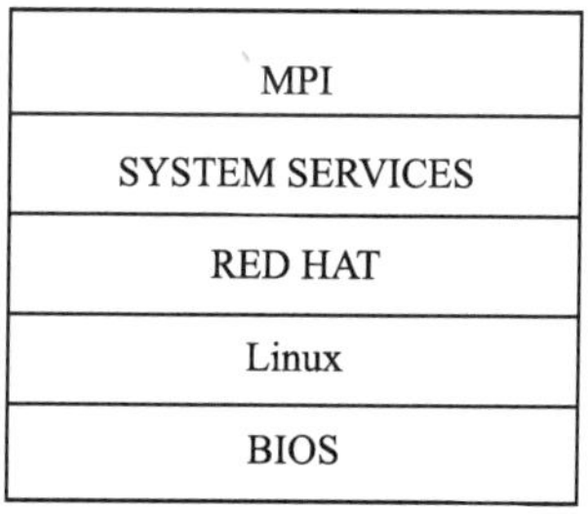

Fig. 13.3 For a traditional cluster configuration, each node is a fully loaded independent system

| MPI |
| SYSTEM SERVICES |
| RED HAT |
| Linux |
| BIOS |

problems can be overcome by redesigning the cluster architecture from low-level machine setup to programming support level. By modifying the key components of the cluster and adding vital functionality, the reliability and efficiency of the cluster can be increased with a decrease in autonomy [8, 9].

This cluster architecture design replaces legacy mechanisms for booting (LinuxBIOS) and runs an operating system that provides a *single system image* of the entire cluster (BProc) (Fig. 13.2). It is interesting to compare this method with the traditional cluster architecture which is a loose coupling of many individual single user workstations (Fig. 13.3).

13.7 Clustermatic

Clustermatic is a collection of new technologies being developed specifically for our new cluster architecture and is expected as the complete cluster solution of the future. Each technology can be used separately, and it does not prohibit integration with other clustering efforts or even other types of computing environments. For example, BProc is being used in several production-grade clusters; LinuxBIOS is being sold in products such as web content caching appliances, DVD players, and fiber channel analyzers [10].

13.8 LinuxBIOS

LinxBIOS replaces the normal BIOS bootstrap mechanism with a Linux kernel that can be booted from a cold start. Cluster nodes can now be as simple as they need to be – perhaps as simple as a CPU and memory, without any disk, floppy, and file system. As a consequence of this, the nodes are up and fit for running in less than two or three seconds.

13.9 BProc

The Beowulf Distributed Process Space (BProc) provides a single system image of the entire cluster. LinuxBIOS cluster nodes come up autonomously and contact the "front end" node which sends them to a BProc kernel to boot and register them as part of the cluster. Users run programs on the front end, which will be carried to other cluster nodes. BProc itself consists of a small set of kernel modifications, utilities, and libraries which allow a user to start processing on other machines in a cluster (including reboot). Remote processes started with this mechanism appear in the process table of the front end. It allows remote process management using the normal UNIX process control facilities. Signals are transparently forwarded to remote processes and exit status is received using the usual "wait" mechanisms. Clusters with thousands of nodes may experience failures very frequently. Programs will need to be much more resilient and *run-through* to completion despite failures [11].

13.10 Configuration

The cluster can contain any number of nodes as we wish. The decision will be based on how much processing capacity we need. It can be a simple cluster with a server and two nodes or a bigger one containing a server and 10 client nodes. They can be connected using a simple switch over the UTP cabling in an Ethernet environment. The sample configuration steps given below is with one machine acting as a server and two other machines acting as the clients for this server. To start with this configuration is pretty good, and once you are able to configure and use this as mentioned, you can go on adding more number of machines to get a better performance. However, please do not think that by just adding several nodes you will get a high performance machine. Since we are communicating over a Ethernet network, and the system works as a cluster of workstations exchanging the data and the results over the network, due to the network traffic overheads sometimes the performance of the cluster may go down also. So, we need to find a optimum number of clients in the cluster [12].

The overall configuration outlook can be summarized as follows. ClusterNFS software allows minimizing system administration overheads for the cluster. Most configuration files are shared amongst the client nodes. Because of the shared root, any package installed on the server is automatically available on clients. In the designed cluster, the client nodes are not supposed to be used as an independent workstations. Therefore, most of the network services are switched off [13].

13.11 Setup

This document has been written *after* the actual installation. Therefore, some minor but important points may be missed. Also, some of package versions used in building are no longer available. This can be upgraded from the net. There is no master node from the MOSIX point of view. However, one node (server) plays a special role by booting the rest of the cluster nodes, running their root directories via NFS, connecting the cluster with the external network, and providing disk space.

13.12 The Steps to Configure a Cluster

The steps for the configuration of the cluster are given below.

1. Physically, connect the machines through a switch so that they will be able to communicate one another once they are configured as a cluster.
2. Select the machine which is to act as the master node. In this machine, go to the BIOS setup and configure the boot sequence to point to the CD-ROM as the first boot device.
3. Now, identify the machines which will be acting as the clients or nodes for the master node. Configure these machines with the CD-ROM to be the first boot device so that we will be able to boot with the Linux boot CD-ROM. In addition, we need to configure some hardware settings also on the clients. This is because when the client nodes are working as a cluster, they will not be having any monitor, keyboard or mouse connected to it and only the network card will have a connection going out from the system box.

However, when the computer boots up, during the post routine, it usually gives an error message and halts if the peripheral devices such as the keyboard or mouse are not found, since they were present at the time of installation and are removed only after the installation and prior to connecting to the switch to act as a cluster [14, 15].

So, in the BIOS, we can set up the passwords for the user and also for the system, and this disables the peripheral device checking. In the "Advanced Options" of the BIOS, we can set "Post mode" as Quick boot and the "Post messages" value to be set as Disable. Under "Device Options" set Monitor Tracking as Disable and Integrated Video as Disable. If the BIOS type is a generic one, set the "Halt On" option with the value "No errors" so that even if the peripheral devices are not found, it will not create an error. Once these settings are effected in the BIOS, save the settings and come out.

Put the Linux Boot CD-ROM in the CD drive and start the machine. The boot process starts and the machine starts reading from the drive and displays the "boot prompt." At this prompt, type "linux text" to denote that you want to boot from the Linux kernel and want to opt for a text mode of installation as opposed to the graphics mode.

Now, the installation proceeds and the cluster will be configured. We have used RedHat Linux 9.0 in our sample setup and the hardware included the Pentium III class of machines from Compaq; we found that even a normal assembled PC was working fine as a cluster.

The step-by-step installation procedure involves the following:

1. Installation of the LAM.
2. Configuration of the NIS server.
3. Configuration of the NIS clients.
4. Network configuration of the server node.
5. Creation of a network file system.
6. Clustermatic installation.

Details of the installation with a suitable example have been included in the text URL.

13.13 Clustering Through Windows

Windows mainly supports three cluster technologies to provide high availability, reliability, and scalability [16,17]. These technologies are described in the following sections.

13.13.1 Network Load Balancing Clusters

Network load balancing (NLB) clusters provide failover support for IP-based applications. They are ideally suited for Web-tier and front-end services. NLB clusters can make use of multiple adapters and different broadcast methods to assist in the load balancing of TCP, UDP, and GRE traffic requests.

13.13.2 Server Clusters

Server clusters are suited for back-end applications and services, such as database servers. Server clusters can use various combinations of active and passive nodes to provide failover support for mission critical applications and services.

13.13.3 Component Load Balancing

Component load balancing (CLB) provides dynamic load balancing of middle-tier application components that use COM+ and is ideally suited for application servers. CLB clusters use two clusters. The routing cluster can be configured as a routing list on the front-end web servers or as separate servers that run server cluster.

13.14 Installing the Windows Cluster

When we install the Windows 2003 Server, the Cluster Administrator is installed by default along with installing the Windows server (e.g., the Windows 2003 server – WS2K3). We need to launch the Cluster Administrator to start the configuration of the cluster by going to Start-Administrative Tools-Cluster Administrator. When installing a new cluster, we do not have to reboot the system, which is a great time saver. The major advantages of WS2K3 server are:

1. *Larger clusters:* The Enterprise Edition supports up to 8-node clusters. Previous editions only supported 2-node clusters. The Datacenter Edition supports 8-node clusters as well. In Windows 2000, it supported only 4-node clusters.
2. *64-bit support:* This feature allows clustering to take advantage of the 64-bit version of Windows Server 2003, which is especially important to optimize the SQL Server 2000 Enterprise Edition.
3. *High availability:* With this update to the clustering service, the Terminal Server directory service can now be configured for failover.
4. *Cluster Installation Wizard:* A completely redesigned wizard allowing us to join and add nodes to the cluster, and providing an additional troubleshooting facility to view logs and details if things go wrong. It helps us to save some trips to the Add/Remove Programs applet.
5. *Microsoft Distributed Transaction Coordinator* (MSDTC) configuration: We can now configure MSDTC once and it is replicated to all nodes, which eliminates the requirement to run the comclust.exe utility on each node.

13.15 Grid Computing

Grid computing, most simply stated, is distributed computing taken to the next evolutionary level. The goal is to create the illusion of a simple yet large and powerful self-managing virtual computer out of a large collection of connected heterogeneous systems, sharing various combinations of resources.

The standardization of communications between heterogeneous systems have created the Internet explosion. The emerging standardization for sharing resources, along with the availability of higher bandwidth, is the driving force behind the evolutionary step in grid computing. The basic principle of grid computing are summarized in the next sections.

13.15.1 Exploiting Underutilized Resources

The minimum use of grid computing is to run an existing application on a different machine. The machine on which the application is normally run might be unusually busy due to an unusual peak in activity. The job could be run on an idle machine elsewhere on the grid.

There are at least two prerequisites for this scenario. The application must be executable in a remote site without undue overhead of any system. The remote machine must be in a position to meet any special hardware, software, or resource requirements imposed by the application.

For example, a batch job that spends a significant amount of time for processing a set of huge input data to produce an output set is perhaps the most ideal and simple use of a grid. If the size of the input and output are large, proper planning might be required to efficiently use the grid. It would usually not make sense to use a word processor remotely on a grid because there would probably be greater delays and more potential points of failure.

In most organizations, there are large amounts of underutilized computing resources. Most desktop machines are busy less than 5 percent of the time. In some organizations, even the server machines can often be relatively idle. Grid computing provides a framework for exploiting these underutilized resources and thus has the possibility of substantially increasing the efficiency of resource usage.

The processing resources are not the only ones that may be underutilized. Often, machines may have enormous unused disk drive capacity. Grid computing, more specifically, a "data grid", can be used to aggregate this unused storage into a much larger virtual data store, possibly configured to achieve improved performance and reliability over that of any single machine.

13.15.2 Parallel CPU Capacity

The potential for massive parallel CPU capacity is one of the most attractive features of a grid. In addition to pure scientific needs, such computing power is driving a new evolution in industries such as the biomedical field, financial modeling, oil exploration, motion picture animation, and many others.

The common attribute among such uses is that the applications have been written to use algorithms that can be partitioned into independently running parts. A CPU intensive grid application can be thought of as many smaller "subjobs," each executing on a different machine of the grid. If the subjobs do not need to communicate with each other, then the application becomes more "scalable." A perfectly scalable application will, for example, finish 10 times faster if it uses 10 times the number of processors.

Barriers often exist to perfect scalability. The major barrier depends on the algorithms used for splitting the application among many CPUs. If the algorithm can only be split into a limited number of independently running parts, then that itself becomes a scalability barrier. The second barrier appears if the parts are not completely independent, which causes contention, limiting the scalability.

For example, if all the subjobs need to read and write from one common file or database, the access limits of that file or database will become a limiting factor in the application's scalability. Other sources of interjob contention in a parallel grid application include message communications latencies among the jobs, network communication capacities, synchronization protocols, input-output bandwidth to devices, and storage devices and latencies interfering with real-time requirements.

13.16 Types of Resources Required to Create a Grid

A grid is a collection of machines referred to as nodes, resources, members, donors, clients, hosts, engines, and so on. They all contribute any combination of resources to the grid as a whole. Some resources may be used by all users of the grid while others may have specific restrictions.

13.16.1 Computational Resources

The most common resource is computing cycles provided by the processors of the machines on the grid. The processors can vary in speed, architecture, software platform, and other associated factors, such as memory, storage, and connectivity. There are three primary ways to exploit the computation resources of a grid. The first and the most common way is to run an existing application on an available machine on the grid rather than locally. The second is to use an application designed to split its work in such a way that the separate parts can execute the job in parallel on different processors. The third way is to run an application that needs to be executed many times, on many different machines in the grid.

Scalability is a measure of how efficiently the multiple processors on a grid are used. If twice as many processors makes an application complete in one half of the time, then it is said to be perfectly scalable. However, there may be limits to scalability when applications can only be split into a limited number of separately running parts or if those parts experience some other contention for resources of some kind.

13.16.2 Storage Resources

The second resource used in a grid is data storage. A grid providing an integrated view of data storage is sometimes referred to as a data grid. Each machine on the grid usually provides some quantity of storage facility for grid use, even if it is temporary. Storage can be memory attached to the processor, secondary storage using hard disk drives, or other permanent storage media. Memory attached to a processor usually has very fast access, but it is highly volatile. It would best be used to cache data or to serve as temporary storage for running applications.

Secondary storage in a grid can be used in an effective manner to increase the capacity, the performance, the sharing, and the reliability of data. Many grid systems use mountable "networked file systems," such as the Andrew File System (AFS®), the Network File System (NFS), the Distributed File System (DFS™), or the General Parallel File System (GPFS). These offer varying degrees of performance, security features, and reliability features.

Capacity on the grid can be increased by using the storage on multiple machines with a unifying file system. Any individual file or database can span several storage devices and machines, eliminating maximum size restrictions. This may often anchor with the operating system imposed by file systems. A unifying file system can also provide a single uniform name space for grid storage. This makes it easier for the users to access the reference data residing in the grid. In a similar way, special database software can amalgamate an assortment of individual databases and file to form a larger, more comprehensive database, which are accessible using database query functions.

More advanced file systems on a grid can automatically duplicate sets of data, to provide redundancy for increased reliability and increased performance. An intelligent grid scheduler can help to select the appropriate storage devices to hold data, based on usage patterns. Jobs can be assigned closer to the data, preferably on the machines directly connected to the storage devices holding the required data.

Data striping can also be implemented by grid file systems. When there are sequential or predictable access patterns to data, we can create the virtual effect of having storage devices that can transfer data at a faster rate than any individual disk drive. This effect is very important either for multimedia data streams, or while collecting large quantities of data at extremely high rates from CAT scans, or particle physics experiments, for example.

A grid file system can also implement journaling so that data can be recovered in a reliable manner, even after certain kinds of unexpected failures. In addition, some file systems implement advanced synchronization mechanisms to reduce contention while sharing and updating the data by a number of users.

13.16.3 Communications Mechanisms

The rapid development of communication capacity among machines today makes grid computing practical, compared to the limited bandwidth available when distributed computing was first emerged. Hence, another important resource of a grid is data communication capacity. This includes communications within the grid and external to the grid. Communications within the grid are required for sending jobs and their required data to points within the grid. Some jobs require a large amount of data to be processed and it may not always reside on the machine running the job. The bandwidth available for such communications can often be a critical resource that can limit the utilization of the grid.

External communication access to the Internet, for example, can be a valuable factor while building search engines. Machines on the grid may have connections to the external Internet besides the connectivity among the grid machines. When these connections do not share the same communication path, then that may be added to the total available bandwidth for accessing the Internet.

Redundant communication paths are sometimes needed to handle the potential network failures and excessive data traffic. In some cases, higher speed networks must be provided to meet the demands of jobs transferring larger amounts of data. A grid management system can better show the topology of the grid and highlight the communication bottlenecks. This information can in turn be used to plan for hardware upgrades.

13.16.4 The Software and Licenses Required
to Create the Grid

The grid may have software installed that may be too expensive to install separately on every grid machine. Using a grid, the jobs requiring this software can be sent to the particular machines on which this software happens to be installed. When the licensing fees are significant, this approach can save significant expenses for an organization.

Some software licensing arrangements permit the software to be installed on all of the machines of a grid but may limit the number of installations that can be simultaneously used at any given instant. *License management software* keeps track of how many concurrent copies of the software are being used, and it prevents the users from executing the job simultaneously. The grid job schedulers can be configured to take software licenses into account, optionally balancing them against other priorities or policies.

13.17 Grid Types – Intragrid to Intergrid

There have been attempts to formulate a precise definition for what a "grid" is. In fact, the very concept of grid computing is still evolving. We will be pragmatic in this regard. We do not claim to make any complete definition of a grid. Therefore, the following descriptions of various kinds of "grids" must be considered in that spirit.

Grids can be built in all sizes, ranging from just a few machines in a department to groups of machines organized as a hierarchy spanning the world. In this section, we will describe some examples in this range of grid system topologies.

The simplest grid consists of just a few machines, all with the same hardware architecture and the same operating system, connected on a local network. This kind of grid uses homogeneous systems. The machines may be in one department of an organization, and their use as a grid may not require any special policies or security concerns. As the machines have the same architecture and operating system, choosing application software for the grid is usually simple. Some people would call this a cluster implementation rather than a "grid."

The next progression would be to include heterogeneous machines. In this configuration, more types of resources are available. The grid system is likely to include some scheduling components. File sharing may still be accomplished using networked file systems. Machines participating in the grid may include multiple departments still within the same organization. Such a grid is referred to as an *intragrid*.

As the grid expands to many departments, policies may be set up for the use of the grid. For example, there may be policies for the kind of work allotted to the grid and even the duration completion of work. There may be a set prioritization for each department regarding the users, applications, and resources of the grid.

The security element becomes more important if more or different organizations are involved. Sensitive data in one department may need to be protected from access by jobs running for other departments. Dedicated grid machines may be added to increase the quality of service for grid computing.

The grid may grow geographically in an organization that has facilities in different cities. Dedicated communications connections may be used among these facilities and the grid. In some cases, VPN tunneling or other technologies may be used over the Internet to connect the different parts of the organization. The grid may grow to be hierarchically organized to reduce the contention implied by central control, increasing scalability.

Over time, a grid may grow to cross organization boundaries, and may be used to collaborate on projects of common interest. This is known as an *intergrid*. The highest levels of security are usually required in this configuration to prevent possible attacks and spying. The intragrid offers the prospect for trading or brokering resources over a much wider audience. Resources may be purchased as a utility from trusted suppliers.

13.18 The Globus Toolkit

The Globus Toolkit (GT) is a joint initiative of the University of Southern California, the Argonne National Lab, and the University of Chicago. It provides an open-source set of services addressing fundamental grid issues, such as security, information discovery, resource management, data management, and communication.

The GT is described by its authors as being made up of three pillars of *resource management* (RM), allocating resources provided by the grid to the respective consumer, *information services* (IS), providing information about available resources and their attributes, and *data management* (DM), dealing with accessing and managing data in a grid (e.g., it provides a more robust and high-performance ftp, customized to grid needs). Each pillar embeds core services given by Globus Security Infrastructure (GSI). GSI ensures fundamental security services such as authentication, confidentiality, and integrity.

The GT supports Red Hat Linux on xSeries, AIX on pSeries and SuSE Linux Enterprise Server 8 (SLES 8) on zSeries, containing the pre-compiled binary distribution of the Globus 2.0 code for Linux on zSeries. We can find out more about the GTPL at: http://www.globus.org/toolkit/download/license.html. For platform specific system requirements for the GT 2.2, please refer to the following Web site: http://www.globus.org/gt2.2/platform.html.

13.19 Bundles and Grid Packaging Technology

Grid packaging technology (GPT) is a package used for installation and distribution, which includes libraries, files, and modules to support package creation and installation. It supports the installation of GT bundles. The package contains the executable files, script files, and configuration files. There are two types of bundles, *source bundles* (Table 13.1) and *binary bundles* (Table 13.2). The binary bundles contain the binary executable files that have been precompiled for specific platforms.

Other platform-specific binary bundles are available at the following Globus FTP site: ftp://ftp.globus.org/pub/gt2/2.2/2.2-latest/bundles/bin/. The installation of the grid involves the following steps:

1. Installing the GPT.
2. Installing the source and binary bundles.

Table 13.1 Source bundle

	Client bundle	Server bundle	SDK bundle
Resource management	globus-resourcemanagement-client-2.2.2-src_bundle.tar.gz	globus-resourcemanagementserver-2.2.2-src_bundle.tar.gz	globus-resourcemanagement-sdk-2.2.2-src_bundle.tar.gz
Information services	globus-informationservices-client-2.2.2-src_bundle.tar.gz	globus-informationservices-server-2.2.2-src_bundle.tar.gz	globus-informationservices-sdk-2.2.2-src_bundle.tar.gz
Data management	globus-datamanagement-client-2.2.2-src_bundle.tar.gz	globus-datamanagementserver-2.2.2-src_bundle.tar.gz	globus-datamanagement-sdk-2.2.2-src_bundle.tar.gz

Table 13.2 Binary bundles

Binary bundle	Contents
globus-all-2.2.2-i686-pclinux-gnu-bin.tar.gz	Client and server packages: resource management, information services and data management
globus-all-server-2.2.2-i686-pc-linux-gnu-bin.tar. gz	Server packages
globus-all-client-2.2.2-i686-pc-linux-gnu-bin.tar.gz	Client packages
globus-all-sdk-2.2.2-i686-pc-linux-gnu-bin.tar.gz	SDK packages
globus-data-managementserver-2.2.2-i686-pc-linuxgnu- bin.tar.gz	Server packages for the data management
globus-data-managementclient-2.2.2-i686-pc-linuxgnu-bin.tar.gz	Client packages for the data management
globus-data-managementsdk-2.2.2-i686-pc-linuxgnu-bin.tar.gz	SDK bundles for the data management
globus-informationservices-server-2.2.2-i686-pc-linux-gnu-bin.tar.gz	Server packages for the information service
globus-informationservices-client-2.2.2-i686-pc-linux-gnu-bin.tar.gz	Client packages for the information service
globus-informationservices-sdk-2.2.2-i686-pc-linux-gnu-bin.tar.gz	SDK packages for the information service
globus-resourcemanagement-server-2.2.2-i686-pc-linux-gnu-bin.tar.gz	Server packages for the resource management
globus-resourcemanagement-client-2.2.2-i686-pc-linux-gnu-bin.tar.gz	Client packages for the resource management
globus-resourcemanagement-sdk-2.2.2-i686-pc-linux-gnu-bin.tar.gz	SDK packages for the resource management

3. Installation of the grid node and the certificate authority.
4. Setting up of the grid environment.
5. Creating the certificate authority.

6. Creating the file to be distributed.
7. Requesting and signing the gate keeper certificates for servers.
8. Requesting and signing the user certificates.
9. Setting up gate keepers.
10. Setting up the Monitory Discovery Service (MDS).
11. Setting up the Grid Information Index Service (GIIS) in the alpha machine, which collects the data reported by the Grid Resource Information Servers (GRIS).
12. Setting up the GRIS on beta, gamma, and delta.
13. Starting the MDS on all servers.
14. Setting up the MDS client.
15. Setting up a secure MDS.
16. Requesting and signing certificates for each server machine.
17. Checking the installation.

An illustrative example showing all these steps is included in the URL.

13.20 The HPC for Computational Chemistry

13.20.1 The Valence-Electron Approximation

In the modeling formulation of a molecule containing an n-electron, the first step is to write the Slater determinant of orbitals which will be of the dimension $n \times n$. If the molecule has a very large number of electrons, the computation becomes really difficult. One of the methods to simplify the calculation is to make the valence-electron approximation. In this approximation, core (inner) electrons are considered as point charges coinciding with the nucleus. As for example, for the system Na_2, a 22×22 determinant can be reduced to a 2×2 determinant. The Hamiltonian for the system becomes identical with that of H_2. Here, we make a constraint to avoid collapsing of valence electrons into the inner orbital, which is supposed to be vacant in this approximation. One way to overcome this difficulty is by making the variational functions of the valence electrons orthogonal to the orbitals of the core electrons.

13.20.2 The Effective Core Potential

Another approach is to treat the core electrons as an imaginary sphere of dense charge distribution providing a high repulsive potential and preventing the valence electrons to collapse into the inner orbital. This potential is referred to as the effective core potential (ECP) or pseudopotential.

The ECP is a one-electron operator that replaces the two-electron Coulomb and exchange operators of the HF equation in the computation of the Hamiltonian of valence electrons. For compounds of the main-group elements, calculations with the ECP gives results comparable with all-electron ab initio calculations. However for transition metals, accurate results with the ECP is harder [19].

13.20.3 The Direct SCF Method

Another suggestion in this regard is to calculate all the integrals and sore them properly so that they can be recalled in any SCF iteration. Here, the problem is the storing difficulty especially for calculations with higher basis sets. If an external disk is storing the integrals, then the iteration may become very slow.

To avoid the use of external storage memory, Almlof developed a method known as "the direct SCF method". In this method, the integrals are calculated and used immediately on each iteration and are never stored. This requires more CPU time, but much less disk space. Three improvements on the direct self-consistent field method are proposed by Marco Häser and Reinhart Ahlrichs, which together increase CPU efficiency by about 50%: (1) the selective storage of costly integral batches, (2) the improved integral bond for prescreening, and (3) the decomposition of the current density matrix into a linear combination of previous density matrices – for which the two-electron contributions to the Fock matrix are available – and a remainder ΔD, which is minimized; the construction of the current Fock matrix only requires processing of the small ΔD which enhances prescreening.

13.20.4 The Partially Direct SCF Method

The partially direct SCF method was developed to improve the computing efficiency by parallelization using a PC cluster without secondary storage on each processor (Table 13.3). Some of the electron repulsion integrals are stored in the buffer (unused memory) with their four indices at the first SCF cycle, and they are reused at the later SCF cycles. This simple method achieved super-linear scalability, for example, the parallelization efficiency became ca. 1.13 in the Fock matrix generation of the Crambin molecule (1974 basis functions), equipped by the 128 Xeon processors (2.8 GHz) with 16 GB buffer area. This algorithm is suitable for the special purpose

Table 13.3 Efficiency of parallelization: a comparative study with direct SCF

Type of computation	2 Proc	4 Proc	8 Proc	16 Proc	32 Proc	64 Proc	128 Proc
Direct SCF	0.992	0.980	0.988	0.976	0.981	0.980	0.978
PDSCF 16	0.980	0.982	0.976	0.982	0.988	0.992	0.998
PDSCF-32	0.981	0.984	0.995	0.993	0.991	1.01	1.02
PDSCF-64	0.983	0.986	0.996	0.992	1.006	1.021	1.052
PDSCF-128	0.989	0.989	0.999	1.005	1.02	1.059	1.131

computers for fast evaluation of the electron repulsion integrals because the recent special purpose processor has usually no secondary storage and has a relatively large main memory.

13.21 The Pseudopotential Method

13.21.1 The Block-Localized Wavefunction Method

The block-localized wave function (BLW) method was developed to circumvent the delocalized nature of molecular orbitals in the HF theory to study properties of localized, or valence bond-like, electronic structures. Although the ab initio valence bond (VB) method can be used to study the resonance effect and to define electronic localized states, its computational costs can quickly become intractable and thus prevent applications to large molecular systems. The BLW method provides a convenient approach to define valence bond-like resonance configurations at the computational cost comparable to HF molecular orbital calculations.

We have seen that DFT-based methods employing non-hybrid exchange-correlation functionals are more accurate than standard HF methods. They are applicable to a much wider class of chemical compounds and are faster by orders of magnitudes compared to HF implementations. This remarkable feature arises from the separate treatment of the Coulomb and exchange contributions to the KS matrix, which allows exploiting more efficient techniques for their evaluation. With DFT, employing hybrid exchange-correlation functionals this advantage is lost and only the (slower) traditional direct HF procedures are applicable. Thus, non-hybrid DFT is the natural choice for electronic structure calculations on much-extended systems, which are otherwise intractable by quantum mechanical methods. However, as the exchange-correlation functional is unknown, DFT suffers from the distinct disadvantage that, in contrast to more traditional quantum chemistry methods, there is no systematic way to improve and to assess the accuracy of a calculation. Fortunately, extensive experience shows which classes of chemical compounds can be modeled with good success.

Serial linear algebra routines have to be replaced in many cases by parallel versions, either because the size of the matrices enforces distributed data or due to the cubic scaling with the problem size. In some cases, the replacement by alternative algorithms is more advantageous either due to better parallel scalability or more favorable cache usage. The evaluation of a pairwise potential over a large number of particles is a rather widespread problem in the natural sciences. One way to avoid the quadratic scaling with the number of particles is the fast multipole method (FMM) which treats a collection of distant charges as a single charge by expanding this collection of charges in a single multipole expansion. The FMM is a scheme to group the particles into a hierarchy of boxes and to manage the necessary manipulation of the associated expansions such that linear scaling is achieved. An improved version of the FMM employing more stable recurrence relations for the Wigner rotation matrices and an improved error estimate has been implemented. The implementation is essentially parameter free: for a given requested accuracy, the FMM specific parameters are determined automatically such that the computation time is minimized. The achieved accuracy is remarkable and competitive. In addition, the continuous fast multipole method (CFMM), a generalization of the FMM for continuous charge distributions, has been implemented and incorporated into the DSCF module of the TURBOMOLE quantum chemistry package. The treatment of solute-solvent interactions in quantum chemical calculations is an important field of application, since most practical problems are dealing with liquid phase chemistry. The explicit treatment of the solvent by placing a large number of solvent molecules around the solute requires, apart from the electronic relaxation, also the geometric relaxation of the complete solvent-solute system, yielding this approach rather impractical. Continuum solvation models replace the solvent by a continuum which describes the electrostatic behavior of the solvent. The response of the solvent upon the polarization by the solute is represented by screening charges appearing on the boundary surface between continuum and solute. They, however, cannot describe orientation dependent interactions between solute and solvent. The particular advantage of the conductor-like screening model (COSMO) formalism over other continuum models are the simplified boundary conditions. Within the HPC-Chem project, COSMO has been implemented for the HF and DFT methods (including energies, gradients, and numerical second derivatives) as well as for the MP2 energies.

13.22 Exercises

1. Compile and run a simple MPI program.
2. Compile and run simple serial programs.

For a simple exercise on hpc, please refer to the book URL.

References

1. Feller D (1997) The EMSL Ab Initio Methods Benchmark Report: A Measure of Hardware and Software Performance in the Area of Electronic Structure Methods, Pacific Northwest National Laboratory technical report PNNL-10481 (Version 3.0). http://www.emsl.pnl.gov:2080/docs/tms/abinitio/cover.html
2. Center of Excellence in Space Data and Information Sciences (CESDIS), NASA Goddard Space Flight Center Beowulf Project at CESDIS, http://www.beowulf.org/
3. Becker DJ et al. (1995) BEOWULF: A Parallel Workstation for Scientific Computation. Proc Int Conf on Parallel Processing, Aug 1995 1:11–14
4. Ridge D et al. (1997) Beowulf: Harnessing the Power of Parallelism in a Pile-of-PCs. Proc. IEEE Aerospace
5. Beowulf Project, Beowulf Consortium, http://www.beowulf.org/consortium.html

6. Tirado-Rives J, Jorgensen WL (1996) Viability of Molecular Modeling with Pentium based PCs. J Comp Chem 17:11 pp 1385–86

7. Nicklaus MC, Williams RW, Bienfait B, Billings ES, Hodoscek M (1998) Computational Chemistry on Commodity-Type Computers. J Chem Info Comp Sci 38:5 pp 893–905

8. Windus TL, Schmidt MW, Gordon MS (1995) Parallel Processing With the Ab Initio Program GAMESS. In: Toward Teraflop Computing and New Grand Challenge Applications, Kalia RJ, Vashishta, eds., Nova Science Publishers, New York

9. Oak Ridge National Laboratory PVM: Parallel Virtual Machine, http://www.epm.ornl.gov/pvm/pvm_home.html

10. Anderson TE et al. (1995) A Case for NOW (Networks of Workstations). IEEE Micro Feb pp 54-64

11. Supercomputer Research Institute, Florida State University DQS - Distributed Queueing System, http://www.scri.fsu.edu/~pasko/dqs.html

12. Litzkow M, Livny M, Mutka MW (1988) Condor – A Hunter of Idle Workstations. Proc 8th Int Conf Distr Comp Sys, June 1988, pp 104–111

13. Distributed and High-Performance Computing Group, University of Adelaide Perseus: A Beowulf for computational chemistry, http://www.dhpc.adelaide.edu.au/projects/beowulf/perseus.html

14. Karplus M, Chemistry at HARvard Macromolecular Mechanics (CHARMM), http://yuri.harvard.edu/charmm/charmm.html

15. Gaussian, Inc. and Scientific Computing Associates Highly Efficient Parallel Computation of Very Accurate Chemical Structures and Properties with Gaussian and the Linda Parallel Execution Environment, http://www.gaussian.com/wp_linda.htm

16. Hockney RW, Jesshope CR (1988) Parallel Computers 2. Adam Hilger, Bristol

17. Guest MF, Sherwood P, Nichols JA, Massive Parallelism: The Hardware for Computational Chemistry?, http://www.dl.ac.uk/CFS/parallel/MPP/mpp.html

18. Center of Excellence in Space Data and Information Sciences (CESDIS), NASA Goddard Space Flight Center BPROC: Beowulf Distributed Process Space, http://www.beowulf.org/software/bproc.html

19. Krauss M, Stevens WJ (1984) Ann Rev Phys Chem 35:357

Chapter 14
Research in Computational Chemistry and Molecular Modeling

14.1 Introduction

We have seen in Sect. 1.10 some research topics connected with computational chemistry. In this chapter we shall specifically mention some of the research methodologies adopted in this discipline with some examples.

14.2 Molecular Interaction

Molecular interaction is a property to be exploited, which helps to quantitatively and qualitatively compute molecular-level aspects related to the orientation, conformation, and activity. The adsorption and diffusion of a carbon (C) atom on several low-index metal surfaces can be studied based on first-principles calculations. The method can be quantum mechanical or density-functional under plane wave formalism, preferably with ultra soft pseudopotentials. The adsorption energies and diffusion barriers of a C atom on metal surfaces can be calculated. The interactions between a pair of C atoms at different separations on these surfaces can also be investigated.

The adsorption of atomic oxygen and carbon can be studied with plane wave density functional theory on Ni surfaces. Various adsorption sites on these surfaces can be examined in order to identify the most favorable adsorption site for each atomic species. The dependence of surface bonding on the adsorbate can be investigated. Adsorption energies and structural information are obtained and can be compared with existing experimental results. In addition, activation barriers to CO dissociation can be determined on Ni by locating the transition states for these processes [1].

The method can be extended to biomolecules. A study of antibody-antigen interactions can be undertaken. Antigen-contacting residues and combining site shape in the antibody crystal structures are available in the Protein Data Bank. Antigen-contacting propensities are presented for each antibody residue, allowing a new definition for the complementarity determining regions to be proposed based on ob-

K. I. Ramachandran et al., *Computational Chemistry and Molecular Modeling*   297
DOI: 10.1007/978-3-540-77304-7, ©Springer 2008

served antigen contacts. An objective means of classifying protein surfaces by gross topography can be developed and applied to the antibody combining site surfaces.

The prediction of secondary structural class and architecture from sequence composition analysis can also be investigated. Modifications to a well established geometric prediction algorithm to improve accuracy and the estimation of reliability may be tried. The hierarchical prediction of fold architectures may be made based on the computational studies [2].

To complement the ab initio approach of class and architecture prediction, a novel sequence alignment algorithm employing direct comparisons of predicted secondary structure and sequence-derived hydrophobicity may be developed, and applied to fold recognition.

The catalytic growth of carbon (C) nanotubes on clusters of transition metal catalysts is of much significant current interest. The elemental energetics for the atomistic rate processes involved in the initial stages of the growth can be made by a computational study of the C atom on a nickel (Ni) magic cluster (Ni_{38}), which preserves fcc geometry. The same analysis may be carried out to "low-index extended Ni surfaces." Related topics of interest are:

1. Parameterization of peptide-metal surface or water-metal surface interactions.
2. Molecular dynamics simulations of peptide adsorption at the interface between water and model hydrophobic/hydrophilic surfaces.
3. Dynamics and thermodynamics of polymer/penetrant systems.
4. Solvent interaction with beta-sheeted crystalline polymers.

14.3 Shape Selective Catalysts

Molecular dynamics and a quantum chemical investigation of partially amorphous material derived from zeolite is important for technological and industrial applications such as catalysis, ion-exchange, and ceramic chemistry. Zeolite is a shape-selecive catalyst, which changes its catalytic activity on changing its shape. The ZSM-5 developed from zeolite can convert methyl and ethyl alcohol into petrol. Properties of such catalysts need proper investigation. In the computational procedure [3], initially a modeling is done to predict catalytic properties. We can even set up a mathematical model correlating molecular shape and catalytic activity. Partial amorphization as is seen in zeolites can be used to tune specific properties. We can apply molecular dynamics using classical interaction potentials and canonical ensembling to excavate the required property.

In order to generate partially amorphous structures the silicious crystalline configuration will be heated to high temperatures, equilibrated, and finally quenched to 300 K. The expected (local) minimum configurations will be stored and then quenched to zero temperature using a combined steepest-descent-conjugate-gradient algorithm. The extent of amorphization can be estimated as the percentage of energy crystallinity (PEC):

$$PEC = \frac{\left(E_{amorphous} - E_{configuration}\right) \times 100}{\left(E_{amorphous} - E_{crystalline}\right)} \quad (14.1)$$

For the detected local minima the dynamic matrices will be calculated and diagonalized in order to obtain eigenvalues (squares of eigenfrequencies) and eigenvectors (types of motion).The structural properties of the partially amorphous materials can be analyzed by means of pair-distribution functions and bond angle distributions. A comparison to the crystalline ZSM-5 may be made.

An important quantitative term for zeolites is the internal surface area (ISA). For its determination the system is modeled as an ensemble of intersecting hard spheres with radii R_{coord} depending on the coordination number (CN) [4]. The ISA can be determined using the so-called probe-atom model:

$$ISA = \frac{1}{M} \left(\sum_{i=1}^{N} 4\pi \left[R_{coord}(i) + r_{prob} \right]^2 \frac{p_i}{p} \right) \tag{14.2}$$

Here, r_{prob} denotes the probe-atom radius, p the total number of sample points homogenously distributed on the surfaces of the spheres, and p_i the number of points on sphere i not being inside other spheres. Computational studies of the partial amorphization of zeolite ZSM-5 made by Atashi Basu Mukhopadhyay, Christina Oligschleger, and Michael Dolg revealed the following results:

1. For large probe radii the ISA decreases due to the reduction of the number of large pores, whereas for small probe radii the ISA increases due to the increase in under-coordination and an increasing tendency to convert large rings into smaller rings.
2. The relative contributions of the motions of structural subunits to the total vibrational density of states (VDOS) was analyzed by projecting the eigenvectors onto the vibrational modes of the isolated structural subunits Si−O−Si and SiO4.
3. For structures with PEC of above/below 60% the intensity of the so-called Boson peak decreases/increases. The effect is associated with a decrease of the concentration of 10-fold rings and a general lowering of symmetry by the puckering of large rings. The latter behavior is related to an increasing participation of under-coordinated centers in the relevant low-frequency motions.
4. Finally, the structure and relative stability of edge-sharing SiO4 tetrahedra vs. the common corner-sharing SiO_4 tetrahedra was investigated by quantum chemical ab initio techniques for the model systems W-silica and alpha-quartz.

14.4 Optimized Basis Sets for Lanthanide and Actinide Systems

Ab initio calculations of the electronic structure of lanthanide and actinide elements and their molecules are very demanding due to the large relativistic and electron correlation effects. The ab initio energy-consistent pseudopotential approach proved to be a reliable approximate relativistic scheme for calculations of the valence electron structure of lanthanide and actinide systems when a small core is used. Polarized valence basis sets of roughly quadruple-zeta quality have to be used for both the $4f$ and $5f$ series. An atomic natural orbital-based generalized contraction scheme can be applied, which allows to reduce the basis set size to triple- or double-zeta quality by omitting the outermost contractions corresponding to the least occupied atomic natural orbitals. The contractions coefficients need to be optimized for the $f^n d^1 s^2$ and $f^{n+1} s^2$ configurations simultaneously, by averaging the corresponding density matrices. As an alternative, segmented contracted basis sets may also be derived. Both sets can be successfully tested in atomic and molecular calibration calculations (e.g., for some monohydrides, monoxides, and monofluorides) and are available, e.g., through the Internet (URL: http://www.theochem.uni-stuttgart.de/pseudopotentiale). As an application, the electronic structure of selected lanthanide dimers (La_2, Ce_2, Eu_2, Gd_2, Yb_2, Lu_2) were investigated in large-scale considering correlated electronic structure calculations by Xiaoyan Cao and Michael Dolg. It was concluded that e.g., the ground state configurations of La_2 and Lu_2 differ (mainly) due to an increase of relativistic effects and (partially) shell structure effects. The vibrational frequency of the La_2 system is most likely affected by the rare gas matrix much more than the one of the Lu_2 system, thus explaining remaining differences with recent experimental data. Gd_2 is confirmed to have 18 unpaired electrons in the ground state, 14 of them in the two $4f$ shells [5].

The higher lanthanide and actinide ionization potentials exhibit very large differential electron correlation effects, since the f occupation number of the involved electronic states changes. In order to come to reliable estimates for the higher ionization potentials, computations were performed at the CASSCF/ACPF and partially at the CCSD(T) level (including spin-orbit correlations) basis set extrapolation studies using uncontracted valence basis sets with up to i-type functions. The results are in good agreement with the experimentally better known values for the lanthanides and provide (in our opinion) the best and most complete theoretical set of values for the actinides. Similar techniques have been recently used to calculate the electron affinity of the Ce atom. Here, we obtained excellent agreement with all-electron ab initio calculations as well with as earlier experimental results, whereas the most recent experiment was interpreted to lead to a substantially higher value. Finally, using large-core ($4f$-in-core) pseudopotentials they selected lanthanide(III)texaphyrin complexes, which are important for cancer theraphy.

14.5 Designing Biomolecular Motors

Molecular motors can be considered as "nanomachines" that consume energy in one form and convert it into motion or mechanical work. In fact, they are the ultimate nanomachines, providing maximum efficiency. There are a number of biopolymers which can function as efficient molecular (bio) motors. For example, many protein-based molecular motors make use of the chemical free energy (Gibbs free energy) released by the hydrolysis of ATP (Adenosine tri phosphate, the energy currency) in order to perform mechanical work. In terms of thermodynamic efficiency, these types of motors will be superior to currently available man-made motors. Hence, designing molecular motors of this type is of much research interest. A computational analysis of biopolymers to identify this mechano-chemical property is of much research interest. The property can be analyzed through quantum mechanical and molecular mechanics computational techniques by taking biomotors such as myosinV (actin) and kinesin (microtubule), etc. The computational technique involved in designing new biomotors is comprised of the following steps.

1. Modeling the control of the patterning of motor raceways as functioning tracks for the motion of motor proteins.
2. Studying the two of the main classes of proteins actin/myosin and microtubule/kinesin to understand their relative merits towards nanotechnology applications.
3. Making suitable computational studies to model structures, molecular orbitals, electrostatic potential, densities, vibrational frequencies, NMR shielding tensors, and reaction pathways.
4. Predicting the thermodynamics of the process, through computational modeling, which is of much importance in designing molecular motors.
5. Studying the application of single motors and collections of motor proteins.
6. Studying the coupling of nanotubes to the electrical circuit through electro/dielectrokinesis at the nanometer scale.
7. Understanding a processing methodology for incorporating nanometer scale e-beam lithography, nanotube placement/growth, patterned chemical functionalization, and motor binding and motility. These capabilities and fundamental characterizations will be applied to new force-sensing analyzing devices and multiplexing arrays.

14.6 Protein Folding and Distributed Computing

Protein folding is the current poster child of the distributed computing world. This is because figuring out the folding order of a protein and obtaining its final structure is an extremely complicated molecular dynamics problem. To put it in perspective, the individual structural units move around their bonds on a time scale in the 10 to 100 picoseconds range (10^{-12} s) but the protein might take anywhere from a few microseconds to a few minutes to reach its final structure. This implies that at least 10,000 moves per structural unit are required for a small protein that obtains its structure, while more complicated proteins are likely to involve around 600 billion moves per structural unit [5, 6].

Speeding up the process appears to be exactly what M. Sega and P. Faccioli et. al have done. They have found a way to quickly calculate the most probable path from the unfolded state (or any other state) to any stable, folded state. They use a form of the diffusion equation, which is the same equation that describes how a drop of liquid sugar will spread out through water. Using this equation, the probability of finding a protein in a particular state at a particular time can be calculated. It is also trivial to determine if that state is stable by minimizing a potential energy function. Hence, the time and path from a denatured (e.g., unfolded) protein to the folded state can be found by minimizing a potential energy function and performing an integration, which supplies the path and time taken to traverse the path.

The potential energy function that is minimized is found by a combination of more traditional molecular dynamics and experimental knowledge. For most proteins, a stable structure can be determined using experimental techniques. Performing a short molecular dynamics simulation with the protein configured in its stable form determines the potential energy function for the stable form. Then similar simulations on several unstable forms (e.g., unfolded) are used to determine a background potential for this minimized potential to sit in. According to the researchers, these simulations are short enough that the entire calculation can be performed on a normal desktop computer.

Using this surface, the researchers can calculate the most probable path between any two locations on the surface. That can then be mapped to time and, through the entropy of the protein, the structures it passes through on the way. An additional advantage of this approach is what it tells us about the stability of the stable state and the presence of other stable states, and how likely it is to make a transition between states. Since structure is very important to protein function, this seems like it could be a useful tool.

14.7 Computational Drug Designing and Biocomputing

The cellular targets (or receptors) of many drugs used for medical treatment are proteins. By binding to the receptor, drugs either enhance or inhibit its activity. Basically, there are two major groups of receptor proteins: proteins that "float" around in the cytoplasm of the cell, and proteins that are incorporated into the cell membrane. In the latter case, a drug does not even need to enter the cell; it can bind simply to an extracellular binding site of the protein and control intracellular reactions from the outside. An important criterion to determine the medical value of a drug is specificity: the physiological effect of the drug should be as clearly defined as possible. It has to specifically bind to the target protein in order to minimize undesired side effects. Undesired side effects, however, are not always an indication for insufficient specificity of drugs, as these effects might also result from a reaction of our body to the desired and therefore successful regulation of the malfunctioning biochemical process. On the molecular level specificity includes two more or less independent mechanisms; firstly, the drug has to bind to its receptor site with a suitable affinity (better binding means lower doses) and secondly, it has to either stimulate or inhibit certain movements of the receptor protein in order to regulate its activity. Both mechanisms are mediated by a variety of interactions between the drug and its receptor site. Usually, tens of thousands of compounds have to be screened to find a promising new drug and only very few of these candidates will make their way through the final clinical tests. Looking for help from powerful computers seems straightforward. So, how can they help?

The input of biocomputing in drug discovery is twofold: firstly, the computer may help to optimize the pharmacological profile of existing drugs by guiding the synthesis of new and "better" compounds. Secondly, as more and more structural information on possible protein targets and their biochemical role in the cell becomes available, completely new therapeutic concepts can be developed. The computer helps in both steps: to find out about possible biological functions of a protein by comparing its amino acid sequence to databases of proteins with known functions, and to understand the molecular workings of a given protein structure. Understanding the biological or biochemical mechanism of a disease then often suggests the types of molecules needed for new drugs.

In all cases, the aim of using the computer for drug design is to analyze the interactions between the drug and its receptor site and to "design" molecules that give an optimal fit. The central assumption is that a good fit results from structural and chemical complementarity to the target receptor. The techniques provided by computational methods include computer graphics for visualization and the methodology of theoretical chemistry. By means of quantum mechanics the structure of small molecules can be predicted to experimental accuracy. Statistical mechanics permits molecular motion and solvent effects to be incorporated.

The best possible starting point is an X-ray crystal structure of the target site. If the molecular model of the binding site is precise enough, one can apply docking algorithms that simulate the binding of drugs to the respective receptor site.

Even if the structure of the receptor site is unknown, the computer may help to figure out how it might look by comparing the chemical and physical properties of drugs that are known to act at a specific site. Moreover, if the amino acid sequence of the receptor site is known, one can try to predict the structure of the unknown site. This can either be done "from scratch" or by using a known structure of a related protein as template. If about 25 to 30% of the amino acid residues are identical in two proteins, one may assume that the three-dimensional structure of these two proteins is very similar. The technique used for this approach is called "homology modeling." The folding pattern of the template protein is maintained and the side chain atoms of the template protein are replaced by the side chain atoms of the unknown protein. Basically, the three-dimensional structure of a protein is represented by the three-dimensional organization of the backbone atoms. The side chain atoms, which are different for all 20 amino acids, define the specific interactions with ligands or other protein domains. Replacing the side chains while maintaining the backbone therefore allows to keep the general structure of the protein and to evaluate the specific properties of the unknown protein with respect to ligand interactions.

A prominent example is the design of potent HIV protease inhibitors [7]. The design was based on knowledge of the target structure.

14.8 Artificial Photo Synthesis

In the photosynthetic reaction centers of plants, light energy is converted into chemically useful energy and oxygen is produced. This photochemical reaction is initiated by a charge separation process in the reaction center (RC) complex. Major research in this regard is to analyze the light-driven electron transfer (ET) and to study the response of the protein in which the RC is embedded, stabilizing the charge separation process in photosynthesis. Several computational tools including Density Functional Theory (DFT), Car-Parrinello molecular dynamics simulations, hybrid QM/MM approaches, and topological analysis of the electron density based on the "Atoms in molecule (AIM)" theory can be used for the computation. These methods enable us to calculate the electronic structure, absorption energies, NMR chemical shifts, and dynamical properties of the model system within the same framework. The long-term goal is not only to complement and interpret available spectroscopic data, but also to predict properties of artificial photosynthetic systems.

14.9 Quantum Dynamics of Enzyme Reactions

Many enzyme reactions involve proton or hydride transfer and can be expected to proceed by quantum mechanical tunneling. Although great progress has been made in incorporating quantum effects into gas-phase reactions, most simulations of processes involving proteins have involved classical mechanics, and therefore they have been unable to properly model proton and hydride transfer processes. This has been particularly frustrating because kinetic isotope effects are very sensitive to tunneling, and kinetic isotope effects are often the best means for learning about transition state structure. Recently simulation of the reaction rates and kinetic isotope effects of the hydride transfer for benzyl alcoholate anion to the coenzyme NAD^+, catalyzed by the enzyme liver alcohol dehydrogenase has been reported.

The calculation was made possible by two advances in simulation methods. First is the treatment of the force field, which involves a combination of semiempirical molecular orbital theory, semiempirical valence bond terms, and molecular mechanics. Second is the treatment of atomic motions, which is based on variational transition state theory with quantized vibrations and multidimensional tunneling contributions along optimized tunneling paths.

The calculations agree very well with kinetic isotope effects measured by Professor Judith Klinman and coworkers at the University of California, Berkeley, and they provide an interpretation of the highly nonclassical kinetic isotope effects that they observed in terms of the rehybridization at the donor carbon atom. The hybridization of this carbon atom, caught in the process of releasing the tunneling hydride atom, is clearly intermediate between sp^2 and sp^3.

14.10 Other Important Topics

1. The development of relativistic energy-consistent ab initio pseudopotentials (known as Stuttgart-Cologne pseudopotentials), effective core-polarization potentials, as well as corresponding optimized valence basis sets.
2. The development of a new multi-reference coupled cluster approach.
3. The development of a Hartree-Fock-Wigner approach for periodic systems.
4. A quantum chemical investigation of the haptotropic rearrangement of $Cr(CO)_3$ templates on condensed polyaromatic systems.
5. A quantum chemical investigation of TiCp2-based catalysts.
6. A quantum chemical investigation of the structure and stability of various borate containing crystalline solids.
7. A quantum chemical investigation of the structure and stability of P$-$N containing oligomers and polymers.
8. A quantum chemical investigation of C$-$S containing solids.
9. A quantum chemical investigation of polycations containing As, Sb, Bi, Se, and Te.
10. Performance modeling of HPC applications on computational grids.
11. Quantum mechanical dynamics.
    A critical focus area in computational chemistry is quantum mechanical dynamics. The linear algebraic variational method for calculating converged quantum mechanical transition probabilities for reactive collisions has been introduced. At present, the main application area is quantum photochemistry, i.e., the utilization of electronic excitation energy to promote chemical reactions.
12. Electronically adiabatic reactions.
    Electronically adiabatic reactions are those that take place entirely in the ground electronic state, i.e., thermally activated reactions on a single potential energy surface. Variational transition-state theory with multidimensional semi-classical tunneling contributions (VTST) can be used to study such systems. VTST involves finding the free energy bottleneck for over barrier processes and the optimal tunneling paths for through-barier processes. This theory has been developed for reactions in the gas phase, in a liquid solution, on metallic surfaces, and in enzyme active sites. The role of tunneling and quantum mechanical vibrational energy on rate constants, kinetic isotope effects, and state-selective chemistry needs to be excavated. Application areas include combustion, atmospheric chemistry, environmental chemistry, clusters (from microhydrated species to nanoparticles), and catalysis (heterogeneous, organometallic, and biological).
13. Electronically nonadiabatic collisions.
    Another research area is semi-classical trajectory methods for reactive collisions involving coupled potential energy surfaces. Two types of semi-classical methods are under study: trajectory surface hopping (also called molecular dynamics for quantum transitions) and self-consistent potential methods (also called time-dependent self-consistent-field methods). We can even combine these two methods to make use of the best features of both of these approaches into a single formalism. This technique is called decay of mixing with coherent switches,

and it is more accurate than previously available methods for the whole range of problems encountered in photochemistry. Furthermore, it is practical to apply this method to both simple and complex photochemical reactions such as calculations for ammonia, OH...HH, bromoacetyl chloride, and Na...HF.

14. One area of active work is the extension of molecular mechanics force fields to be able to treat reactive systems that involve bond breaking. An approach called multi-configuration molecular mechanics (MCMM) has been developed for this purpose, and it is very promising.

15. Another area of special concentration is in the interface of electronic structure theory and dynamics. We are developing a variety of single-level and dual-level methods for direct dynamics calculations, where direct dynamics denotes the calculation of rate constants or other dynamical quantities directly from electronic structure calculations without the intermediacy of fitting a potential energy function. In such a case the potential energy surface is implicit but is never actually constructed.

16. A very exciting recent development is the parameterization of multi-coefficient methods for scaling components of the correlation energy and extrapolating electronic structure calculations to an infinite basis set. These methods allow one to calculate accurate gas-phase heats of formation, atomization energies, and potential energy surfaces for large systems at an affordable cost. These methods have better scaling properties than pure ab initio calculations, and they often yield more accurate results with far less computer time. We have now shown how these methods can be improved by adding static correlation with density function theory for even great performance-to-cost ratios.

17. The direct calculation of free energies from potential energy surfaces, without first calculating the energy spectrum, is also of great interest, and we are developing improved Monte Carlo sampling methods for doing this by the Feynman path integral method.

18. Solvation effects.
    Solvation effects are important for several physical, chemical, and biological properties. Energetics and dynamics in the condensed phase are to be made as accurate as their treatment for gas-phase species and processes. The role of the solvent in polarizing the solute is especially interesting. Solvation models for both aqueous and organic solvents can be developed. A variety of applications of compounds to structure and reactivity in solution are underway.

19. Biochemical applications.
    Many enzymatic reactions involve proton and hydride transfer, but until recently, techniques for simulating the dynamics of these processes were usually based entirely on classical mechanics. We can incorporate quantum effects in biological simulations. This includes tunneling, zero point effects, and the effect of quantization on thermally averaged quantities. Proton transfers catalyzed by enolase and hydride transfer catalyzed by liver alcohol dehydrogenase are dominated by quantum mechanical events, and these can be well modeled by semi-classical dynamics methods.

An important application of solvation modeling is the calculation of the partitioning of organic and biological molecules between aqueous and cell membranes. This has an important effect on the bioavailability of drugs.

20. Nanomaterials.

Nanotechnology is the art of manipulating materials on a scale of the order of a nanometer, to build molecular scale devices or to take advantage of the unique chemical, physical, and material properties of nanostructured materials. The major research in this area focuses on computational studies of nanoparticle growth and dynamics. We are concerned with the development and implementation of new methods for the modeling and simulation of nanoparticles and their elementary processes, including nucleation, deposition, melting, and surface reactions. Nanoscale systems present a challenge to computation because they display properties that are not well modeled by methods developed for use in bulk simulations, and because they are expensive to treat using methods developed for molecular systems. The development of new techniques for extending the time and length scales of simulations and their application to problems involving semiconductor nanoparticles and metal nanoparticles is of much concern. To study the importance of quantum effects in nanoparticle reactivity, for example, the reaction of metal particles with hydrocarbons and hydrocarbon fragments, we can develop multilevel methods, such as QM/MM methods, that combine quantum mechanics (QM) and molecular mechanics (MM). The efficiency of these methods potentially allows one to perform accurate calculations for large reactive systems over long time scales. For the simulation of systems with non-localized active areas, it is necessary to adaptively redefine the region to be treated by quantum mechanics. For such systems, we can develop new methods for combining multilevel methods with modern sampling schemes, such as our molecular dynamics code, ANT, or Monte Carlo codes.

21. Integrated tools for computational chemical dynamics.

The goal of this research is to develop more powerful simulation methods and incorporate them into a user-friendly high-throughput integrated software suite for chemical dynamics. Recent advances in computer power and algorithms have made possible accurate calculations of many chemical properties for both equilibria and kinetics. Nonetheless, applications to complex chemical systems, such as reactive processes in the condensed phase, remain problematic due to the lack of a seamless integration of computational methods that allow modern quantum electronic structure calculations to be performed with state-of-the-art methods for electronic structure, chemical thermodynamics, and reactive dynamics. These problems are often exacerbated by invalidated methods, non-modular and non-portable computer codes, and inadequate documentation that drastically limit software reliability, throughput, and ease of use. The goal of the Integrated Tools consortium is to develop an integrated software suite that combines electronic structure packages with dynamics codes and efficient sampling algorithms for the following kinds of condensed-phase modeling problems:

1. Thermochemical kinetics and rate constants
2. Photochemistry and spectroscopy

3. Chemical and phase equilibria
4. Computational electrochemistry
5. Heterogeneous catalysis

The photochemical creation of excited states offers a means to control chemical transformations, because different wavelengths of light can be used to create different vibrational states, thereby directing chemical reactions along different pathways. It is crucial to understand how energy deposited into the system is used; this is particularly complicated in condensed phase systems where many channels lead to dissipation of excess energy. Similar opportunities and challenges present themselves in the areas of electrochemistry and catalysis.

22. Research on theories and the application of electronic structure.
23. Molecular mechanics studies of compounds and introduction of new force fields.
24. Research on condensed matter physics, nanobiospectroscopy and biological molecules.
25. The computational modeling of carbohydrates, drugs, and macromolecules.
26. Applying theoretical chemistry, structure, and the reactivity of clusters and molecules
27. The theory application, computer models, and related data about non-covalent binding and molecular recognition
28. Research on organic quantum mechanical methods and systems.
29. Computational studies and the reactivity of biomacromolecules tested solutions.
30. Computer-assisted methods for studies on physicochemical properties, pharmaceutical activity, and chemical and genetic toxicity.
31. Simulating solvent properties of solutions, proteins, and membranes.
32. Investigating in areas of reaction mechanisms and molecular electronic structures.
33. Computational study of DNA repair.
34. Theoretical and computational methods for application in broad chemical interests.
35. Investigating sources in stability, structures and properties of different macromolecules.
36. Computational electrochemistry: the prediction of environmentally important redox potentials.
    Single-electron transfer steps are often involved as the rate-determining step in reaction pathways that lead to the transformation of certain classes of anthropogenic organic compounds in the environment. A key molecular descriptor in modeling electron-transfer kinetics is the one-electron redox potential.

Pure computational techniques (involving ab initio or semiempirical electronic structure theory and quantum mechanical continuum solvation models) and of certain kinds of linear free energy relationships can be used for predicting the 1-electron oxidation potentials of substituted anilines. Mean accuracies from 20 to 90 mV over 21 different substituted anilines were achieved with different approaches by professors Eric Patterson, Cramer and Truhlar. Figure 14.1 illustrates the use of

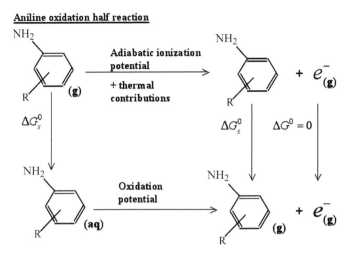

Aniline oxidation half reaction

Normal hydrogen electrode half reaction

$$H^+_{(Aqouos)} + e^- \xrightarrow{\Delta G^0 = -4.44eV} \frac{1}{2}H_{2(g)}$$

Fig. 14.1 Use of a free energy cycle to compute such an oxidation potential in an aqueous solution

a free energy cycle to compute such an oxidation potential in aqueous solution. They have applied this same technology to characterize the reaction path by which hexachloroethane (a common contaminant of drinking water) is transformed in the environment to tetrachloroethylene.

References

1. Zhang M, Wells JC, Gong XG, Zhang Z (2004) Adsorption of a carbon atom on the Ni38 magic cluster and three low-index nickel surfaces: A comparative first-principles study. Phys Rev B 69:205413
2. Li T, Bhatia B, Sholl DS (2004) First-principles study of C adsorption, O adsorption, and CO dissociation on flat and stepped Ni surfaces. J Chem Phys 121:20
3. Cao X, Dolg M (2001) Valence basis sets for relativistic energy-consistent small-core lanthanide pseudopotentials. J Chem Phys 115 pp 7348–7355
4. Mukhopadhyay AB, Oligschleger C, Dolg M (2003) Molecular dynamics investigation of structural properties of zeolite ZSM-5 based amorphous material. Phys Rev B 67 pp 014106–1014107

5. Mukhopadhyay AB, Oligschleger C, Dolg M (2003) Molecular dynamics investigation of vibrational properties of zeolite ZSM-5 based amorphous material. Phys Rev B 68 pp 24205–24215
6. Cao X, Dolg M (2001) Valence basis sets for relativistic energy-consistent small-core lanthanide pseudopotentials. J Chem Phys 115 pp 7348–7355
7. Tucker, TJ (1994) Science 263:380

Chapter 15

Basic Mathematics for Computational Chemistry

15.1 Introduction and Basic Definitions

A matrix (plural *matrices*) is a rectangular table of elements having rows and columns. The horizontal lines of elements in a matrix are called rows and the vertical lines of elements are called columns. The elements may be numbers or, more generally, any abstract quantities that can be added and multiplied. It is customary to enclose the elements of a matrix in parentheses, brackets, or braces. For example, the following is a matrix:

$$\begin{bmatrix} 6 & 9 & 3 \\ -1 & 0 & 8 \end{bmatrix} \tag{15.1}$$

This matrix has two rows and three columns, so it is referred to as a "2 by 3" matrix. The elements of a matrix are represented in the following way:

$$\begin{bmatrix} X_{11} & X_{12} & X_{13} \\ X_{21} & X_{22} & X_{23} \end{bmatrix} \tag{15.2}$$

That is, the first subscript in a matrix refers to the row and the second subscript refers to the column. It is important to remember this convention when matrix algebra is performed.

A *vector* is a special type of matrix that has only one row (called a *row vector*) or one column (called a *column vector*). Below, a is a column vector while b is a row vector.

$$a = \begin{bmatrix} 8 \\ 3 \\ 4 \end{bmatrix}, \quad b = \begin{bmatrix} -3 & 8 & 5 \end{bmatrix} \tag{15.3}$$

A *scalar* is a matrix with only one row and one column. It is customary to denote scalars by italicized, lower case letters (e.g., x), to denote vectors by bold, lower case letters (e.g., $\mathbf{x}$), and to denote matrices with more than one row and one column by bold, upper case letters (e.g., $\mathbf{X}$). We shall see the application of MATLAB in our computations.

K. I. Ramachandran et al., *Computational Chemistry and Molecular Modeling* 311
DOI: 10.1007/978-3-540-77304-7, ©Springer 2008

MATLAB is software for scientific and technical computing from Mathworks Inc., USA. The name MATLAB originated from MATrix LABoratory and matrices are the building blocks of MATLAB. MATLAB has inbuilt functions for doing all matrix computations. It is one of the most popular software packages for scientific computing. In this chapter, all the matrix computations are illustrated using MATLAB [1].

15.1.1 Example 1

To enter the matrix

```
1   2
3   4
```

and store it in a variable a, type

```
>> a = [1 2; 3 4]
```

To redisplay the matrix, just type its name:

```
>> a
```

A *square matrix* has as many rows as it has columns. Matrix **A** is square but matrix **B** is not square:

$$A = \begin{bmatrix} 2 & 7 \\ 4 & 1 \end{bmatrix}, \quad B = \begin{bmatrix} 1 & 9 \\ 0 & 2 \\ 7 & -3 \end{bmatrix} \tag{15.4}$$

A *symmetric matrix* is a square matrix in which $x_{ij} = x_{ji}$ for all i and j. Matrix **A** is symmetric; matrix **B** is not symmetric.

$$A = \begin{bmatrix} 9 & 1 & 5 \\ 1 & 6 & 2 \\ 5 & 2 & 7 \end{bmatrix}, \quad B = \begin{bmatrix} 9 & 1 & 5 \\ 2 & 6 & 2 \\ 5 & 1 & 7 \end{bmatrix} \tag{15.5}$$

A *diagonal matrix* is a symmetric matrix where all the off diagonal elements are 0. Matrix **A** is diagonal.

$$A = \begin{bmatrix} 8 & 0 & 0 \\ 0 & 5 & 0 \\ 0 & 0 & 3 \end{bmatrix}$$

15.1.2 Example 2 Using MATLAB

This example illustrates how MATLAB can be used to display the diagonal of a matrix **A**.

```
A =
     3 0 0
     0 2 0
     0 0 2
>> B = diag(A)
```

Answer:

```
B =
     3
     2
     2
```

An *identity matrix* (also called a *unit matrix*) is a diagonal matrix with all its elements on the diagonal as unity (one or 1) and zeros elsewhere. The identity matrix is usually denoted as **I**. For example, a 3-by-3 identity can be written as follows:

$$\mathbf{I} = \begin{bmatrix} 1 & 0 & 0 \\ 0 & 1 & 0 \\ 0 & 0 & 1 \end{bmatrix} \tag{15.6}$$

15.2 Matrix Addition and Subtraction

To add two matrices, they both must have the same number of rows and columns (i.e., they should have the same dimensions). The elements of the two matrices are simply added together, element by element, to produce the results. That is, for $\mathbf{R} = \mathbf{A} + \mathbf{B}$, then $r_{ij} = a_{ij} + b_{ij}$ for all i and j. Thus:

$$\begin{bmatrix} 9 & 5 & 1 \\ -4 & 7 & 6 \end{bmatrix} = \begin{bmatrix} 1 & 9 & -2 \\ 3 & 6 & 0 \end{bmatrix} + \begin{bmatrix} 8 & -4 & 3 \\ -7 & 1 & 6 \end{bmatrix}$$

Matrix subtraction works in the same way, except that elements are subtracted instead of added.

15.2.1 *Example 3: Matrix Addition Using MATLAB*

```
>> A = [1 9 -2; 3 6 0]
   A =
        1   9  -2
        3   6   0

>> B = [8 -4 3; -7 1 6]
   B =
        8  -4   3
       -7   1   6

>> C = A + B
   C =
        9   5   1
       -4   7   6
```

15.3 Matrix Multiplication

There are several rules for matrix multiplication. The first concerns the multiplication between a matrix and a scalar. Here, each element in the product matrix is simply the scalar multiplied by the element in the matrix. That is, for $\mathbf{R} = a\mathbf{B}$, then $r_{ij} = ab_{ij}$ for all i and j. Thus:

$$8 \begin{bmatrix} 2 & 6 \\ 3 & 7 \end{bmatrix} = \begin{bmatrix} 16 & 48 \\ 24 & 56 \end{bmatrix} \tag{15.7}$$

Matrix multiplication involving a scalar is commutative. That is, $a\mathbf{B} = \mathbf{B}a$. The next rule involves the multiplication of a row vector by a column vector. To perform this, the row vector must have as many columns as the column vector has rows [2,3]. For example:

$\begin{bmatrix} 1 & 7 & 5 \end{bmatrix} \begin{bmatrix} 2 \\ 4 \\ 1 \end{bmatrix}$ is legal. However $\begin{bmatrix} 1 & 7 & 5 \end{bmatrix} \begin{bmatrix} 2 \\ 4 \\ 1 \\ 7 \end{bmatrix}$

is not legal because the row vector has three columns while the column vector has four rows. The row vector multiplied by a column vector (i.e., the dot product) will be a scalar. This scalar is simply the sum of the first element of the row vector multiplied by the first element of the column vector, plus the second element of the row vector multiplied by the second element of the column vector, plus the third element of the row vector multiplied by the third element of the column vector, and so on. In linear algebra, this can be represented as $r = \mathbf{ab}$ or

$$r = \sum_{i=1}^{n} a_i b_i \tag{15.8}$$

Thus:

$$\begin{bmatrix} 2 & 6 & 3 \end{bmatrix} \begin{bmatrix} 8 \\ 1 \\ 4 \end{bmatrix} = 2*8+6*1+3*4 = 34$$

All other types of matrix multiplication involve the multiplication of a row vector and a column vector. Specifically, in the expression $\mathbf{R} = \mathbf{AB}$:

$$r_{ij} = a_{i.}b_{.j} \tag{15.9}$$

where $a_{i.}$ is the i-th row vector in matrix $\mathbf{A}$ and $b_{.j}$ is the j-th column vector in matrix $\mathbf{B}$. Thus, if:

$$A = \begin{bmatrix} 2 & 8 & 1 \\ 3 & 6 & 4 \end{bmatrix} \text{ and } B = \begin{bmatrix} 1 & 7 \\ 9 & -2 \\ 6 & 3 \end{bmatrix}$$

$$r_{11} = a_{1.}b_{.1} = \begin{bmatrix} 2 & 8 & 1 \end{bmatrix} \begin{bmatrix} 1 \\ 9 \\ 6 \end{bmatrix} = 2*1+8*9+1*6 = 80$$

$$r_{12} = a_{1.}b_{.2} = \begin{bmatrix} 2 & 8 & 1 \end{bmatrix} \begin{bmatrix} 7 \\ -2 \\ 3 \end{bmatrix} = 2*7+8*(-2)+1*3 = 1$$

$$r_{21} = a_{2.}b_{.1} = \begin{bmatrix} 3 & 6 & 4 \end{bmatrix} \begin{bmatrix} 1 \\ 9 \\ 6 \end{bmatrix} = 3*1+6*9+4*6 = 81$$

$$r_{22} = a_{2.}b_{.2} = \begin{bmatrix} 3 & 6 & 4 \end{bmatrix} \begin{bmatrix} 7 \\ -2 \\ 3 \end{bmatrix} = 3*7+6*(-2)+4*3 = 21$$

Thus:

$$\begin{bmatrix} 2 & 8 & 1 \\ 3 & 6 & 4 \end{bmatrix} \begin{bmatrix} 1 & 7 \\ 9 & -2 \\ 6 & 3 \end{bmatrix} = \begin{bmatrix} 80 & 1 \\ 81 & 21 \end{bmatrix}$$

For matrix multiplication to be legal, the first matrix must have as many columns as the second matrix has rows. This, of course, is the requirement for multiplying a row vector by a column vector. The resulting matrix will have as many rows as the first matrix and as many columns as the second matrix. Because $\mathbf{A}$ has 2 rows and 3 columns while $\mathbf{B}$ has 3 rows and 2 columns, the matrix multiplication may legally proceed and the resulting matrix will have 2 rows and 2 columns. Because of these requirements, matrix multiplication is usually not commutative. That is, usually $\mathbf{AB} \neq \mathbf{BA}$. And even if $\mathbf{AB}$ is a legal operation, there is no guarantee that $\mathbf{BA}$ will also be legal. For these reasons, the terms *premultiply* and *postmultiply* are often encountered in matrix algebra, while they are seldom encountered in scalar algebra.

15.3.1 Example 4: Matrix Multiplication Using MATLAB

```
A = [2 8 1;3 6 4]
A =
    2   8   1
    3   6   4

>> B = [1 7;9 -2;6 3]
   B =
         1   7
         9  -2
         6   3

>> C = A * B
   C =
        80   1
        81  21
```

Note: In MATLAB, there is another kind of multiplication involving two matrices of the same dimensions, wherein each element of the product matrix is the product of the corresponding elements of the matrices (i.e., element by element multiplication) involved in multiplication. This is represented as C = A.*B. For example:

```
>> A = [1 2 3; 4 5 6; 7 8 9]
   A =
         1 2 3
         4 5 6
         7 8 9

>> B = [3 5 7;2 6 8; 4 7 3]
   B =
         3 5 7
         2 6 8
         4 7 3

>> C = A.* B
   C =
         3 10 21
         8 30 48
        28 56 27
```

15.4 The Matrix Transpose

If A is a m-by-n matrix with elements a_{ij}, the n-by-m matrix obtained from A by interchanging the rows and columns is called the transpose of A and is written as prime (A') or a superscript t or T (A^t or A^T) Thus:

$$A = \begin{bmatrix} 2 & 7 & 1 \\ 8 & 6 & 4 \end{bmatrix} \text{ and } A^T = \begin{bmatrix} 2 & 8 \\ 7 & 6 \\ 1 & 4 \end{bmatrix} \tag{15.10}$$

The transpose of a row vector will be a column vector, and the transpose of a column vector will be a row vector. The transpose of a symmetric matrix is simply the original matrix.

15.4.1 Example 5: The Transpose of a Matrix Using MATLAB

```
>> A = [2 7 1; 8 6 4]
   A =
        2  7  1
        8  6  4

>> B = A'
   B =
        2  8
        7  6
        1  4
```

The properties of the transpose are as follows: The transposition operation is reflective; i.e., $\left(A^T\right)^T = A$. The transpose of the product of two matrices is equal to the product of their transposes in the reserve order; i.e., $(AB)^T = B^T A^T$.

15.5 The Matrix Inverse

In scalar algebra, the inverse of a number is that number which, when multiplied by the original number, gives a product of 1. Thus, the inverse of x is $1/x$ and is denoted as x^{-1}. In matrix algebra, the inverse of a matrix is that matrix which, when multiplied by the original matrix, gives an identity matrix. The inverse of a matrix is denoted by the superscript "-1". Hence:

$$AA^{-1} = A^{-1}A = I \tag{15.11}$$

If A has an inverse, it is said to be invertible. If an n-by-n matrix A is invertible, the elements of the inverse of matrix A can be computed using its determinant and the transpose of the matrix of its cofactors. Firstly, we form the matrix composed of the cofactors of A with the elements. The cofactor of a matrix is the signed minor of the matrix which can be formed by the elements of the matrix that do not fall in the same row and column of the minor element and taking the determinant of the resulting matrix.

$$B = b_{ij} = (-1)^{i+j} \det(M_{ij}) \tag{15.12}$$

Secondly, the adjoint of $\mathbf{A}$ is defined as the transpose of the matrix $\mathbf{B}$. Thus, $\text{adj}(A)$ is the matrix B^T. The inverse is then:

$$A^{-1} = \frac{1}{|A|} \text{adj}(A) \tag{15.13}$$

The determinant of a square matrix A (denoted by $|A|$) is a single number associated with every square matrix which can be calculated by using all the elements of the matrix. The calculation of determinant of a matrix is very useful in the determination of the matrix inverse and the analysis and solution of systems of linear systems of equations. For a two-dimensional matrix, the determinant is given by:

$$|A| = \begin{vmatrix} a_1 & b_1 \\ a_2 & b_2 \end{vmatrix} = a_1 b_2 - a_2 b_1$$

For a three-dimensional matrix, the determinant is given by:

$$|A| = \begin{vmatrix} a_1 & b_1 & c_1 \\ a_2 & b_2 & c_2 \\ a_3 & b_3 & c_3 \end{vmatrix} = a_1 \begin{vmatrix} b_2 & c_2 \\ b_3 & c_3 \end{vmatrix} - b_1 \begin{vmatrix} a_2 & c_2 \\ a_3 & c_3 \end{vmatrix} + c_1 \begin{vmatrix} a_2 & b_2 \\ a_3 & b_3 \end{vmatrix}$$

The determinant has the following important properties, which include invariance under elementary row and column operations: (a) Switching two rows or columns changes the sign (b) Scalars can be factored out from rows and columns (c) Multiples of rows and columns can be added together without changing the determinant's value (d) Scalar multiplication of a row by a constant c multiplies the determinant by c (e) A determinant with a row or column of zeros has value 0 and (f) Any determinant with two rows or columns equal has value 0.

15.5.1 Example 6

Let

$$A = \begin{bmatrix} 1 & 4 & 8 \\ 1 & 0 & 0 \\ 1 & -3 & -7 \end{bmatrix}$$

$$\det(A) = -1 \begin{vmatrix} 4 & 8 \\ -3 & -7 \end{vmatrix} = -1(-28+24) = 4$$

B, the matrix composed of the cofactors of **A** is given by:

$$B = \begin{bmatrix} 0 & 7 & -3 \\ 4 & -15 & 7 \\ 0 & 8 & -4 \end{bmatrix}$$

$$A^{-1} = \frac{1}{4}B^T = \begin{bmatrix} 0 & 1 & 0 \\ 7/4 & -15/4 & 2 \\ -3/4 & 7/4 & -1 \end{bmatrix}$$

15.5.2 MATLAB Implementation

```
>> A = [1 4 8;1 0 0 ;1 -3 -7]
   A =
       1   4   8
       1   0   0
       1  -3  -7

>> det(A)
   ans =

          4

>> Ainv = inv(A)
   Ainv =
             0   1.0000        0
        1.7500  -3.7500   2.0000
       -0.7500   1.7500  -1.0000
```

To get the matrix in rational form (in some cases, it may be approximate), one can use the function *rats* in MATLAB as follows:

```
>> rats (Ainv)
   ans =
          0      1   0
        7/4  -15/4   2
       -3/4    7/4  -1
```

To check the property of inverse matrix that $AA^{-1} = I$, enter:

```
>> Ainv* A
   ans =
       1.0000        0        0
            0   1.0000        0
            0  -0.0000   1.0000
```

```
>> At = A'
   At =
          1   1   1
          4   0  -3
          8   0  -7

>> det (At)
   ans =
          4
```

For a matrix that is singular, the determinant is zero and it does not have an inverse. The determinant of a matrix close to zero indicates that the matrix is near singular and there may be numerical difficulties in calculating the inverse of such matrices [4–6].

15.6 Systems of Linear Equations

A system of equations is just a list of equations in one or more unknowns (also called variables). It turns out that many situations in life can be described by systems of equations of various sorts.

For example, one of the primary functions of air traffic control is to make sure that airplanes do not crash each other in the air. How do they do this? The path of each airplane is tracked and described by an algebraic equation. Then the equations are compared to see if there are any points at which they intersect. That is, one tries to find a solution for the system of equations that describe the routes of a set of airplanes – if there is one solution, then it means that the airplanes are on a collision course. The equations that arise may be linear (if a plane is flying in a straight line) or of other types such as quadratic (if a plane is circling the airport, for example).

This example and other situations give rise to possibly very complicated systems of equations. The reason for this name is that this type of equations describes straight lines – in 2-space, 3-space, or higher dimensional space. A solution is a set of numbers once substituted for the unknowns will satisfy the equations of the system

15.6.1 Example 7

Consider the system of equations:

$$2x + y = 10$$
$$x - y = 5$$

The values $x = 5, y = 0$ yield a solution for the system, since $2(5) + 0 = 10$ and $5 - 0 = 5$.

The solution to the system is the pair of values $(x,y) = (5,0)$.
Not every system of equations has a solution.

15.6.2 Example 8

Consider the system of equations:

$$2x + y = 10$$
$$2x + y = 20$$

Clearly, this system has no solutions, since whatever values we pick for x and y can satisfy at most one of these equations. In a case like this, we say that the system is inconsistent. A third possibility is that a given system has infinitely many solutions! When will this happen? In general, if the system has more unknowns than equations, and if there is a solution, then there will be infinitely many solutions. Alternatively, if it can be transformed into such a system, then it also has infinitely many solutions. We will discuss this transformation in the section on solutions of systems of linear equations. An example of such a system is Example 9.

15.6.3 Example 9

Consider the equation: $2x - y + z = 1$.
For any values one picks for y and z, there is a corresponding value for x which satisfies this equation. Of course there are infinitely many values one could choose for y and z, and so infinitely many solutions to the system. Because these are linear equations, their graphs will be straight lines. This can help us visualize the situation graphically. There are three possibilities.

15.6.3.1 Independent Equations

In this case (Fig. 15.1) the two equations describe lines that intersect at one particular point. Clearly, this point is on both lines, and therefore its coordinates (x,y) will satisfy the equation of either line. Thus, the pair (x,y) is the one and only solution to the system of equations.

15.6.3.2 Dependent Equations

Sometimes two equations might look different but actually describe the same line. For example, in:

$$2x + 3y = 1$$
$$4x + 6y = 2$$

Fig. 15.1 Independent equations

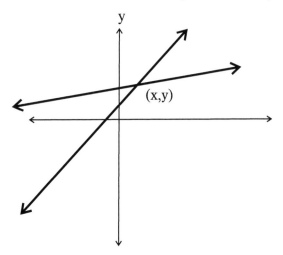

the second equation is just two times the first equation, so they are actually equivalent and would both be equations of the same line. Because the two equations describe the same line, they have *all* their points in common; hence, there are an infinite number of solutions to the system (Fig. 15.2). If you try to solve a dependent system by algebraic methods, you will eventually run into an equation that is an *identity*. An identity is an equation that is always true, independent of the value(s) of any variable(s). For example, you might get an equation that looks like $x = x$, or $3 = 3$. This would tell you that the system is a dependent system, and you could stop right there because you will never find a unique solution.

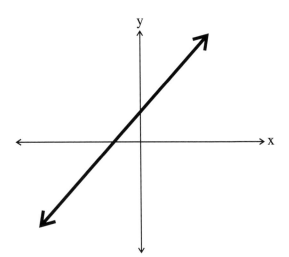

Fig. 15.2 Dependent equations

Fig. 15.3 Inconsistent equations

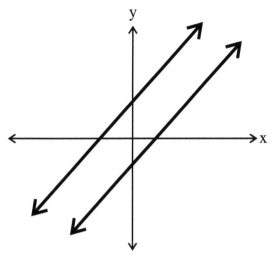

15.6.3.3 Inconsistent Equations

If two lines happen to have the same slope, but are not identically the same line, then they will never intersect. There is no pair (x, y) that could satisfy both equations, because there is no point (x, y) that is simultaneously on both lines. Thus, these equations are said to be *inconsistent*, and there is no solution (Fig. 15.3). The fact that they both have the same slope may not be obvious from the equations, because they are not written in one of the standard forms for straight lines. The slope is not readily evident in the form we use for writing systems of equations. (If you think about it you will see that the slope is the negative of the coefficient of x divided by the coefficient of y).

By attempting to solve such a system of equations algebraically, you are operating on a false assumption – namely, that a solution exists. This will eventually lead you to a *contradiction*: a statement that is obviously false, regardless of the value(s) of the variable(s). At some point in your work you would get an obviously false equation like $3 = 4$. This would tell you that the system of equations is inconsistent, and there is no solution [7].

15.6.4 Example 10: A MATLAB Solution of the Linear System of Equations

Consider the following cases of linear systems of equations:

1. An inconsistent system

$$2x - y + z = 1,$$

$$x+y-z=2,$$
$$3x-y+z=0;$$

2. An undetermined system:

$$-x+y+3z=-2$$
$$y+2z=4$$

3. A consistent system with a unique solution:

$$x-2y=-1,$$
$$2x+3y=7;$$

To solve the equations in the first case using MATLAB, enter the following commands:

```
a = [2 -1 1;1 1 -1;3 -1 1] ;
b = [1 2 0]' ;
x = inv(a)*b
```

On entering this, a message will be displayed: *Warning: Matrix is close to singular or badly scaled. Results may be inaccurate.* In this case, the determinant of the matrix is zero very close to zero, and hence, there will be difficulties in numerical computations. Hence, the system is inconsistent. For the second case, obviously, there are three variables, namely x, y, and z, but only two equations. Such a system cannot be solved and it is a case of an undetermined system. For the third case, enter the following commands in MATLAB:

```
>> a = [1 -2;2 3] ;
   b = [-1 7]' ;
   x = inv(a)*b
   %x = a\b

x =
     1.5714
     1.2857
```

As displayed, there exists a unique solution and it is the case of the consistent system of equations. The consistency of the system of linear equations to have a unique solution can be checked by using the Gauss-Jordan elimination to reduce the augmented matrix into row-reduced echelon form (rref) by elementary row operations. A matrix is in row-reduced echelon form if the following conditions are satisfied (a) the leading entry in each row (if any) is a one, (b) there are no entries in the column above or below any leading entry and, (c) any leading entries in a row is to the right of a leading entry in a row above. The Gauss-Jordan elimination eliminates the need for a back substitution of the Gauss elimination. In MATLAB, the function *rref(A)* produces the reduced row echelon form of matrix *A* using the Gauss Jordan elimination. For the abovementioned problem, see the following steps.

1. The first case of an inconsistent system produces:

```
>> A = [2 -1 1 1;1 1 -1 2;3 -1 1 0] ;
   rref(A)
   ans =
        1   0   0   0
        0   1  -1   0
        0   0   0   1
```

2. For the second case (an undetermined system):

```
>> A = [-1 1 3 -2; 0 1 2 4; 0 0 0 0] ;
   rref(A)
   ans =
        1   0  -1   6
        0   1   2   4
        0   0   0   0
```

3. For the third case of a consistent system:

```
>> A = [1 -2 -1;2 3 7] ;
   rref(A)
   ans =
        1.0000        0 1.5714
             0 1.0000 1.2857
```

The third case has obviously a unique solution, while the first and second cases have no solutions.

The determinant, the matrix inverse, and the solution to a system of equations are closely related to each other and each of these can be calculated from the LU decomposition of a matrix. After the LU decomposition, the determinant is simply the product of the diagonal elements of the LU decomposed matrix. The LU function expresses a matrix A as the product of two essentially triangular matrices, one of them a permutation of a lower triangular matrix and the other an upper triangular matrix. The factorization is called the LU. In MATLAB, the function $[L,U,P] = LU(A)$ returns unit lower triangular matrix L, upper triangular matrix U, and permutation matrix P so that $P*A = L*U$. Given a matrix equation $Ax = LUx = b$, the equation has to be solved for A and b. Firstly, the equation $Ly = b$ is solved for y and secondly, the equation $Ux = y$ is solved for x.
For example, consider the system of equations:

$$3x + 2y + z = 10$$
$$2x + y + 3z = 13$$
$$x + 3y + 2z = 13$$

To solve using *LU* decomposition, the following MATLAB commands can be run in sequence:

```
A = [3 2 1;2 1 3;1 3 2] ;
b = [10 13 13]' ;
[L U P] = lu(A) ;
%Ly = b using Forward substitution
y(1,1) = b(1,1) ;
y(2,1) = b(2,1) - L(2,1) * y(1,1) ;
y(3,1) = b(3,1) - L(3,1) * y(1,1) - L(3,2) * y(2,1) ;
%Ux = y using backward substitution
x(3,1) = y(3,1)/U(3,3) ;
x(2,1) = (y(2,1) - U(2,3) * x(3,1)) / U(2,2) ;
x(1,1) = (y(1,1) - U(1,2) * x(2,1) - U(1,3) * x(3,1))/U(1,1) ;
```

This produces the following result which are the values of x, y, and z:

```
x =
      1.0000
      2.0000
      3.0000
```

It can be also seen that product of the diagonal elements of the *LU* decomposed matrix is the same as the determinant.

15.7 The Least-Squares Method

As an example for the application of the matrix algebra in the solution of system of simultaneous equations, we discuss here the least square method for regression. The least square method is a statistical approach to estimate an expected value or function with the highest probability from the observations with random errors. The highest probability is replaced by minimizing the sum of square of residuals in the least square method, where the residual is defined as the difference between the observation and an estimated value of a function. The least-squares line uses a straight line:

$$y = a + bx \tag{15.14}$$

to approximate the given set of data, $(x_1, y_1), (x_2, y_2), \ldots (x_n, y_n)$ where $n \geq 2$. The best fitting curve $f(x)$ has the least square error, i.e.,

$$\Pi = \sum_{i=1}^{n} [y_i - f(x_i)]^2 = \sum_{i=1}^{n} [y_i - (a + bx_i)]^2 = \min. \tag{15.15}$$

Please note that a and b are unknown coefficients while all x_i and y_i are given. To obtain the least square error, the unknown coefficients a and b must yield zero first derivatives.

$$\frac{\partial \Pi}{\partial a} = 2 \sum_{i=1}^{n} [y_i - (a + bx_i)] = 0$$

$$\frac{\partial \Pi}{\partial b} = 2 \sum_{i=1}^{n} x_i [y_i - (a + bx_i)] = 0 \tag{15.16}$$

Expanding the above equations, we have:

$$\sum_{i=1}^{n} y_i = a \sum_{i=1}^{n} 1 + b \sum_{i=1}^{n} x_i$$

$$\sum_{i=1}^{n} x_i y_i = a \sum_{i=1}^{n} x_i + b \sum_{i=1}^{n} x_i^2 \tag{15.17}$$

The unknown coefficients a and b can therefore be obtained:

$$a = \frac{\left(\sum\limits_{i=1}^{n} y\right)\left(\sum\limits_{i=1}^{n} x^2\right) - \left(\sum\limits_{i=1}^{n} x\right)\left(\sum\limits_{i=1}^{n} xy\right)}{n \sum\limits_{i=1}^{n} x^2 - \left(\sum\limits_{i=1}^{n} x\right)^2} \tag{15.18}$$

$$b = \frac{n \sum\limits_{i=1}^{n} xy - \left(\sum\limits_{i=1}^{n} x\right)\left(\sum\limits_{i=1}^{n} y\right)}{n \sum\limits_{i=1}^{n} x^2 - \left(\sum\limits_{i=1}^{n} x\right)^2} \tag{15.19}$$

From Eq. 15.16, the matrix form becomes:

$$\begin{bmatrix} N & \sum\limits_{i=1}^{n} x_i \\ \sum\limits_{i=1}^{n} x_i & \sum\limits_{i=1}^{n} x_i^2 \end{bmatrix} \begin{bmatrix} a \\ b \end{bmatrix} = \begin{bmatrix} \sum\limits_{i=1}^{n} y_i \\ \sum\limits_{i=1}^{n} x_i y_i \end{bmatrix} \tag{15.20}$$

The left-hand side of Eq. 15.19 can be written as a product $\left(A^T A\right) X$ if the product is defined as:

$$\left(A^T A\right) X = \begin{bmatrix} 1 & 1 & 1 & \ldots & 1 \\ x_1 & x_2 & x_3 & \ldots & x_n \end{bmatrix} \begin{bmatrix} 1 & x_1 \\ 1 & x_2 \\ 1 & x_3 \\ \cdot & \cdot \\ \cdot & \cdot \\ \cdot & \cdot \\ 1 & x_n \end{bmatrix} \begin{bmatrix} a \\ b \end{bmatrix} \tag{15.21}$$

The right-hand side of Eq. 15.19 can be written as a product $A^T b$:

$$A^T b = \begin{bmatrix} 1 & 1 & 1 & \ldots & 1 \\ x_1 & x_2 & x_3 & \ldots & x_n \end{bmatrix} \begin{bmatrix} y_1 \\ y_2 \\ y_3 \\ . \\ . \\ . \\ y_n \end{bmatrix}$$

(15.22)

Thus, the least square equations defined by Eq. 15.16 becomes:

$$\left(A^T A\right) X = A^T b$$

(15.23)

15.7.1 Example 11

Consider the following data.

Table 15.1 x-y data

x	y
1	2
2	3
3	7
4	8
5	9

Data points included here can be plotted to get a 'data point graph' as shown in Fig. 15.4. These data points can be used to make a 'continuous graph' as shown in Fig. 15.5. Evaluated error with different values of x and y are included in Table 15.2. If we choose the line that goes through the points when $x = 1$ and 2, we get the line $y = 1 + x$.

Table 15.2 Error evaluation

x	y	predicated y	error	(error)^2
1	2	2	0	0
2	3	3	0	0
3	7	4	3	9
4	8	5	3	9
5	9	6	3	9

If we choose the line that goes through the points when $x = 3$ and 4, we get the line $y = 4 + x$, for which the data points are tabulated in Table 15.3 and the graph is plotted in Fig. 15.6.

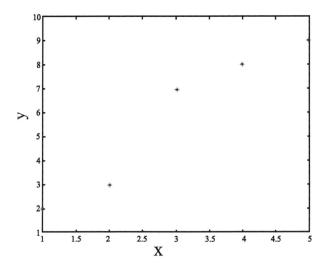

Fig. 15.4 Graphical representation of data

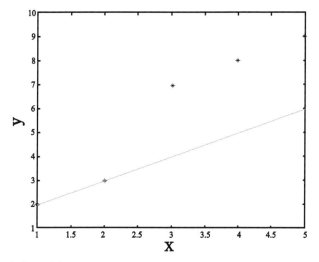

Fig. 15.5 Graph fit to minimum error

Table 15.3 Data for the line passing through 3 and 4

No	y	predicated y	error	(error)^2
1	2	5	−3	9
2	3	6	−3	9
3	7	7	0	0
4	8	8	0	0
5	9	9	0	0

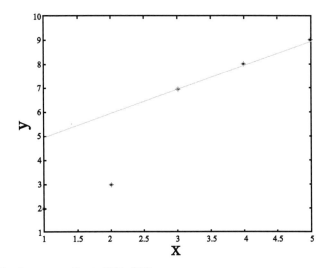

Fig. 15.6 Graph corresponding to Table 15.3

Let us try the line that is halfway between these two lines. The equation would be $y = 2.5 + x$. Data points generated from this equation are included in Table 15.4 and the corresponding graph is as shown Fig. 15.7. Evaluated error and square of error are tabulated in Table 15.5.

Table 15.4 Data for the line that is halfway between the graphs in Figs. 15.5 and 15.6

x	y	predicated y	error	(error)^2
1	2	3.5	−1.5	2.25
2	3	4.5	−1.5	2.25
3	7	5.5	1.5	2.25
4	8	6.5	1.5	2.25
5	9	7.5	1.5	2.25

Using matrix form of least square, we have:

$$\left(A^T A\right) X = A^T b$$

$$\begin{bmatrix} 1 & 1 & 1 & 1 & 1 \\ 1 & 2 & 3 & 4 & 5 \end{bmatrix} \begin{bmatrix} 1 & 1 \\ 1 & 2 \\ 1 & 3 \\ 1 & 4 \\ 1 & 5 \end{bmatrix} \begin{bmatrix} a \\ b \end{bmatrix} = \begin{bmatrix} 1 & 1 & 1 & 1 & 1 \\ 1 & 2 & 3 & 4 & 5 \end{bmatrix} \begin{bmatrix} 2 \\ 3 \\ 7 \\ 8 \\ 9 \end{bmatrix}$$

$$\begin{bmatrix} 5 & 15 & 29 \\ 15 & 55 & 106 \end{bmatrix}_{r_1 \div 5} \Rightarrow \begin{bmatrix} 1 & 3 & 5.8 \\ 15 & 55 & 106 \end{bmatrix}_{-15*r_1+r_2} \Rightarrow$$

$$\begin{bmatrix} 1 & 3 & 5.8 \\ 0 & 10 & 19 \end{bmatrix}_{r_2 \div 10} \Rightarrow \begin{bmatrix} 1 & 3 & 5.8 \\ 0 & 1 & 1.9 \end{bmatrix}_{-3*r_2+r_1} \Rightarrow \begin{bmatrix} 1 & 0 & 0.1 \\ 0 & 1 & 1.9 \end{bmatrix} \Rightarrow$$

$$\begin{bmatrix} a \\ b \end{bmatrix} = \begin{bmatrix} 0.1 \\ 1.9 \end{bmatrix}$$

For the line $y = 0.1 + 1.9x$

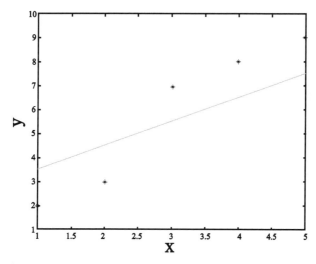

Fig. 15.7 Graph corresponding to Table 15.4

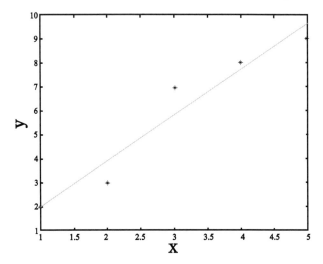

Fig. 15.8 Graph for $y = 0.1 + 1.9x$

The MATLAB implementation of the least square curve fitting for the above example is illustrated below by the sequence of commands.

```
x = [1 2 3 4 5] ;
y = [2 3 7 8 9] ;
A1= [1 1 1 1 1]' ;
A = [A1 x'] ;
U = A'*A ;
V = A'*y' ;
LS1 = U\V ; % solves the equation to find a and b
LS2 = polyfit(x,y,1) ; % Fit a straight line
f1 = polyval (LS2,x) ; % Evaluates the polynomial with x
error = y-f1 ; %Calculates the error
disp(' x y f1 y-f1') ;
disp([x' y' f1' error']) ;
plot(x,y,'o',x,f1,'-') %Plots the graph
axis([1 5 1 10 ]) % set the axis ranges
xlabel('x') % label the x-axis
ylabel('y') % label the y-axis
V = A''*y' ;
LS1 = U\V ; % solves the equation to find a and b (stored in LS1)
LS2 = polyfit(x,y,1) ; % Fit a straight line with coefficients in LS2
f1 = polyval (LS2,x) ; % Evaluates the polynomial with x
error = y-f1 ; % Calculates error
disp(' x y f1 y-f1') ;
disp([x' y' f1' error']) ;
plot(x,y,'o',x,f1,'-') % Plots the graph
axis([1 5 1 10 ]) % set the axis ranges
xlabel('x') % label the x-axis
```

Table 15.5 Error and square of error

x	y predicated y	error	(error)^2
1	2	2.0	0	0
2	3	3.9	−0.9	0.81
3	7	5.8	1.2	1.44
4	8	7.7	0.3	0.09
5	9	9.6	−0.6	0.36

The result is the following graph:

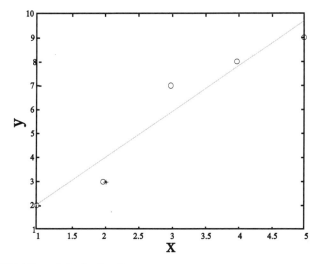

Fig. 15.9 MATLAB graph for the function

15.8 Eigenvalues and Eigenvectors

The eigenvalue problem is a problem of considerable theoretical interest and wide-ranging application. For example, this problem is crucial in solving systems of differential equations, analyzing population growth models, and calculating powers of matrices (in order to define the exponential matrix). Other areas such as physics, sociology, biology, economics, and statistics have focused considerable attention on "eigenvalues" and "eigenvectors" – their applications and their computations. Before we give the formal definition, let us introduce these concepts in an example.

15.8.1 Example 12

Consider the matrix:

$$A = \begin{bmatrix} 1 & 2 & 1 \\ 6 & -1 & 0 \\ -1 & -2 & -1 \end{bmatrix}$$

Consider the three column matrices:

$$C_1 = \begin{bmatrix} 1 \\ 6 \\ -13 \end{bmatrix}, \quad C_2 = \begin{bmatrix} -1 \\ 2 \\ 1 \end{bmatrix}, \quad C_3 = \begin{bmatrix} 2 \\ 3 \\ -2 \end{bmatrix}$$

We have:

$$AC_1 = \begin{bmatrix} 0 \\ 0 \\ 0 \end{bmatrix}, \quad AC_2 = \begin{bmatrix} 4 \\ -8 \\ -4 \end{bmatrix}, \quad AC_3 = \begin{bmatrix} 6 \\ 9 \\ -6 \end{bmatrix}.$$

In other words, we have:

$$AC_1 = 0C_1, AC_2 = -4C_2, AC_3 = 3C_3$$

Next consider the matrix P for which the columns are C_1, C_2, and C_3, i.e.,

$$P = \begin{bmatrix} 1 & -1 & 2 \\ 6 & 2 & 3 \\ -13 & 1 & -2 \end{bmatrix}$$

We have $det(P) = 84$. So, this matrix is invertible. Easy calculations give:

$$P^{-1} = \frac{1}{84} \begin{bmatrix} -7 & 0 & -7 \\ -27 & 24 & 9 \\ 32 & 12 & 8 \end{bmatrix}$$

Next, we evaluate the matrix $P^{-1}AP$. We leave the details to the reader to check that we have:

$$\frac{1}{84} \begin{bmatrix} -7 & 0 & -7 \\ -27 & 24 & 9 \\ 32 & 12 & 8 \end{bmatrix} \begin{bmatrix} 1 & 2 & 1 \\ 6 & -1 & 0 \\ -1 & -2 & -1 \end{bmatrix} \begin{bmatrix} 1 & -1 & 2 \\ 6 & 2 & 3 \\ -13 & 1 & -2 \end{bmatrix} = \begin{bmatrix} 0 & 0 & 0 \\ 0 & -4 & 0 \\ 0 & 0 & 3 \end{bmatrix}$$

In other words, we have:

$$P^{-1}AP = \begin{bmatrix} 0 & 0 & 0 \\ 0 & -4 & 0 \\ 0 & 0 & 3 \end{bmatrix}$$

Using the matrix multiplication, we obtain:

$$A = P \begin{bmatrix} 0 & 0 & 0 \\ 0 & -4 & 0 \\ 0 & 0 & 3 \end{bmatrix} P^{-1}$$

which implies that A is similar to a diagonal matrix. In particular, we have:

$$A^n = P \begin{bmatrix} 0 & 0 & 0 \\ 0 & (-4)^n & 0 \\ 0 & 0 & 3^n \end{bmatrix} P^{-1}$$

for $n = 1, 2, 3, \ldots$. Note that it is almost impossible to find A^{75} directly from the original form of A. This example is so rich with conclusions that many questions impose themselves in a natural way. For example, given a square matrix A, how do we find column matrices which have similar behaviors as the above ones? In other words, how do we find these column matrices which will help find the invertible matrix P such that $P^{-1}AP$ is a diagonal matrix?

From now on, we will call column matrices vectors. So, the above column matrices C_1, C_2, and C_3 are now vectors. We have the following definition:

Definition. Let A be a square matrix. A non-zero vector C is called an eigenvector of A if and only if there exists a number (real or complex) λ such that:

$$AC = \lambda C \tag{15.24}$$

If such a number λ exists, it is called an *eigenvalue* of A. The vector C is called an eigenvector associated to the eigenvalue λ.

15.8.2 Example 13

Consider the matrix:

$$A = \begin{bmatrix} 1 & 2 & 1 \\ 6 & -1 & 0 \\ -1 & -2 & -1 \end{bmatrix}$$

We have seen that:

$$AC_1 = 0C_1, AC_2 = -4C_2, AC_3 = 3C_3$$

where:

$$C_1 = \begin{bmatrix} 1 \\ 6 \\ -13 \end{bmatrix}, \quad C_2 = \begin{bmatrix} -1 \\ 2 \\ 1 \end{bmatrix}, \quad C_3 = \begin{bmatrix} 2 \\ 3 \\ -2 \end{bmatrix}$$

So, C_1 is an eigenvector of A associated to the eigenvalue 0. C_2 is an eigenvector of A associated to the eigenvalue -4, while C_3 is an eigenvector of A associated to the eigenvalue 3.

15.8.3 The Computation of Eigenvalues

For a square matrix A of order n, the number λ is an eigenvalue if, and only if, there exists a non-zero vector C such that:

$$AC = \lambda C$$

Using the matrix multiplication properties, we obtain:

$$(A - \lambda I_n)C = 0 \tag{15.25}$$

This is a linear system for which the matrix coefficient is $A - \lambda I_n$. Since the zero-vector is a solution and C is not the zero vector, then we must have:

$$\det(A - \lambda I_n) = 0 \tag{15.26}$$

In general, for a square matrix A of order n, the above equation will give the eigenvalues of A. This equation is called the characteristic equation or characteristic polynomial of A. It is a polynomial function in λ of degree n. So, we know that this equation will not have more than n roots or solutions. Therefore, a square matrix A of order n will not have more than n eigenvalues.

15.8.4 Example 14

Consider the matrix:

$$A = \begin{bmatrix} 1 & -2 \\ -2 & 0 \end{bmatrix}$$

The equation $\det(A - \lambda I_n) = 0$ translates into:

$$\begin{vmatrix} 1 - \lambda & -2 \\ -2 & 0 - \lambda \end{vmatrix} = (1 - \chi)(0 - \lambda) - 4 = 0$$

which is equivalent to the quadratic equation:

$$\lambda^2 - \lambda - 4 = 0$$

Solving this equation leads to:

$$\lambda = \frac{1 + \sqrt{17}}{2}, \quad \lambda = \frac{1 - \sqrt{17}}{2}$$

In other words, the matrix A has only two eigenvalues.

15.8.5 The Computation of Eigenvectors

Let A be a square matrix of order n and λ one of its eigenvalues. Let X be an eigenvector of A associated to λ. We must have:

$$AX = \lambda X \text{ or } (A - \lambda I_n)X = 0 \tag{15.27}$$

This is a linear system for which the matrix coefficient is $A - \lambda I_n$. Since the zero-vector is a solution, the system is consistent.

15.8.6 Example 15

Consider the matrix:

$$A = \begin{bmatrix} 1 & 2 & 1 \\ 6 & -1 & 0 \\ -1 & -2 & -1 \end{bmatrix}$$

Firstly, we look for the eigenvalues of A. These are given by the characteristic equation $\det(A - \lambda I_3) = 0$, i.e.:

$$\begin{vmatrix} 1-\lambda & 2 & 1 \\ 6 & -1-\lambda & 0 \\ -1 & -2 & -1-\lambda \end{vmatrix} = 0$$

If we develop this determinant using the third column, we obtain:

$$\begin{vmatrix} 6 & -1-\lambda \\ -1 & -2 \end{vmatrix} + (-1-\lambda)\begin{vmatrix} 1-\lambda & 2 \\ 6 & -1-\lambda \end{vmatrix} = 0$$

Using easy algebraic manipulations, we get:

$$-\lambda(\lambda+4)(\lambda-3) = 0$$

which implies that the eigenvalues of A are 0, -4, and 3. Secondly, we look for the eigenvectors.

1. *Case $\lambda = 0$:* The associated eigenvectors are given by the linear system $AX = 0$ which may be rewritten by:

$$x + 2y + z = 0$$
$$6x - y = 0$$
$$-x - 2y - z = 0$$

Many ways may be used to solve this system. The third equation is identical to the first. Since, from the second equations, we have $y = 6x$, the first equation reduces to $13x + z = 0$. So this system is equivalent to:

$$y = 6x$$
$$z = -13x$$

So, the unknown vector X is given by:

$$X = \begin{pmatrix} x \\ y \\ z \end{pmatrix} = \begin{bmatrix} x \\ 6x \\ -13x \end{bmatrix} = x\begin{bmatrix} 1 \\ 6 \\ -13 \end{bmatrix}$$

Therefore, any eigenvector X of A associated to the eigenvalue 0 is given by:

$$X = c \begin{bmatrix} 1 \\ 6 \\ -13 \end{bmatrix}$$

where c is an arbitrary number.

2. *Case 2$\lambda = -4$*: The associated eigenvectors are given by the linear system $AX = -4X$ or $(A + 4I_3)X = 0$ which may be rewritten by:

$$\begin{cases} 5x + 2y + z = 0 \\ 6x + 3y = 0 \\ -x - 2y + 3z = 0 \end{cases}$$

In this case, we will use elementary operations to solve it. Firstly, we consider the augmented matrix, i.e.:

$$\begin{bmatrix} 5 & 2 & 1 & | & 0 \\ 6 & 3 & 0 & | & 0 \\ -1 & -2 & 3 & | & 0 \end{bmatrix}$$

Secondly, we use elementary row operations to reduce it to a upper-triangular form. We interchange the first row with the first one to get:

$$\begin{bmatrix} -1 & 2 & 3 & | & 0 \\ 5 & 2 & 1 & | & 0 \\ 6 & 3 & 0 & | & 0 \end{bmatrix}$$

Next, we use the first row to eliminate the 5 and 6 on the first column. We obtain:

$$\begin{bmatrix} -1 & 2 & 3 & | & 0 \\ 0 & -8 & 16 & | & 0 \\ 0 & -9 & 18 & | & 0 \end{bmatrix}$$

If we cancel the 8 and 9 from the second and third row, we obtain:

$$\begin{bmatrix} -1 & 2 & 3 & | & 0 \\ 0 & -1 & 2 & | & 0 \\ 0 & -1 & 2 & | & 0 \end{bmatrix}$$

Finally, we subtract the second row from the third to get:

$$\begin{bmatrix} -1 & 2 & 3 & | & 0 \\ 0 & -1 & 2 & | & 0 \\ 0 & 0 & 0 & | & 0 \end{bmatrix}$$

Next, we set $z = c$. From the second row, we get $y = 2z = 2c$. The first row will imply $x = -2y + 3z = -c$. Hence:

$$X = \begin{bmatrix} x \\ y \\ z \end{bmatrix} = \begin{bmatrix} -c \\ 2c \\ c \end{bmatrix} = c \begin{bmatrix} -1 \\ 2 \\ 1 \end{bmatrix}$$

Therefore, any eigenvector X of A associated to the eigenvalue -4 is given by:

$$X = c \begin{bmatrix} -1 \\ 2 \\ 1 \end{bmatrix}$$

where c is an arbitrary number.

3. *Case $\lambda = 3$*: The details for this case will be left to the reader. Using similar ideas as the one described above, one may easily show that any eigenvector X of A associated to the eigenvalue 3 is given by:

$$X = c \begin{bmatrix} 2 \\ 3 \\ -2 \end{bmatrix}$$

where c is an arbitrary number.

MATLAB implementation of eigenvalues and eigenvectors:

```
>> A = [1 2 1;6 -1 0; -1 -2 -1] ;
   A =
      1   2   1
      6  -1   0
     -1  -2  -1

[V,D] = eig(A)
     V =
          -0.4082   0.4851  -0.0697
           0.8165   0.7276  -0.4180
           0.4082  -0.4851   0.9058

     D =
          -4.0000        0        0
                0   3.0000        0
                0        0  -0.0000
```

For the applications of eigenvalues and eigenvectors in computational chemistry, refer to Chaps. 4, 5, and 6.

15.9 Exercises

1. Solve the system of the equations:

$$x + y + x = 150$$
$$x + 2y + 3x = 150$$
$$2x + 3y + 4z = 200$$

2. Show that the system is consistent and undetermined:

$$2x_1 + 4x_2 + 5x_3 = 47$$
$$3x_1 + 10x_2 + 11x_3 = 104$$
$$3x_1 + 2x_2 + 4x_3 = 37$$

3. Fit a straight line using the least square method. Check your answer by comparing normal equations and matrix form.

X	0	1.0	2.0	3.0	5.0
Y	0	1.4	2.2	3.5	4.4

Find the eigenvalues and eigenvectors for each of the following:

$$A = \begin{bmatrix} 1 & -1 & 0 \\ 0 & 1 & 1 \\ 0 & 0 & -2 \end{bmatrix} ; \quad B = \begin{bmatrix} 2 & -2 & 3 \\ -2 & -1 & 6 \\ 1 & 2 & 0 \end{bmatrix} ; \quad C = \begin{bmatrix} 8 & 0 & 3 \\ 2 & 2 & 1 \\ 2 & 0 & 3 \end{bmatrix}$$

15.10 Summary

Only a basic treatment of matrix computation is attempted with MATLAB examples in this chapter. Numerical linear algebra is the heart of any computational science and engineering subject such as computational chemistry and deals with matrix multiplications, matrix transformations, matrix factorization, singular value decomposition, solution of systems of equations, computation of eigenvalues and eigenvectors, sparse matrices, etc. While a good working knowledge of the subject is very essential for a computational scientist, an extensive treatment of the subject is beyond the scope of this book. For example, any physical system is generally governed by partial differential equations and such governing partial differential equations in general are nonlinear in themselves, or the domain of the problem where the solution is sought after may be very complex. For such problems involving nonlinear partial differential equations with complex boundary or initial conditions, there are no analytical solutions and one has to resort to numerical methods. The numerical solution of any partial differential equation finally boils down to a system of a large number of simultaneous equations, where one has to employ the computational methods used in numerical linear algebra. For a detailed understanding of the subject, readers can refer to the books on the subject cited in the references below.

References

1. Trefethen LN, Bau D III (1997) Numerical Linear Algebra. SIAM, Philadelphia, PA
2. Demmel JW (1997) Applied Numerical Linear Algebra. SIAM, Philadelphia, PA
3. Golub GH, Van Loan CF (1996) Matrix Computations. Johns Hopkins University Press, Baltimore, MD
4. Strang G (2003) Linear Algebra and Its Applications. Thomson, Cambridge
5. Quarteroni A, Sacco R, Saleri F (2007) Numerical Mathematics, 2nd ed. Springer, New York
6. Yang WY, Cao W, Chung T-S, Morris J (2005) Applied Numerical Methods Using MATLAB. Wiley-Interscience, New York
7. Meyer CD (2007) Matrix Analysis and Applied Linear Algebra. SIAM, Philadelphia, PA

Appendix A
Operators

A.1 Introduction

Levine defines an *operator* as "a rule that transforms a given function into another function." The differentiation operator d/dx is an example. It transforms a differentiable function $f(x)$ into another function $f'(x)$. Other examples include integration, the square root, and so forth. Numbers can also be considered as operators (they multiply a function). McQuarrie gives an even more general definition for an operator: "An *operator* is a symbol that tells you to do something with whatever follows the symbol." Perhaps this definition is more appropriate if we want to refer to the $\hat{C}_3$ operator acting on NH_3, for example.

A.2 Operators and Quantum Mechanics

In quantum mechanics, physical observables (e.g., energy, momentum, position, etc.) are represented mathematically by operators. For instance, the operator corresponding to energy is the Hamiltonian operator:

$$\hat{H} = -\frac{\hbar^2}{2} \sum_i \frac{1}{m_i} \nabla_i^2 + V \tag{A.1}$$

where i is an index over all the particles of the system. We have already encountered the single-particle Hamiltonian. The average value of an observable A represented by an operator $\hat{A}$ for a quantum molecular state $\psi(r)$ is given by the "expectation value" formula:

$$\langle A \rangle = \int \psi^*(r) \hat{A} \psi(r) \tag{A.2}$$

K. I. Ramachandran et al., *Computational Chemistry and Molecular Modeling*   343
DOI: 10.1007/978-3-540-77304-7, ©Springer 2008

A.3 Basic Properties of Operators

Most of the properties of operators are obvious, but they are summarized below for completeness.

The *sum* and *difference* of two operators $\hat{A}$ and $\hat{B}$ are given by:

$$(\hat{A} + \hat{B})f = \hat{A}f + \hat{B}f \tag{A.3}$$

$$(\hat{A} - \hat{B})f = \hat{A}f - \hat{B}f \tag{A.4}$$

The *product* of two operators is defined by

$$\hat{A}\hat{B}f = \hat{A}\left[\hat{B}f\right] \tag{A.5}$$

Two operators are *equal* if:

$$\hat{A}f = \hat{B}f \tag{A.6}$$

for all functions f.

The *identity operator* $\hat{I}$ does nothing (or multiplies by 1):

$$\hat{I}f = f \tag{A.7}$$

A common mathematical trick is to write this operator as a sum over a complete set of states (more on this later).

$$\sum_i |i\rangle\langle i| f = f \tag{A.8}$$

The *associative law* holds for operators:

$$\hat{A}\left(\hat{B}\hat{C}\right) = \left(\hat{A}\hat{B}\right)\hat{C} \tag{A.9}$$

The *commutative law* does *not* generally hold for operators. In general, $\hat{A}\hat{B} \neq \hat{B}\hat{A}$. It is convenient to define the quantity as:

$$\left[\hat{A}, \hat{B}\right] \equiv \hat{A}\hat{B} - \hat{B}\hat{A}, \tag{A.10}$$

which is called the *commutator* of $\hat{A}$ and $\hat{B}$. Note that the order matters, so that $\left[\hat{A}, \hat{B}\right] = -\left[\hat{B}, \hat{A}\right]$. If $\hat{A}, \hat{B}$ happen to commute, then

$$\left[\hat{A}, \hat{B}\right] = 0. \tag{A.11}$$

The *n-th power* of an operator $\hat{A}^n$ is defined as n successive applications of the operator, e.g.:

$$\hat{A}^2 f = \hat{A}\hat{A}f \tag{A.12}$$

The *exponential* of an operator $e^{\hat{A}}$ is defined via the power series

$$e^{\hat{A}} = \hat{I} + \hat{A} + \frac{\hat{A}^2}{2!} + \frac{\hat{A}^3}{3!} + - - - - - \tag{A.13}$$

A.4 Linear Operators

Almost all operators encountered in quantum mechanics are *linear operators*. A linear operator is an operator, which satisfies the following two conditions:

$$\hat{A}(f+g) = \hat{A}f + \hat{A}g \tag{A.14}$$

$$\hat{A}(cf) = c\hat{A}f \tag{A.15}$$

where c is a constant and f and g are functions. As an example, consider the operators d/dx and $()^2$. We can see that d/dx is a linear operator because:

$$(d/dx)[f(x) + g(x)] = (d/dx)f(x) + (d/dx)g(x) \tag{A.16}$$

$$(d/dx)[cf(x)] = c\left(\frac{d}{dx}\right)f(x) \tag{A.17}$$

However, $()^2$ is not a linear operator because:

$$[f(x) + g(x)]^2 \neq [f(x)]^2 + [g(x)]^2 \tag{A.18}$$

The only other category of operators relevant to quantum mechanics is the set of *anti-linear* operators, for which:

$$\hat{A}(\lambda f + \mu g) = \lambda^* \hat{A}f + \mu^* \hat{A}g \tag{A.19}$$

Time-reversal operators are antilinear.

A.5 Eigenfunctions and Eigenvalues

An *eigenfunction* of an operator $\hat{A}$ is a function f such that the application of $\hat{A}$ on f gives f again, times a constant:

$$\hat{A}f = kf \tag{A.20}$$

where, k is a constant called the *eigenvalue*. It is easy to show that if $\hat{A}$ is a linear operator with an eigenfunction g, then any multiple of g is also an eigenfunction of $\hat{A}$.

When a system is in an *eigenstate* of observable A (i.e., when the wavefunction is an eigenfunction of the operator $\hat{A}$) then the expectation value of A is the eigenvalue of the wavefunction. Thus, if:

$$\hat{A}\psi(r) = a\psi(r) \tag{A.21}$$

then:

$$\langle A \rangle = \int \psi^*(r)\hat{A}\psi(r) = \int \psi^*(r)a\psi(r) = a\int \psi^*(r)\psi(r) = a \tag{A.22}$$

assuming that the wavefunction is normalized to 1, as is generally the case. In the event that $\psi(r)$ is not or cannot be normalized (free particle, etc.) then we may use the formula:

$$\langle A \rangle = \frac{\int \psi^*(r)\hat{A}\psi(r)}{\int \psi^*(r)\psi(r)} \tag{A.23}$$

What if the wavefunction is a combination of eigenstates? Let us assume that we have a wavefunction which is a linear combination of two eigenstates of $\hat{A}$ with eigenvalues a and b:

$$\psi = C_a\psi_a + C_b\psi_b \tag{A.24}$$

where $\hat{A}\psi_a = a\psi_a$ and $\hat{A}\psi_b = a\psi_b$. Then, what is the expectation value of A?

$$\langle A \rangle = \int \psi^*\hat{A}\psi$$
$$= \int [C_a\psi_a + C_a\psi_a]^*\hat{A}[C_a\psi_a + C_a\psi_a]$$
$$= \int [C_a\psi_a + C_a\psi_a]^*[aC_a\psi_a + bC_a\psi_a]$$
$$= a|C_a|^2 \int \psi_a^*\psi_a + bC_a^*C_b \int \psi_a^*\psi_b + aC_b^*C_a \int \psi_b^*\psi_a + b|C_b|^2 \int \psi_b^*$$
$$= a|c_a|^2 + b|c_b|^2 \tag{A.25}$$

assuming that ψ_a and ψ_b are orthonormal (shortly, we will show that eigenvectors of Hermitian operators are orthogonal). Thus, the average value of A is a weighted average of eigenvalues, with the weights being the squares of the coefficients of the eigenvectors in the overall wavefunction.

Appendix B
Hückel MO Heteroatom Parameters

Heteroatom parameters (h and k) for common atoms and bonds are listed below (Table B.1).

Table B.1 Heteroatom parameters

Element	Coulomb integral parameter (h_X)	Resonance integral parameter (k_{C-X})
B	$h_B = -1.0$	$k_{C-B} = 0.7$
		$k_{B-N} = 0.8$
C	$h_C = 0.0$	$k_{C-C} = 1.0$
N	$h_{N.} = 0.5$	$k_{C-N} = 0.8$
	$h_{N:} = 1.5$	$k_{C=N} = 1.0$
	$h_{N+} = 2.0$	$k_{N-O} = 0.7$
O	$h_{O.} = 1.0$	$k_{C-O} = 0.8$
	$h_{O:} = 2.0$	$k_{C=O} = 1.0$
	$h_{O+} = 2.5$	
F	$h_F = 3.0$	$k_{C-F} = 0.7$
Cl	$h_{Cl} = 2.0$	$k_{C-Cl} = 0.4$
Br	$h_{Br} = 1.5$	$k_{C-Br} = 0.3$

K. I. Ramachandran et al., *Computational Chemistry and Molecular Modeling*
DOI: 10.1007/978-3-540-77304-7, ©Springer 2008

Appendix C
Using Microsoft Excel to Balance Chemical Equations

C.1 Introduction

A chemical reaction can be represented by an equation, which should be in accordance with the laws of conservation of mass, atoms and charge. Hence for chemical equations, the mass of the reactants should be equal to the mass of products. (The law of conservation of mass) The number of each atom on the reactant side should be equal to that on the product side (the principle of atom conservation, or POAC). The total charge of the reactants should be equal to the charge of products (the conservation of charge).

A number of traditional methods have been introduced for balancing chemical equations, such as the hit and trial method (trial and error), the oxidation number method, the partial equation method and ion-electron method. However, none of these methods proves to be applicable for all types of reactions. To overcome this difficulty an algebraic method was proposed. In this method, a reactant-product system (reaction) is treated as a linear system. The mathematical equations obtained are solved to get the chemical equation balanced. This method was not very popular due to the difficulty in solving simultaneous equations. The development of modern scientific computing techniques helps to overcome the difficulty of solving these equations making the algebraic method again important. The balancing of equations by using Microsoft Excel is explained here.

C.2 The Matrix Method

C.2.1 Methodology

A reactant-product system (equation) to be balanced is treated as a matrix of the form $Ax = b$ where matrix A is a square matrix corresponding to the atomicities of various atoms and "x" is a column vector corresponding to the molar coefficients of reactants and products. The matrix equation set up is solved using Microsoft Excel.

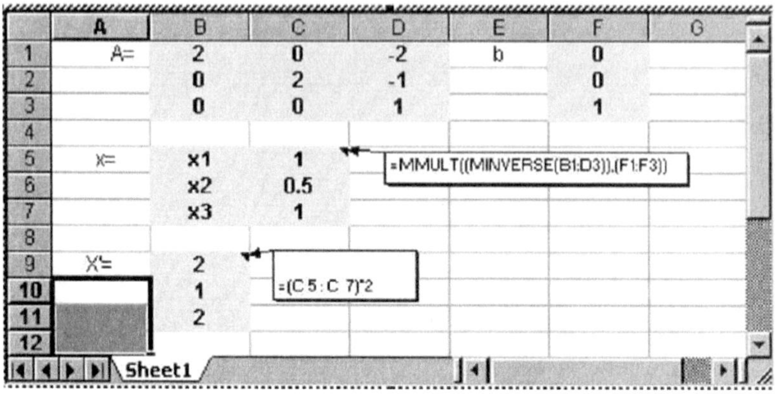

	A	B	C	D	E	F	G
1	A=	2	0	-2	b	0	
2		0	2	-1		0	
3		0	0	1		1	
4							
5	x=	x1	1	=MMULT((MINVERSE(B1:D3)),(F1:F3))			
6		x2	0.5				
7		x3	1				
8							
9	X'=	2					
10		1	=(C5:C 7)*2				
11		2					
12							

Sheet1

Fig. C.1 Microsoft Excel work sheet for Eq. C.1

C.2.2 Example 1

The combustion of hydrogen in oxygen, producing water, can be written as,

$$x_1 H_2 + x_2 O_2 \rightarrow x_3 H_2 O \tag{C.1}$$

We have to determine the unknown coefficients, x_1, x_2, and x_3. In this equation three elements are involved. Make separate equations for each element in the equation:

Hydrogen(H): $2x_1 + 0x_2 = 2x_3$
Oxygen (O): $0x_1 + 2x_2 = x_3$

These equations can been written as:

$$2x_1 + 0x_2 - 2x_3 = 0$$
$$x_1 + 2x_2 - x_3 = 0$$

We have two equations and three unknowns. To complete the system, we define an auxiliary equation by arbitrarily choosing a value (normally one) for one of the coefficients. Here, let us assume x_3 as one. The system can be represented in the matrix form $Ax = b$, where:

$$A = \begin{bmatrix} 2 & 0 & -2 \\ 0 & 2 & -1 \\ 0 & 0 & 1 \end{bmatrix} ; \quad x = \begin{bmatrix} x_1 \\ x_2 \\ x_3 \end{bmatrix} \text{ and } b = \begin{bmatrix} 0 \\ 0 \\ 1 \end{bmatrix}$$

$x =$ Matrix product of the inverse of A and b.

The Microsoft Excel method for solving this equation is illustrated in the excel work sheet (Fig. C.1). If fractional values are obtained as coefficients, they can be changed into whole numbers using Microsoft Excel.

Thus, the balanced equation becomes:

$$2H_2 + O_2 \rightarrow 2H_2O \tag{C.2}$$

The same method is illustrated for an ionic equation.

C.2.2.1 Example 2

Ionic equations conserve mass and charge.
Example:

$$x_1 MnO_4^- + x_2 Fe^{2+} + x_3 H^+ \rightarrow x_4 Mn^{2+} + x_5 Fe^{3+} + x_6 H_2O \tag{C.3}$$

Balancing the equation with respect to atoms:

Manganese: $x_1 + 0x_2 + 0x_3 - x_4 - 0x_5 - 0x_6 = 0$
Oxygen: $\quad 4x_1 + 0x_2 + 0x_3 - 0x_4 - 0x_5 - 1x_6 = 0$
Iron: $\quad\quad 0x_1 + x_2 + 0x_3 - 0x_4 - x_5 - 0x_6 = 0$
Hydrogen: $\quad 0x_1 + 0x_2 + x_3 - 0x_4 - 0x_5 - 2x_6 = 0$
The charge should also be balanced giving one more equation:

$$-x_1 + 2x_2 + x_3 - 2x_4 - 3x_5 - 0x_6 = 0$$

Setting an auxiliary equation by setting the value of one of the coefficients as $1 (x_6 = 1)$ and solving the matrix equation (Fig. C.2):

$$A = \begin{bmatrix} 1 & 0 & 0 & -1 & 0 & 0 \\ 4 & 0 & 0 & 0 & 0 & -1 \\ 0 & 1 & 0 & 0 & -1 & 0 \\ 0 & 0 & 1 & 0 & 0 & -2 \\ -1 & 0 & 1 & -2 & -3 & 0 \\ 0 & 0 & 0 & 0 & 0 & 1 \end{bmatrix}, \quad b = \begin{bmatrix} 0 \\ 0 \\ 0 \\ 0 \\ 0 \\ 1 \end{bmatrix} \text{ and } x = \begin{bmatrix} x_1 \\ x_2 \\ x_3 \\ x_4 \\ x_5 \\ x_6 \end{bmatrix}$$

Hence, the balanced equation for the reaction is:

$$MnO_4^- + 5Fe^{2+} + 8H^+ \rightarrow Mn^{2+} + 5Fe^{3+} + 4H_2O \tag{C.4}$$

C.3 Undermined Systems

A chemical system, where the number of mathematical equations that can be set up, is less than the number variables to be determined, is said to be an *undermined system*. However, such a system can be split up into partial equations. Balance the partial equations by matrix method using Microsoft Excel and combine the partial equations to get the parent equation balanced.

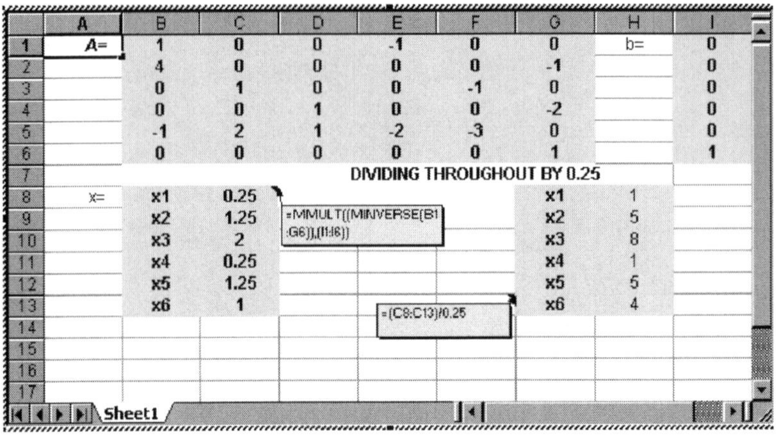

	A	B	C	D	E	F	G	H	I
1	A=	1	0	0	-1	0	0	b=	0
2		4	0	0	0	0	-1		0
3		0	1	0	0	-1	0		0
4		0	0	1	0	0	-2		0
5		-1	2	1	-2	-3	0		0
6		0	0	0	0	0	1		1
7					DIVIDING THROUGHOUT BY 0.25				
8	x=	x1	0.25				x1	1	
9		x2	1.25	=MMULT((MINVERSE(B1			x2	5	
10		x3	2	:G6)),(I1:I6))			x3	8	
11		x4	0.25				x4	1	
12		x5	1.25				x5	5	
13		x6	1		=(C8:C13)/0.25		x6	4	
14									
15									
16									
17									

Sheet1

Fig. C.2 Microsoft Excel work sheet for Example 2

C.4 Balancing as an Optimization Problem

In this method, the chemical equation is treated as a system consisting of n simultaneous linear algebraic equations with m unknowns (molar coefficients). If $n < m$, the chemical equation becomes underdetermined.

C.4.1 Example 3

The reaction between hydrogen peroxide and acidified potassium permanganate to get manganese ions, oxygen, and water is an example of an underdetermined ionic equation and is given in the following equation:

$$x_1 MnO_4^- + x_2 H_2O_2 + x_3 H^+ \rightarrow x_4 Mn^{2+} + x_5 O_2 + x_6 H_2O \qquad (C.5)$$

These variables, x_1, x_2, x_3, x_4, x_5 and x_6 correspond to the molar coefficients of reactants and products. The objective function in the linear optimization problem to be minimized is the sum of these coefficients represented as $\sum_{i=1}^{n} x_i$. While formulating the constraints, the POAC and the charge will only be considered. The constraints are set up on the basis of the POAC with respect to each element in the reaction system as given in Eq. C.6.

$$x_1 n_1 + x_2 n_2 + \ldots = 0 \qquad (C.6)$$

where $x_1, x_2, \ldots$ are the molar coefficients of reactants and products and $n_1, n_2, \ldots$ are the number of atoms of each element in various reactants and products. Obviously, the number of such equations obtained will be equal to the number of elements involved in the reaction. While formulating the constraints for ionic equations, the conservation of charge should also be considered. Constraints set up in the optimization problem can be generalized as follows:

1. $x_1 n_1 + x_2 n_2 + \ldots = 0$ – with respect to each element (the sum product).
2. $\sum C_i x_i = 0$. (where C_i is the charge of the species i and x_i is the molar coefficient of that species.)
3. Molar coefficients should be positive nonzero integers.
4. The problem can be solved using Microsoft Excel Solver.

C.4.1.1 Illustration

Here, the balancing of the underdetermined ionic equation (Eq. C.1) mentioned earlier is illustrated.

In the mathematical form, the equation can be written as:

$$x_1 MnO_4^- + x_2 H_2O_2 + x_3 H^+ - x_4 Mn^{2+} - x_5 O_2 - x_6 H_2O = 0 \qquad (C.7)$$

The objective function to be minimized in this optimization problem is:

$$\sum_{i=1}^{6} x_i \qquad (C.8)$$

subject to the constraints:

(a) $1x_1 + 0x_2 + 0x_3 - 1x_4 - 0x_5 - 0x_6 = 0$ (with respect to manganese (Mn).)
(b) $4x_1 + 2x_2 + 0x_3 - 1x_4 - 2x_5 - 1x_6 = 0$ (with respect to oxygen (O).)
(c) $0x_1 + 2x_2 + 1x_3 - 0x_4 - 0x_5 - 2x_6 = 0$ (with respect to hydrogen (H).)
(d) $-1x_1 + 0x_2 + 1x_3 - 2x_4 - 0x_5 - 0x_6 = 0$ (with respect to the charge.)
(e) x_1, x_2, x_3, x_4, x_5 and x_6 should be positive nonzero integers.

This is solved using Microsoft Excel Solver in the following manner:

1. Construct a worksheet data with the molar coefficients, the elements and the charge, as is given in Fig. C.3.
2. Find the "sum product" with respect to all elements and the charge.
3. Provide space for coefficients to be determined after optimization (row-6).
4. Find the sum of these (Objective function-D8).
5. Set the Solver option with the objective function and the constraints, as is shown in Fig. C.4.
6. Solve the optimization problem to get the molar coefficients, as is shown in Fig. C.5.

Hence the balanced equation is:

$$2MnO_4^- + H_2O_2 + 6H^+ \rightarrow 2Mn^{2+} + 3O_2 + 4H_2O \qquad (C.9)$$

The balancing of some more complex equations by this method is also included.

	A	B	C	D	E	F	G	H
1		x_1	x_2	x_3	x_4	x_5	x_6	sumprod.
2	Mn	1	0	0	-1	0	0	0
3	O	4	2	0	0	-2	-1	0
4	H	0	2	1	0	0	-2	0
5	charge	-1	0	1	-2	0	0	0
6		x_1	x_2	x_3	x_4	x_5	x_6	=SUM PRODUCT
7								(B2:G2,
8				0	=SUM(B7:G7)			B$7:G$7)
9								

Fig. C.3 Worksheet for the example before optimization

	A	B	C	D	E	F	G	H
1		x_1	x_2	x_3	x_4	x_5	x_6	sumprod.
2	Mn	1	0	0	-1	0	0	0
3	O	4	2	0	0	-2	-1	0
4	H	0	2	1	0	0	-2	0
5	charge	-1	0	1	-2	0	0	0
6		x_1	x_2	x_3	x_4	x_5	x_6	=SUM PRODUCT
7		0	0	0	0	0	0	(B2:G2,
8				0	=SUM(B7:G7)			B$7:G$7)

Solver Parameters   **? X**

Set Target Cell: $D$8

Equal To: ○ Max   ● Min   ○ Value of: 0   **Solve**   **Close**

By Changing Cells:

$B$7:$G$7   **Guess**

Subject to the Constraints:

$B$7:$G$7 = integer   **Add**   **Options**
$B$7:$G$7 >= 1
$H$2:$H$5 = 0   **Change**
                    Delete   **Reset All**
                                     Help

Fig. C.4 Worksheet with the solver parameters

	A	B	C	D	E	F	G	H
1		x_1	x_2	x_3	x_4	x_5	x_6	sumprod.
2	Mn	1	0	0	-1	0	0	0
3	O	4	2	0	0	-2	-1	0
4	H	0	2	1	0	0	-2	0
5	charge	-1	0	1	-2	0	0	0
6		x_1	x_2	x_3	x_4	x_5	x_6	=SUM PRODUCT
7		2	1	6	2	3	4	(B2:G2,
8				18	=SUM(B7:G7)			B$7:G$7)
9								

Fig. C.5 Worksheet after optimization

C.4.2 Example 4

$$x_1Cl_2 + x_2NaOH \rightarrow x_3NaCl + x_4NaClO_3 + x_5H_2O \tag{C.10}$$

This system can be written in the form of a mathematical equation as given in Eq. C.9:

$$x_1Cl_2 + x_2NaOH - x_3NaCl - x_4NaClO_3 - x_5H_2O = 0 \tag{C.11}$$

In the optimization procedure, the objective function is:

$$\sum_{i=1}^{5} x_i \tag{C.12}$$

subject to the constraints:

(a) $2x_1 + 0x_2 - 1x_3 - 1x_4 - 0x_5 = 0$ (with respect to chlorine (Cl))
(b) $0x_1 + 1x_2 - 1x_3 - 1x_4 - 0x_5 = 0$ (with respect to sodium (Na))
(c) $0x_1 + 1x_2 - 0x_3 - 3x_4 - 0x_5 = 0$ (with respect to oxygen (O))
(d) $0x_1 + 1x_2 - 0x_3 - 0x_4 - 2x_5 = 0$ (with respect to hydrogen (H))
(e) x_1, x_2, x_3, x_4 and x_5 should be positive nonzero integers.

The balanced equation for the reaction is given in Eq. C.13:

$$3Cl_2 + 6NaOH \rightarrow 5NaCl + NaClO_3 + 3H_2O \tag{C.13}$$

C.4.3 Example 5

$$x_1P_2I_4 + x_2P_4 + x_3H_2O \rightarrow x_4PH_4I + x_5H_3PO_4 \tag{C.14}$$

It can be written in the mathematical form as is given below:

$$x_1P_2I_4 + x_2P_4 + x_3H_2O - x_4PH_4I - x_5H_3PO_4 = 0 \tag{C.15}$$

The objective function for the optimization is:

$$\sum_{i=1}^{5} x_i \tag{C.16}$$

subject to the constraints.

(a) $2x_1 + 4x_2 + 0x_3 - 1x_4 - 1x_5 = 0$ (with respect to phosphorus (P))
(b) $4x_1 + 0x_2 + 0x_3 - 1x_4 - 0x_5 = 0$ (with respect to iodine (I))
(c) $0x_1 + 0x_2 + 2x_3 - 4x_4 - 3x_5 = 0$ (with respect to hydrogen (H))

(d) $0x_1 + 0x_2 + 1x_3 - 0x_4 - 4x_5 = 0$ (with respect to oxygen (O))
(e) x_1, x_2, x_3, x_4 and x_5 should be positive nonzero integers.

The balanced equation is:

$$10P_2I_4 + 13P_4 + 128H_2O \rightarrow 40PH_4I + 32H_3PO_4 \qquad (C.17)$$

The balancing of all types of chemical equations can be effectively carried out through this simple optimization approach. This computational method helps to provide a uniform technique for balancing all types of chemical equations. As Microsoft Excel is familiar to even high school students, the method can be adopted during the introduction of stoichiometric calculations.

Appendix D
Simultaneous Spectrophotometric Analysis

D.1 Introduction

A spectrum is a consequence of interaction of matter with energy. A spectrophotometer is employed to measure the amount of light that a sample absorbs. The instrument operates by passing a beam of light through a sample and measuring the intensity of light reaching a detector.

The spectrophotometric techniques can be used to measure concentration of solutes in solution. To do this, we will measure the amount of light that is absorbed by the solutes in solution in a cuvette in the spectrophotometer. Spectrophotometry takes advantage of the dual nature of light. Namely, light has:

1. a particle nature which gives rise to the photoelectric effect (used in the spectrophotometer)
2. a wave nature which gives rise to the visible spectrum of light.

A spectrophotometer measures the intensity of a light beam after it is directed through and emerges from a solution (Fig. D.1). As an example, let's look at how a solution of copper sulphate ($CuSO_4$) absorbs light.

The red part of the spectrum has been almost completely absorbed by $CuSO_4$ and blue light has been transmitted. Thus, $CuSO_4$ absorbs little blue light and therefore

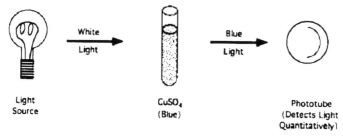

Fig. D.1 Principle of spectrophotometer

K. I. Ramachandran et al., *Computational Chemistry and Molecular Modeling*
DOI: 10.1007/978-3-540-77304-7, ©Springer 2008

appears blue. In spectrophotometry, we can gain greater sensitivity by directing red
light through the solution because $CuSO_4$ absorbs strongest at the red end of the
visible spectrum. But, to do this, we have to isolate the red wavelengths.

The important point to note here is that, colored compounds absorb light differently depending on the λ of incident light.

D.2 The Absorption Spectrum

Different compounds having dissimilar atomic and molecular interactions have
characteristic absorption phenomena and absorption spectra, which differ. The point
(wavelength) at which any given solute exhibits the maximum absorption of light
(the peaks on the curves on the Fig. D.2) is defined as that compounds particular
λ-max.

A spectrophotometric problem in the simultaneous analysis of spectra of solutions is obtained by reacting hydrogen peroxide with Mo, Ti, and V ions in the same
solution to produce compounds that absorb light strongly in overlapping peaks with
absorbances at 330, 410, and 460 nm, respectively is shown in Fig. D.3. These values are included in the matrix.

$$C = \begin{bmatrix} 0.416 & 0.130 & 0.000 \\ 0.048 & 0.608 & 0.148 \\ 0.002 & 0.410 & 0.200 \end{bmatrix} \tag{D.1}$$

The absorbance of light A from a dissolved complex is given by $A = abc$ where
a is the absorptivity, a function of the wavelength, which is characteristic of the
complex, b is the length of the light path through the absorbing solution in centime-
ters, and c is the concentration of the absorbing species in grams per liter. If more
than one complex is present, the absorbance at any selected wavelength is the sum
of contributions of each constituent. Individual solutions of Mo, Ti, and V ions were
made into complexes by hydrogen peroxide, and each spectrum in the visible region

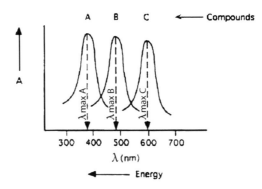

Fig. D.2 Absorption spectrum

was taken with a 1.00-cm cell, with the results shown in Fig. D.3. The absorbance of solutions containing a single complex was recorded at one of the wavelengths shown. The remaining two complexes were measured at the same wavelength, yielding three measurements. This was repeated with the other two complexes, each at its selected wavelength, yielding a total of nine measurements. The concentrations of the metal complex solutions were all the same: $40.0\,\text{mg}\,\text{L}^{-1}$. The absorbance table at L for each of the metal complexes constitutes a matrix with rows of absorbances, at one wavelength, of Mo, Ti, and V complexes, in that order. Each column comprises absorbances for one metal complex at 330, 410, and 460 nm, in that order:

$$C = \begin{bmatrix} 0.416 & 0.130 & 0.000 \\ 0.048 & 0.608 & 0.148 \\ 0.002 & 0.410 & 0.200 \end{bmatrix} . \tag{D.2}$$

Dividing throughout by 0.04 to convert C to $\text{Lg}^{-1}\text{cm}^{-1}$:

$$C = \begin{bmatrix} 10.4 & 3.25 & 0.000 \\ 1.20 & 15.2 & 3.70 \\ 0.05 & 10.25 & 5.00 \end{bmatrix} \tag{D.3}$$

Notice that the matrix has been arranged so that it is as nearly a diagonal dominant as the data permits.

Now, an unknown solution containing Mo, Ti, and V ions was treated with hydrogen peroxide, and its absorbance was determined with a 1.00-cm cell at the three wavelengths, in the same order (lowest to highest), that were used to generate the absorbance matrix for the single complexes. The absorbance of the unknown solution at the three wavelengths was 0.284, 0.857, and 0.718. The ordered set of absorbances of any mixture of the complexes constitutes a b vector, in this case:

$$b = \begin{bmatrix} 0.284 \\ 0.857 \\ 0.718 \end{bmatrix} . \tag{D.4}$$

Let M, T, and V be concentrations of three solutions involved; then, the concentration vector (x) is given as:

$$x = \begin{bmatrix} M \\ T \\ V \end{bmatrix}$$

This is solved with Microsoft Excel as shown in Fig. D.4. Hence:

1. The concentration of Mo $= 0.014641\,\text{Lg}^{-1}\text{cm}^{-1}$
2. The concentration of Ti $= 0.040532\,\text{Lg}^{-1}\text{cm}^{-1}$
3. The concentration of V $= 0.060363\,\text{Lg}^{-1}\text{cm}^{-1}$

Absorbance
0.7
0.6
0.5
0.4
0.3
0.2
0.1
0.0
320 340 360 380 400 420 440 460 480
Wavelength

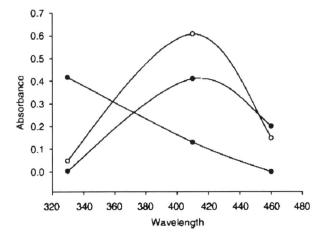

Fig. D.3 Absorption spectrum of mixture

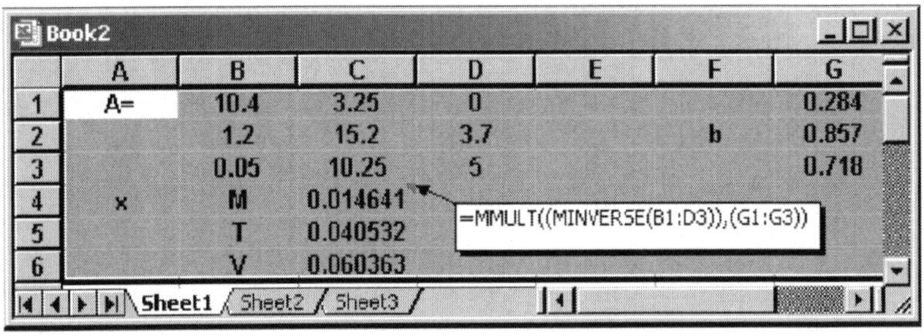

Book2
A B C D E F G
1 A= 10.4 3.25 0 0.284
2 1.2 15.2 3.7 b 0.857
3 0.05 10.25 5 0.718
4 x M 0.014641
5 T 0.040532
6 V 0.060363
=MMULT((MINVERSE(B1:D3)),(G1:G3))
Sheet1 Sheet2 Sheet3

Fig. D.4 Microsoft Excel worksheet for finding the concentration

Appendix E
Bond Enthalpy of Hydrocarbons

The derivation of bond enthalpies from thermo-chemical data involves a system of simultaneous equations in which the sum of unknown bond enthalpies, each multiplied by the number of times the bond appears in a given molecule, is set equal to the enthalpy of atomization of that molecule (Atkins, 1998). Taking a number of molecules equal to the number of bond enthalpies to be determined, one can generate an $n \times n$ set of equations in which the matrix of coefficients is populated by the (integral) number of bonds in the molecule and the set of n atomization enthalpies in the b vector. (Obviously, each bond must appear at least once in the set.)

Carrying out this procedure for propane and butane, $CH_3 - CH_2 - CH_3$ and $CH_3 - CH_2 - CH_2 - CH_3$ yield the bond matrix.

$$\begin{bmatrix} 2 & 8 \\ 3 & 10 \end{bmatrix}.$$

The bond energy data is taken from a chemical database such as the NIST database (http://webbook.nist.gov/chemistry/).

The simultaneous equations obtained are:

$$2(C{-}C) + 8(H{-}C) - \text{propane}$$
$$3(C{-}C) + 10(C{-}H) - \text{butane}$$

We can substitute from the above table to get the enthalpy of atomization of hydrocarbons. Here, the enthalpy vector is as follows:

$$\begin{bmatrix} 3994 \\ 5166 \end{bmatrix}$$

From the enthalpy of atomization of constituent elements, the enthalpy of atomization of the compound is computed based on the equation:

Enthalpy of atomization (bond enthalpy) of compound $= \Sigma$ Enthalpy of atomization of constituents $-$ enthalpy of formation of the compound.

K. I. Ramachandran et al., *Computational Chemistry and Molecular Modeling*   361
DOI: 10.1007/978-3-540-77304-7, ©Springer 2008

Table E.1 Bond energy table

Bond	Bond energy ($kj\,mol^{-1}$)	Bond	Bond energy ($kj\,mol^{-1}$)
H–H	435.4	N–H	389
H–F	565	C–H	413
H–Cl	431	C–Cl	328
H–Br	364	C–O	335
H–I	297	C=O	707
F–F	155	C–N	293
Cl–Cl	242	C=N	616
Br–Br	190	C≡N	879
I–I	149	C–Br	275.6
O=O	494	O–O	138
N≡N	941	N–N	159
C=C	619	N=N	418
C≡C	812	C–C	347
O–H	463		

Here, for propane, enthalpy of atomization is obtained by subtracting the enthalpy of formation of the alkane from the sum of atomic atomization enthalpies (C: 716; H: 218 kJ mol^{-1}) for the molecule. For example, the molecular atomization enthalpy of propane is:

$$3(716) + 8(218) - (-104) = 3996\,kJ\,mol^{-1}$$

Benson, in seeking group additivity values for different kinds of $(CH)_n$ groups defines primary P, secondary S, tertiary T, and quaternary Q carbons and then sets up the simultaneous equations to obtain energetic contributions for P, S, T, and Q.

$$\Delta_f H^{298}(ethane) = -83.81 = 2P$$
$$\Delta_f H^{298}(propane) = -104.7 = 2P+S$$
$$\Delta_f H^{298}(isobutane) = -134.2 = 3P+T$$
$$\Delta_f H^{298}(neopentane) = -168.1 = 4P+Q$$

The b vector in this equation set has been converted from kilocalories per mole to kilojoules per mol. Computing P, S, Q and T:

$$P = -41.905$$
$$S = -20.89$$
$$T = -8.485$$
$$Q = -0.48$$

Appendix F
Graphing Chemical Analysis Data

We can plot and analyze data using a spreadsheet. Guidelines (heuristics) for creating a good graph are reviewed.

F.1 Guidelines

1. Enter and format data in an Microsoft Excel spreadsheet in a form appropriate for graphing.
2. Create a scatter plot from spreadsheet data.
3. Insert a linear regression line (trend line) into the scatter plot
4. Use the slope/intercept formula for the regression line to calculate an x value for a known y value.
5. Explore curve fitting to scatter plot data:

   a. Create a connected point (line) graph.
   b. Place a reference line in a graph.

F.2 Example: Beer's Law Absorption Spectra Tools

F.2.1 Basic Information

This exercise is primarily designed to give students basic skills in creating scatter plots in Microsoft Excel, and then adding either a regression line or a fitted curve to the data points. These techniques are very good for labs in fields such as chemistry and physics where the data collected by the students needs to be interpreted in relation to some theoretical model. For example, does the slope of the regression line, fitted to the collected data, match the theoretical slope calculated from the equation?

K. I. Ramachandran et al., *Computational Chemistry and Molecular Modeling*
DOI: 10.1007/978-3-540-77304-7, ©Springer 2008

In addition to these basic skills, some principles of good graph design are demonstrated with the somewhat modest graph formatting options allowed in Microsoft Excel.

F.2.2 Beer's Law Scatter Plot and Linear Regression

F.2.2.1 Introduction

Beer's Law states that there is a linear relationship between the concentration of a colored compound in the solution and the light absorption of the solution. This fact can be used to calculate the concentration of unknown solutions, given their absorption readings. Firstly, a series of solutions of a known concentration are tested for their absorption level. Secondly, a scatter plot is made of this empirical data and a linear regression line is fitted to the data. This regression line can be expressed as a formula and used to calculate the concentration of unknown solutions. Finally, some techniques are demonstrated as to how to make the plot more readable using the formatting options available in Microsoft Excel.

F.2.2.2 Entering and Formatting the Data in Microsoft Excel

Your data will go in the first two columns in the spreadsheet (Fig. F.1).

1. Title the spreadsheet page in cell A1.
2. Label Column A as the concentration of the known solutions in cell A3.
3. Label Column B as the absorption readings for each of the solutions in cell B3.

Begin by formatting the spreadsheet cells so the appropriate number of decimal places is displayed (see Fig. F.1).

1. Click and drag over the range of cells that will hold the concentration data (A5 through A10 for the sample data).
2. Choose Format > Cells... (this is shorthand for choosing Cells...> from the Format menu at the top of the Microsoft Excel window).
3. Click on the Number tab.
4. Under Category choose Number and set Decimal places to 5.
5. Click OK.
6. Repeat for the absorbance data column (B5 through B10 for the sample data), setting the decimal places to 4.

Let us take data from Fig. F.2.

1. Enter the data below the column titles.
2. We can also place the absorption readings for the unknown solutions below the other data.

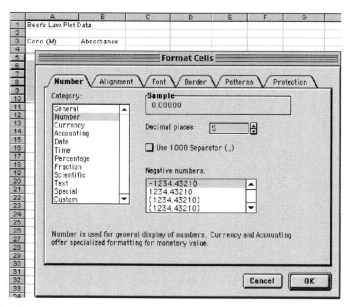

Fig. F.1 Beer's law

Fig. F.2 Data for Beer's law plot

The concentration data is probably better expressed in scientific notation.

1. Highlight the concentration data and choose Format > Cells...
2. Choose the Scientific Category and set the Decimal places to 2.
3. Highlight the data in both the concentration and absorbance columns (but not the unknown data) by selecting them.

With the data you want graphed, start the Chart Wizard.

1. Choose the Chart Wizard icon from the tool bar. If the Chart Wizard is not visible, you can also choose Insert > Chart...
   The first dialogue of the wizard comes up.
2. Choose XY (Scatter) and the unconnected points icon for the Chart sub-type.
3. Click Next > The Data Range box should reflect the data you highlighted in the spreadsheet. The Series option should be set to Columns, which is how your data is organized (Fig. F.3).
4. Click Next > The next dialogue in the wizard is where you label your chart (Fig. F.4)
5. Enter Beer's Law for the Chart Title.
6. Enter Concentration (M) for the Value X Axis.
7. Enter Absorbance for the Value Y Axis.
8. Click on the Legend tab.
9. Click off the Show Legend option (Fig. F.5).
10. Click Next > Keep the chart as an object in the current sheet (Fig. F.6). Note: Your current sheet is probably named with the default name of "Sheet 1".
11. Click Finish.

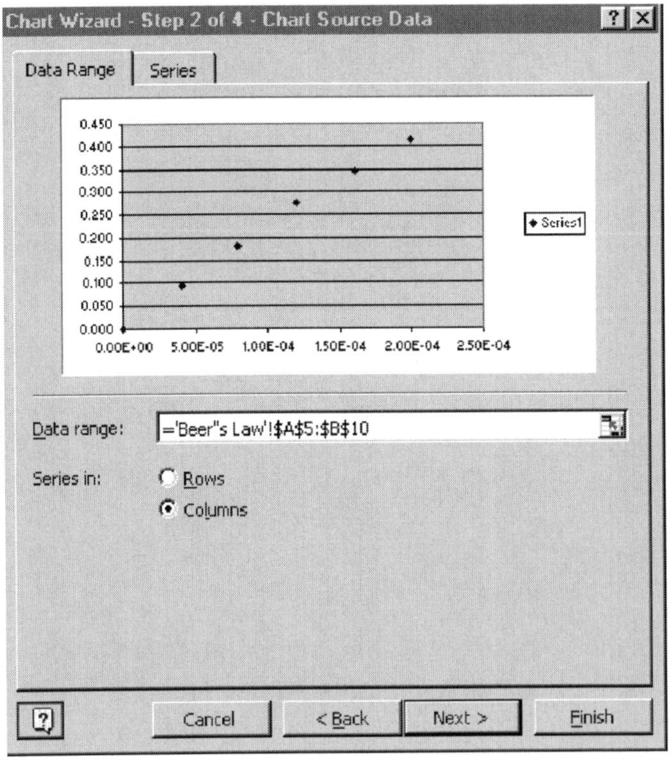

Fig. F.3 Graph plotting from data

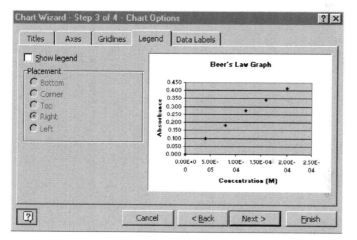

Fig. F.4 Chart wizard

Fig. F.5 Step 3

Fig. F.6 Step 4

The initial scatter plot is now finished and should appear on the same spreadsheet page as your original data. Your chart should look like Fig. F.7.

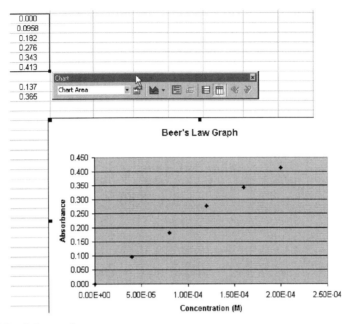

Fig. F.7 Beer's law graph

A few items to be noted:

1. The data should look as though it falls along a linear path.
2. Horizontal reference lines were automatically placed in your chart, along with a gray background.
3. The chart is highlighted with square handles on the corners. When your chart is highlighted, a special chart floating palette should also appear, as is seen in Fig. F.7. If the chart floating palette does not appear, go to Tools > Customize..., click on the Toolbars tab, and then click on the Chart checkbox. If it still doesn't show up as a floating palette, it may be "docked" on one of your tool bars at the top of the Microsoft Excel window.
4. With your graph highlighted, you can click and drag the chart to wherever you would like it located on the spreadsheet page. Grabbing one of the four corner handles allows you to resize the graph. Note: the graph will automatically adjust a number of chart properties as you resize the graph, including the font size of the text in the graph. You may need to go back and alter these properties. At the end of the first part of this tutorial, you will learn how to do this.

F.3 Creating a Linear Regression Line (Trendline)

When the chart window is highlighted, besides having the chart floating palette appear, a chart menu also appears. From the chart menu, you can add a regression line to the chart.

1. Choose Chart > Add trendline...
   A dialogue box appears (Fig. F.8).
2. Select the Linear Trend/Regression type.
3. Choose the Options tab and select Display equation on chart (Fig. F.9).
4. Click OK to close the dialogue.

The chart now displays the regression line (Fig. F.10),

F.4 Using the Regression Equation to Calculate Concentrations

The linear equation shown on the chart represents the relationship between Concentration (x) and Absorbance (y) for the compound in the solution. The regression line can be considered an acceptable estimation of the true relationship between concentration and absorbance. We have been given the absorbance readings for two solutions of unknown concentration.

Using the linear equation (labeled A in Fig. F.11), a spreadsheet cell can have an equation associated with it to do the calculation for us. We have a value for

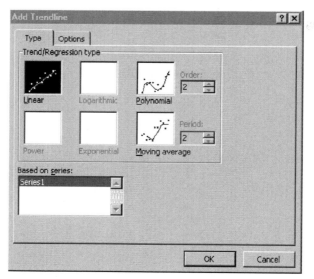

Fig. F.8 Adding trendlines

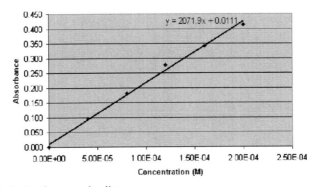

Fig. F.9 Selected display equation on chart

Fig. F.10 Displaying the regression line

y (Absorbance) and need to solve for x (Concentration). Below are the algebraic equations working out this calculation:

$$y = 2071.9x + 0.111$$
$$y - 0.0111 = 2071.9x$$
$$(y - 0.0111)/2071.9 = x$$

Now, we have to convert this final equation into an equation in a spreadsheet cell. The equation associated with the spreadsheet cell will look like what is labeled C in Fig. F.8. B12 in the equation represents y (the absorbance of the unknown). The solution for x (Concentration) is then displayed in cell C12.

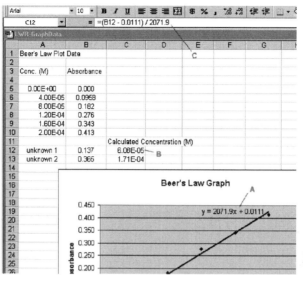

Fig. F.11 Beer's law graph

1. Highlight a spreadsheet cell to hold x, the result of the final equation (cell C12, labeled B in Fig. F.11).
2. Click in the equation area (labeled C, Fig. F.11).
3. Type an equal sign and then a parentheses.
4. Click in the cell representing y in your equation (cell B12 in Fig. F.11) to put this cell label in your equation.
5. Finish typing your equation.

Note: If your equation differs for the one in this example, use your equation
    Duplicate your equation for the other unknown.

1. Highlight the original equation cell (C12 in Fig. F.11) and the cell below it (C13).
2. Choose Edit > Fill > Down.

F.4.1 Adjusting the Chart Display

The readability and display of the scatterplot can be further enhanced by modifying a number of the parameters and options for the chart. Many of these modifications can be accessed through the Chart menu, the Chart floating palette, and by double-clicking the element on the chart itself. Let's start by creating a better contrast between the data points and regression line and the background.

1. Double-click in the gray background area of the chart or by selecting Chart Area in the Chart floating palette and then clicking on the Format icon (Fig. F.12).

In the Chart Area Format dialogue, set the border and background colors.

1. Choose None for a Border.
2. Choose the white square from the color palette for an Area color.
3. Click OK.

Now, delete the horizontal grid lines.

1. Click on the horizontal grid lines in the chart and press the Delete key.

Now, adjust the color and line weight of the regression line and the color of the data points.

1. Double-click on the regression line (or choose Series 1 Trendline 1 from the Chart floating palette and then click the Format icon).
2. Choose a thinner line for the Line Weight.
3. Click on the word Automatic next to Line Color and the color palette appears. Choose dark blue from the color palette.
4. Click OK.
5. Double-click on one of the data points (or choose Series 1 and click the Format icon).
6. Choose dark red from the color palette for the Marker Foreground and Background.
7. Click OK.

Finally, you can move the regression equation to a more central location on the chart

1. Click and drag the regression equation.

If necessary, resize the font size for text elements in the graph.

1. Either double click the text element or choose it from the floating palette.
2. Click on the Fonts tab.
3. Choose a different font size.

The results can be seen in Fig. F.13.

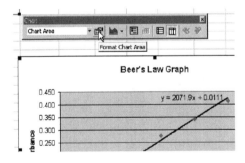

Fig. F.12 Formatting chart area

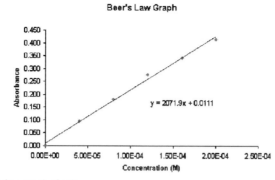

Fig. F.13 Beer's law graph (final)

Appendix G
Titration Data Plotting

G.1 Creating a Scatter Plot of Titration Data

In this next part of the tutorial, we will work with another set of data. In this case, it is of a strong acid-strong base titration (Table G.1). With this titration, a strong base (NaOH) of known concentration is added to a strong acid (also of known concentration, in this case). As the strong base is added to solution, its OH^- ions bind with the free H^+ ions of the acid. An equivalence point is reached when there are no free OH^- nor H^+ ions in the solution. This equivalence point can be found with a color indicator in the solution or through a pH titration curve. This part of the tutorial will show you how to do the latter.

In the last part of the tutorial, the axis scale is manipulated on the plot in order to get a closer look at the most critical part of the plot: the equivalence point.

Table G.1 pH variation with acid-base neutralization

Titration of 50 mL of 0.1 M HCl with 0.1 M NaOH.	
Volume of NaOH added (in mL)	pH
0.00	1.00
10.0	1.17
25.0	1.48
45.0	2.28
49.5	3.30
49.75	3.60
50.0	7.00
50.25	10.40
55.0	11.68
60.0	11.96

K. I. Ramachandran et al., *Computational Chemistry and Molecular Modeling*
DOI: 10.1007/978-3-540-77304-7, ©Springer 2008

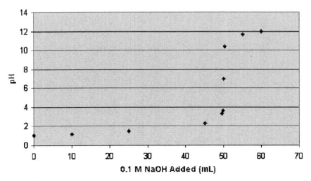

Fig. G.1 Titration graph

Note that there should be two columns of data in your spreadsheet:
Column A: mL of 0.1 M NaOH added
Column B: pH of the 0.1 M HCl/0.1M NaOH mixture

1. Using a new page in the spreadsheet, enter your titration data.

Highlight the titration data and the Column headers.

1. Click on the Chart wizard icon.
2. Choose XY (Scatter) and the scatter Chart sub-type. Continue through steps 2 through 4 of the Chart wizard.
3. The defaults for step 2 should be fine if you properly highlighted the data.
4. In step 3 enter the chart title and x and y axis labels and turn off the legend.
5. In step 4, leave as an object in the current page.
6. The resulting plot should look like Fig. G.1.

G.2 Curve Fitting to Titration Data

The next logical question that you might ask is whether a linear regression line or a curved regression line might help us interpret the titration data. You may remember that our goal with this plot is to calculate the equivalence point, that is, what amount of NaOH is needed to change the pH of the mixture to 7 (neutral).

Create a linear regression line:

1. Choose Chart > Add Trendline…
2. Pick Linear sub-type.

Looking at the data (Fig. G.2), it is clear that the first 45 ml of NaOH do little to alter the pH of the mixture. Then between 45 ml and 55 ml, there is a sharp rise in pH before leveling off again. The data trend does not seem linear at all and, in fact, a linear regression line does not fit the data well at all.

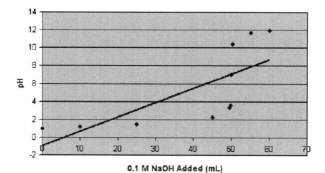

Fig. G.2 Linear regression

The next approach might be to choose a different type of trendline (Fig. G.3):

1. Click on the linear regression line in the plot and press the delete key to delete the line.
2. Choose Chart > Add Trendline...
3. Pick Polynomial subtype.
4. Set the Order of the curve to 2.

You can see that a second order polynomial curve does not capture the steep rise of the data well. A higher order curve might be tried (Fig. G.4):

1. Double-click on the curved regression line.
2. Set the Order of the curve to 3.

Still, the third order polynomial does not capture the steep part of the curve where it passes through a pH of 7. Even higher order curves could be created to see if they

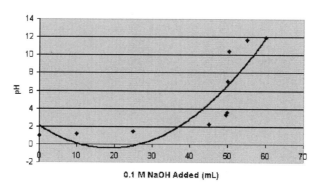

Fig. G.3 Polynomial regression

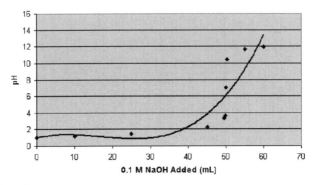

Fig. G.4 Higher order curve

fit the data better. Instead, a different approach will be taken for this data. Go ahead and delete the regression curve:

1. Click on the curved regression line in the plot and press the delete key.

G.3 Changing the Scatter Plot to a Line Graph

Instead of adding a curved regression line, all of the points of the titration data are connected with a smooth curve. With this approach, the curve is guaranteed to go through all of the data points. This is both good and bad. This option can be used if you have only one pH reading per amount of NaOH added. If you have multiple pH readings for each amount added on the scatter plot, you will not end up with a smooth curve. To change the scatter plot is a (smoothed) line graph (Fig. G.5):

1. Choose Chart > Chart Type...
2. Select the Scatter connected by smooth lines Chart subtype.

The result should look like Fig. G.6.

This smooth, connected curve helps locate where the steep part of the curve passes through pH 7.

G.4 Adding a Reference Line

The chart can be enhanced by adding a reference line at pH 7. This clearly marks the point where the curve passes through this pH.

1. A set of drawing tools should be visible at the bottom of the window. If not, click on the Draw icon two to the right of the Chart wizard icon.

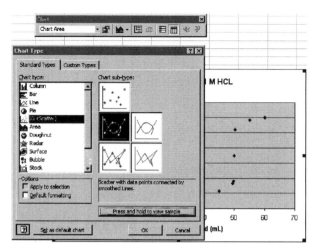

Fig. G.5 Changing the scatter plot

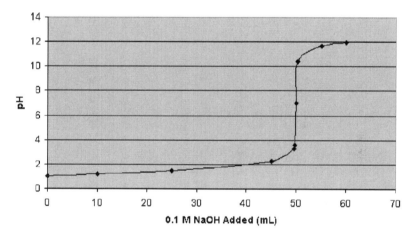

Fig. G.6 Scatter plot changed

2. Make sure your chart is highlighted.
3. Choose the line tool at the bottom of the window.
4. Draw a horizontal line at pH 7 across the width of the chart by clicking and dragging a line across the chart area.
5. With the horizontal line still highlighted, choose a 3/4 pt line thickness and a dashed line type at the bottom of the window.
6. Remove the other horizontal grid lines.
7. Turn off the border.
8. Change the chart colors.

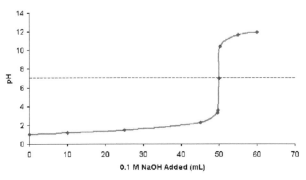

Fig. G.7 Refined graph

9. Thicken the curve and shrink the data points, emphasizing the fitted curve over the individual data points.

The result should look like Fig. G.7.

G.5 Modifying the Chart Axis Scale

The above chart gives a good overview of the entire titration. If you would like to focus exclusively on the steep part of the curve between 45 and 55 ml of added NaOH, a new chart can be created which limits the x axis range. Start by making a copy of the current chart:

1. Select the current chart by clicking near its border.
2. Choose Edit > Copy.
3. Click a spreadsheet cell about 10 rows below the current chart.
4. Choose Edit > Paste.

With the new chart highlighted (Fig. G.8):

1. Choose Value (X) Axis from the Chart floating palette.
2. Click on the Format icon.
3. Set the Minimum to 45, Maximum to 55.
4. Set the Major unit to 1 and Minor unit to 0.25.
5. Click OK.

Next, both vertical and horizontal gridlines can be added to more accurately locate the equivalency point (Fig. G.9):

1. Choose Chart > Chart Options...
2. Click on the Gridlines tab.
3. Select X axis Major gridlines and Y axis Major gridlines.
4. Click OK.

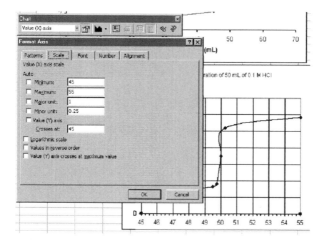

Fig. G.8 New chart highlighted

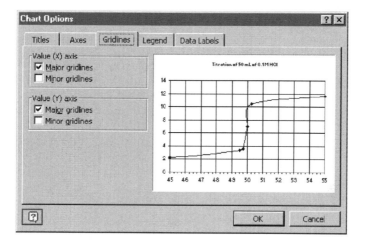

Fig. G.9 Locating the equivalency point

With enhancements similar to what you did to the other chart, the result will look like Fig. G.10.

Even with this smooth curve passing through all of the data points, it is still an estimation of what intermediate mL added/pH data points would be. A clear inaccuracy is where the curve moves in a negative X direction between the 50 and 51 mL data points. More data points collected between 49 and 51 mL would both better smooth the curve and give a more accurate estimation of the equivalency point.

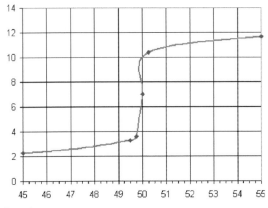

Fig. G.10 Modified graph

G.6 Extensions

Possible extensions include making charts and graphs of other chemical reactions
carried out in the lab. This type of graphing also lends itself to physics and technology education labs where data is collected, graphed, and compared to some theoretical equation. Examples might be a lab on Ohm's law or velocity of a toy car on
a downhill track. Make sure if experiments are carried out, that the lab and students
are properly outfitted with safety equipment.

Appendix H
Curve Fitting in Chemistry

H.1 Membrane Potential

Whenever an ionic conductor separates two electrolyte solutions of different composition, it is possible in principle to observe all or part of that difference in composition as a difference in potential, which obeys the Nernst equation. This can be done experimentally if one electrode is placed on each side of the ionic conductor so that one is in each of the two different electrolyte solutions. These two electrodes usually are identical reference electrodes so that the measured cell potential difference is only the potential difference across the ionic conductor.

If all substances could move through the ionic conductor equally well the cell potential difference would be zero, but if only some can move through or into the conductor (or if not all of them move equally well) then the cell potential difference will not be zero. Natural biological cell membranes act in this way, and so do synthetic polymer membranes; these membranes are called *ion*-selective membranes. Thin glass membranes and crystals of some slightly soluble salts can also act as ion-selective membranes. The cell potential difference observed across an ion-selective membrane is called a membrane potential.

Membrane potentials are responsible for the operation of the nervous systems of living organisms. Chemists make use of them to construct chemical sensors for various ions in aqueous solutions. These sensors routinely determine hydrogen, sodium, potassium, and fluoride ions. We will consider here only one of them, the glass electrode, which is the most common chemical sensor for the hydrogen ion. As such, it is by far the most common method used to determine the pH of aqueous solutions. The glass electrode cell is usually a two-electrode cell containing two silver/silver chloride reference electrodes arranged as follows:

$$Ag/AgCl(s), Cl^-(aq), H^+(aq)/glass/test\ soln.//Cl^-(aq), AgCl(s)/Ag$$

The reference electrode and electrolyte on the left are contained within the thin glass electrode membrane. The reference electrode on the right is connected by a salt bridge to the test solution, which contains an unknown concentration of the

K. I. Ramachandran et al., *Computational Chemistry and Molecular Modeling*
DOI: 10.1007/978-3-540-77304-7, ©Springer 2008

hydrogen ion. The membrane potential is the cell potential difference. A saturated calomel reference electrode sometimes replaces the silver/silver chloride electrode on the right.

For a glass membrane of the type used in these electrodes, only the aqueous hydrogen ion can move into the membrane to any significant extent. The hydrogen ions do not move through the membrane, but only into the hydrated layers on each side of the glass where it touches the inner and test (outer) electrolyte solutions. At 25 °C, the cell potential difference is the membrane potential and it follows the Nernst equation in the form:

$$DE = DE' - 0.05915\,\text{pH}$$

In this equation DE' is a small constant potential difference depending on the reference electrodes, the salt bridge, and the inner electrolyte solution; the pH is that of the test solution.

Over the aqueous pH range 2 to 12, the membrane potential of a glass electrode can accurately track the pH of a test solution in accordance with the Nernst equation. At more extreme values of pH, some response to other species in solution begins to become apparent. This can be improved somewhat by choices of different glasses, so that glass electrodes can be used in aqueous solutions from pH 1 to pH 13. Using still different glasses, electrodes which respond to sodium ion rather than hydrogen ion can be fabricated.

The response of glass electrodes to differences in solution pH was first observed in 1906. Systematic studies of glass composition led to the selection of a soft soda-lime glass (72% SiO_2, 22% Na_2O, and 6% CaO) as the most suitable composition. The glass electrode did not come into general use until about 1935, when electronic voltmeters were first used with it. The commercial pH meter developed by Dr. Arnold O. Beckman established Beckman Instruments as a major supplier of chemical instrumentation. A pH meter is a high-input-impedance electronic voltmeter whose scale is calibrated in pH units (one pH unit is 59.15 mV). Buffers of known pH are used as standards to calibrate the pH meter.

H.2 The Determination of the E^0 of the Silver-Silver Chloride Reference Cell

From the theory of the electrochemical cell, the potential in volts E of a silver-silver chloride-hydrogen cell is related to the molarity m of HCl by the equation:

$$E + \frac{2RT}{F}\ln m = E^0 + \frac{2.343RT}{F}m^{\frac{1}{2}} \tag{H.1}$$

where R is the gas constant, F is the Faraday constant (9.648×10^4 coulombs mol^{-1}), and T is 298.15 K. The silver-silver chloride half-cell potential E^0 is of critical importance in the theory of electrochemical cells and in the measurement of pH. We can measure E at known values of m, and it would seem that simply solving the above equation would lead to E^0. So it would, except for the influence of non-ideality on E. Inter-ionic interference gives us an incorrect value of E^0 at any nonzero value of m. But if m is zero, there are no ions to give a voltage E. The way out of this dilemma is to make measurements at several (non-ideal) molarities m and extrapolate the results to a hypothetical value of E at $m = 0$. In so doing we have "extrapolated out" the non-ideality because at $m = 0$ all solutions are ideal. Rather than ponder the philosophical meaning of a solution in which the solute is not there, it is better to concentrate on the error due to inter-ionic interactions, which becomes smaller and smaller as the ions become more widely separated (Fig. H.1). At the extrapolated value of $m = 0$, ions have been moved to an infinite distance where they cannot interact. Plotting the left side of the equation as a function of $m^{\frac{1}{2}}$ gives a curve with (2.342RT F) as the slope and E^0 as the intercept (Fig. H.2).

From the graph equation, the value of E^0 can be read as 0.2225, which is very close to the modern value of 0.2223 V.

	A	B	C	D	E	F	G
1	Electrode potential of Silver-Silver chloride reference electrode						
2	Trial no	Molarity	sqrt(molarity)	Potential	Calculated Eo		
3	1	0.003214890000	0.056700	0.225600	0.222188248		
4	2	0.005619001600	0.074960	0.226300	0.221789507		
5	3	0.009137448100	0.095590	0.227300	0.221548159		
6	4	0.013409640000	0.115800	0.228200	0.221232083		
7	5	0.025632010000	0.160100	0.230000	0.220366463		
8	6	0.053916840000	0.232200	0.232200	0.218228062		
9	7	0.123833610000	0.351900	0.234600	0.213425474		
10					1.538777996	Mean	0.219825
11		0.256925694444					

Fig. H.1 Electrode potential of silver-silver chloride reference electrode

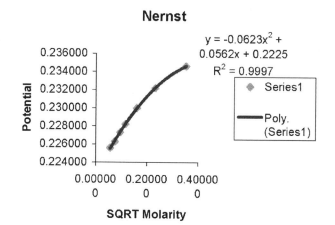

Fig. H.2 Nernst law application

Appendix I
The Solvation of Potassium Fluoride

Linear extrapolation of the experimental behavior of a real gas to zero pressure or a solute to infinite dilution is often used as a technique to "get rid" of molecular or ionic interactions so as to study some property of the molecule or ion to which these interferences are considered extraneous. Emsley (1971) studied the heat (enthalpy) of solutions of potassium fluoride KF and the monosolvated species KF.HOAc in glacial acetic acid at several concentrations. A known weight of the anhydrous salt KF was added to a known weight of glacial acetic acid in a Dewar flask fitted with a heating coil, a stirrer, and a sensitive thermometer. The temperature change on each addition was recorded. The heat capacity C of the flask and its contents was determined by supplying a known amount of electrical energy Q to the flask and noting the temperature rise ΔT in kelvins (K) $Q\,(\text{joules}) = C\Delta T$. The experiment was repeated for the solvated salt KF.HOAc, where the molecule of solvation is acetic acid, HOAc. Some experimental results calculated are included in Table I.1 and the corresponding graphs are given in Figs. I.1 and I.2.

Table I.1 Variation of molality with temperature

KF: $C = 4.168\ \text{kJ K}^{-1}$				
Molality	0.194	0.590	0.821	1.208
Temperature change K	1.592	4.501	5.909	8.115
KF: HOAc: $C = 4.203\ \text{kJ K}^{-1}$				
Molality	0.280	0.504	0.910	1.190
Temperature change K	−0.227	−0.432	−0.866	−1.189

K. I. Ramachandran et al., *Computational Chemistry and Molecular Modeling*
DOI: 10.1007/978-3-540-77304-7, ©Springer 2008

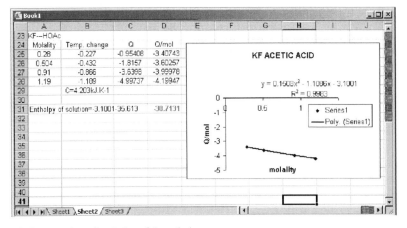

KF			
Molality	Temp. change	Q=C*Delta	Q/mol
0.194	1.592	6.635456	34.20338
0.59	4.502	18.76434	31.80396
0.821	5.909	24.62971	29.99843
1.208	8.115	33.82332	27.99944
	C=4.168kJ.K-1		

KF solvation

$y = 0.7005x^2 - 7.1845x + 35.613$

$R^2 = 0.9965$

Fig. I.1 Molality change in temperature

KF---HOAc			
Molality	Temp. change	Q	Q/mol
0.28	-0.227	-0.95408	-3.40743
0.504	-0.432	-1.8157	-3.60257
0.91	-0.866	-3.6398	-3.99978
1.19	-1.109	-4.99737	-4.19947
	C=4.203kJ.K-1		
Enthalpy of solution= 3.1001-35.613			-30.7131

KF ACETIC ACID

$y = 0.1608x^2 - 1.1006x - 3.1001$

$R^2 = 0.9903$

Fig. I.2 Computation of enthalpy of the solution

Appendix J
Partial Molal Volume of $ZnCl_2$

In general, the volume of a solution, say $ZnCl_2$ in water, is dependent on the number of moles of each of the components. For a binary solution, $V = f(n_1, n_2)$. The change in volume dV on adding a small amount dn_1 of water or dn_2 of $ZnCl_2$ is:

$$dV = \left(\frac{\partial V}{\partial n_1}\right) dn_1 + \left(\frac{\partial V}{\partial n_2}\right) dn_2 \tag{J.1}$$

where we stipulate that pressure, P, and temperature, T are constant for the process and we adopt the usual subscript convention, 1 for solvent and 2 for solute. If we specify 1 kg as the amount of water, n_2 is the molality of $ZnCl_2$. We expect that the volume of the solution will be greater than $1000\,cm^3$ by the volume taken up by the $ZnCl_2$. It may seem reasonable to take the volume of one mole of $ZnCl_2$ in the solid state Vm and add it to $1000\,cm^3$ to get the volume of a 1 molal solution. One-half the molar volume of solute would, by this scheme, lead to the volume of a 0.5 molar solution, and so on. This does not work. The volume of $1000\,g$ of water in the solution is not exactly $1000\,cm^3$, and it is dependent of the temperature. Nor are the volumes additives. Indeed, some solutes cause contraction of the solution to less than $1000\,cm^3$. Interactions at the molecular or ionic level cause an expansion or contraction of the solution so that, in general:

$$V \neq 1000 + V_m \tag{J.2}$$

We define a partial molar volume $\overline{V}_i$ such that $V = n_1\overline{V}_1 + n_2\overline{V}_2$ for a binary solution or, in general:

$$V = \sum_{i=1}^{N} n_i\overline{V}_i \tag{J.3}$$

for a solution of N components. It can be shown (Alberty, 1987) that

$$\overline{V}_i = \left(\frac{\partial V}{\partial n_i}\right)_j \tag{J.4}$$

where the subscript j indicates that all components in the solution other than i are

K. I. Ramachandran et al., *Computational Chemistry and Molecular Modeling*
DOI: 10.1007/978-3-540-77304-7, ©Springer 2008

held constant. If the solution is a binary solution of n_2 moles of solute in 1 kg of water, $\overline{V}_2$ is the partial molal volume of component 2. A partial molal volume is a special case of the partial molar volume for 1 kg of solvent. Refer to Fig. J.1 and Fig. J.2.

The computed slope $= 163.2217$.

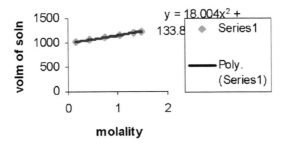

	A	B	C	D	E	F
1	g ZnCl2 per kg of water	Density	Molality	Volm of solution	Volm of water	dV/dm
2	20	1.0167	0.14771	1020.334	1000	145.0743
3	60	1.0532	0.443131	1063.192	1000	154.7554
4	100	1.0891	0.738552	1108.91	1000	165.6619
5	140	1.1275	1.033973	1157.85	1000	176.4262
6	180	1.1665	1.329394	1209.97	1000	185.1595
7	200	1.1866	1.477105	1237.32	1000	
8					Sum	827.0774
9					Average	165.4155

Fig. J.1 Partial molal volume

Fig. J.2 Graphical computation of the partial molal volume

Index